VOLUME EIGHTY FIVE

ADVANCES IN BOTANICAL RESEARCH

Plastid Genome Evolution

ADVANCES IN BOTANICAL RESEARCH

Series Editors

Jean-Pierre Jacquot
Professor, Membre de L'Institut Universitaire de France, Unité Mixte de Recherche INRA, UHP 1136 "Interaction Arbres Microorganismes", Université de Lorraine, Faculté des Sciences, Vandoeuvre, France

Pierre Gadal
Honorary Professor, Université Paris-Sud XI, Institut Biologie des Plantes, Orsay, France

VOLUME EIGHTY FIVE

ADVANCES IN BOTANICAL RESEARCH

Plastid Genome Evolution

Volume Editors

SHU-MIAW CHAW

Biodiversity Research Center, Academia Sinica; Biodiversity Program, Taiwan International Graduate Program, Academia Sinica and National Taiwan Normal University, Taipei, Taiwan

ROBERT K. JANSEN

Department of Integrative Biology, University of Texas at Austin, Austin, TX, United States; Genomics and Biotechnology Research Group, Faculty of Science, King Abdulaziz University, Jeddah, Saudi Arabia

ELSEVIER

ACADEMIC PRESS

An imprint of Elsevier

Academic Press is an imprint of Elsevier
125 London Wall, London, EC2Y 5AS, United Kingdom
The Boulevard, Langford Lane, Kidlington, Oxford OX5 1GB, United Kingdom
50 Hampshire Street, 5th Floor, Cambridge, MA 02139, United States
525 B Street, Suite 1800, San Diego, CA 92101-4495, United States

First edition 2018

ISBN: 978-0-12-813457-3
ISSN: 0065-2296

Publisher: Zoe Kruze
Acquisition Editor: Sam Mahfoudh
Editorial Project Manager: Joanna Collett
Production Project Manager: James Selvam
Cover Designer: Greg Harris

Typeset by SPi Global, India

CONTENTS

CONTRIBUTORS

Abdulrahman L. Al-Malki
Department of Biochemistry, Faculty of Science, King Abdulaziz University, Jeddah, 21589, Saudi Arabia

Alawiah M. Alhebshi
Genomics and Biotechnology Research Group, Faculty of Science, King Abdulaziz University, Jeddah, 21589, Saudi Arabia

John M. Archibald
Dalhousie University, Halifax, NS; Program in Integrated Microbial Biodiversity, Canadian Institute for Advanced Research, Toronto, ON, Canada

Matt P. Ashworth
Department of Integrative Biology, University of Texas at Austin, Austin, TX, United States

Shu-Miaw Chaw
Biodiversity Research Center, Academia Sinica; Biodiversity Program, Taiwan International Graduate Program, Academia Sinica and National Taiwan Normal University, Taipei, Taiwan

Jan de Vries
Dalhousie University, Halifax, NS, Canada

Francisco Figueroa-Martinez
CONACyT-Universidad Autónoma Metropolitana, México City, México

Matthew A. Gitzendanner
Department of Biology, University of Florida, Gainesville, FL, United States

Nahid H. Hajrah
Genomics and Biotechnology Research Group, Faculty of Science, King Abdulaziz University, Jeddah, 21589, Saudi Arabia

Christopher Jackson
School of Biosciences, University of Melbourne, Melbourne, VIC, Australia

Robert K. Jansen
Department of Integrative Biology, University of Texas at Austin, Austin, TX, United States; Genomics and Biotechnology Research Group, Faculty of Science, King Abdulaziz University, Jeddah, Saudi Arabia

Mohammad A. Khiyami
King Abdulaziz City for Science and Technology, Riyadh, 11442, Saudi Arabia

Claude Lemieux
Université Laval, Québec, QC, Canada

De-Zhu Li
Germplasm Bank of Wild Species, Kunming Institute of Botany, Chinese Academy of Sciences, Kunming, PR China

Jeffrey P. Mower
University of Nebraska, Lincoln, NE, United States

Sergio A. Muñoz-Gómez
Centre for Comparative Genomics and Evolutionary Bioinformatics, Dalhousie University, Halifax, NS, Canada

Julia Naumann
Institute for Botany, Technische Universität Dresden, Dresden, Germany

Adrian Reyes-Prieto
University of New Brunswick, Fredericton, NB, Canada

Tracey A. Ruhlman
Department of Integrative Biology, University of Texas at Austin, Austin, TX, United States

Sarah Russell
University of New Brunswick, Fredericton, NB, Canada

Jamal S.M. Sabir
Genomics and Biotechnology Research Group, Faculty of Science, King Abdulaziz University, Jeddah, 21589, Saudi Arabia

Mumdooh J. Sabir
Faculty of Computing and Information Technology, King Abdulaziz University, Jeddah, 21589, Saudi Arabia

Claudio H. Slamovits
Centre for Comparative Genomics and Evolutionary Bioinformatics, Dalhousie University, Halifax, NS, Canada

David R. Smith
University of Western Ontario, London, ON, Canada

Douglas E. Soltis
Department of Biology; Florida Museum of Natural History; Genetics Institute; Biodiversity Institute, University of Florida, Gainesville, FL, United States

Pamela S. Soltis
Florida Museum of Natural History; Genetics Institute; Biodiversity Institute, University of Florida, Gainesville, FL, United States

Edi Sudianto
Biodiversity Research Center, Academia Sinica; Biodiversity Program, Taiwan International Graduate Program, Academia Sinica and National Taiwan Normal University; National Taiwan Normal University, Taipei, Taiwan

Edward C. Theriot
Department of Integrative Biology, University of Texas at Austin, Austin, TX, United States

Monique Turmel
Université Laval, Québec, QC, Canada

Trisha L. Vickrey
University of Nebraska, Lincoln, NE, United States

Susann Wicke
Institute for Evolution and Biodiversity, University of Münster, Münster, Germany

Chung-Shien Wu
Biodiversity Research Center, Academia Sinica, Taipei, Taiwan

Ting-Shuang Yi
Germplasm Bank of Wild Species, Kunming Institute of Botany, Chinese Academy of Sciences, Kunming, PR China

Mengjie Yu
Department of Integrative Biology, University of Texas at Austin, Austin, TX, United States

PREFACE

In plants the plastids are almost certainly to be regarded as differentiations of protoplasmic substance

K. Mereschkowsky 1905; English translation by Martin and Kowallik (1999)

Prior to the invention of electron microscope in 1931, a prominent Russian biologist and botanist named Konstantin Sergeevich Mereschkowski proposed the endosymbiotic origin of cell organelles in a 1905 publication. More than 60 years passed before Lynn Margulis revisited Mereschkowsky's unprecedented hypothesis, and eventually biologists began to accept that the plastid descended from an ancestor of extant cyanobacteria. To manufacture carbohydrates, and concomitantly release oxygen, photosynthetic plastids (i.e. the chloroplasts) capture energy from sunlight and combine that energy with carbon dioxide and water. Plastid genomes, referred to as plastomes in this book, encode many key proteins that are not only vital for regulation of photosynthesis but also play fundamental roles in the synthesis of nucleotides, amino acids, fatty acids and numerous primary metabolites as well as secondary compounds. Therefore, plastomes are crucial to the development of photosynthetic eukaryotes and to their interactions with the environment. In the past decade, advances in high-throughput sequencing technologies have expedited the accumulation of plastome sequences for examining their evolution, the function of plastid-encoded genes and their interaction with nuclear genes. Information about the organization and evolution of plastomes has also played a crucial role in plastid genetic engineering to enhance crop species and to produce pharmaceuticals. Finally, plastome data have played a pivotal role in resolving the phylogeny of photosynthetic organisms.

This book brings together expert contributors who have been working on plastid genome variation and evolution across photosynthetic eukaryotes. The topics range from the diversified plastome architecture of single-celled photosynthetic eukaryotes to seed plants; from the causes and consequences of genomic diversity to the phylogenetic utility of plastomic sequences for resolving relationships across the photosynthetic tree of life. Presently over 41,000 complete plastome sequences (as of April 27, 2017) are available in the National Center for Biotechnology Information (NCBI) organelle genome database. Novel software and comparative plastomics have modified our views on plastome architecture and made tremendous contributions to the resolution of evolutionary relationships within many clades. In terms of plastome organization it is now widely recognized that plastomes are not predominantly circular but instead occur as linear and/or branched molecules that can form more complex multisubunit structural variants that can recombine. Moreover, dominant and subdominant forms have been detected in a number of seed plant lineages. There have been several recent examples of invasion of foreign DNA from the mitochondrion into the plastome, a phenomenon that is likely to be more prevalent as additional plastomes are sequenced. Plastome data have also been instrumental in identifying the earliest diverging flowering plant as *Amborella*, a problem that had vexed plant scientists since Darwin referred to it as an abominable mystery. This volume presents novel insights into this exciting field from leading experts in plastome evolution, including a comprehensive coverage of plastid genome variation in a broad range of taxonomic groups, from protists and multicellular algae to the major clades of land plants. Plant scientists and students in the fields of molecular biology, biotechnology, evolution, phylogenetics, horticulture and agriculture will be prospective readers. We hope that readers find this volume a useful summary of up-to-date work on plastome evolution.

The 11 chapters of this volume have been written with the goal of illuminating plastome evolution across a wide diversity of photosynthetic eukaryotes since their endosymbiotic origin approximately 1.5 billion years ago. We made an effort to provide coverage of all major photosynthetic lineages, but due to page limitations some groups have not been included. Although the topics are technical in nature, each chapter was written in an attempt to be as comprehensible as possible by nonspecialists and students.

We thank all of the authors for their time and effort in contributing to this volume. We also acknowledge the 23 reviewers listed below who assisted in producing high-quality chapters.

Reviewers: John F. Allen, Andrew Alverson, Craig Barrett, Shu-Miaw Chaw, Philippe Delavault, Joshua P. Der, Richard Dorrell, Chris Blazier, Romain Gastineau, Wenhu Guo, Robert K. Jansen, Patrick Kociolek, Shao-Lun Liu, Wolfgang Löffelhardt, Michael Moore, Jeffrey P. Mower, Vincent Savolainen, Thomas Pfannschmidt, David R. Smith, Monique Turmel, Mao-Lun Weng, Chung-Shien Wu and Hwan Su Yoon.

SHU-MIAW CHAW and ROBERT K. JANSEN
TAIPEI and AUSTIN, DECEMBER 2017

REFERENCE

Martin, W., & Kowallik, K. V. (1999). Annotated English translation of Mereschkowsky's 1905 paper 'Über Natur und Ursprung der Chromatophoren imPflanzenreiche'. *European Journal of Phycology, 34*, 287–295.

CHAPTER ONE

Plastid Autonomy vs Nuclear Control Over Plastid Function

Jan de Vries*[,1], John M. Archibald*[,†,1]
*Dalhousie University, Halifax, NS, Canada
[†]Program in Integrated Microbial Biodiversity, Canadian Institute for Advanced Research, Toronto, ON, Canada
[1]Corresponding authors: e-mail address: jan.devries@dal.ca; john.archibald@dal.ca

Contents

Abstract

Plastids stem from free-living cyanobacteria. The transition from endosymbiont to organelle involved strong reductive evolution. Modern-day plastid genomes possess only a small fraction of the genes present in their cyanobacterial progenitors. In addition to genome reduction, plastids underwent modifications that facilitated recruitment of host-derived proteins and metabolites; both processes contributed to organellogenesis and a shift in control over plastid function from the organellar genome to that of the host. It is likely that most of the modifications to the early plastid happened before the major radiations that led to today's algae and plants. Plastids nevertheless exhibit substantial variation in form and function. In this chapter, we highlight some of the evolutionary implications of the differences in the genetic capacities of plastids across the breadth of plant and algal diversity. We focus on the transition from genetic semiautonomy, which is of relevance in the context of the endosymbiotic spread of plastids and kleptoplasty, to the high degree of nuclear control over plastid function seen in land plants. Genomic and transcriptomic investigations of diverse plants and algae have revealed important differences in the coding capacity of plastid genomes in different lineages, raising questions about how the plastid's own genetic capabilities impact its physiology as well as that of its host.

Advances in Botanical Research, Volume 85
ISSN 0065-2296
https://doi.org/10.1016/bs.abr.2017.11.011

1. INTRODUCTION: ENDOSYMBIOSIS AND THE GENOMIC REMNANTS OF CYANOBACTERIA IN ARCHAEPLASTIDA

Prior to the evolution of plastids, photosynthesis was an exclusively prokaryotic trait (for review, see, e.g. Bryant & Frigaard, 2006). In a sense it still is, given that eukaryotes never evolved the ability to photosynthesize de novo. Eukaryotes acquired oxygenic photosynthesis through endosymbiosis, whereby a phototrophic cyanobacterium was assimilated by a plastid-lacking protist (Mereschkowsky, 1905; reviewed by Archibald, 2015a; Cavalier-Smith, 1982; Keeling, 2013; Zimorski, Ku, Martin, & Gould, 2014). It is commonly accepted that there was a single primary endosymbiosis that gave rise to the Archaeplastida (Jackson & Reyes-Prieto, 2014; Rodríguez-Ezpeleta et al., 2005; but see also critical discussions in Larkum, Lockhart, & Howe, 2007 or Mackiewicz & Gagat, 2014); the Archaeplastida circumscribe the group of photosynthetic eukaryotes with 'primary' plastids (cf. Adl et al., 2012), i.e., those stemming directly from a prokaryote. When this landmark event took place is still unclear. Based on the finding of a fossilized red alga (coined *Bangiomorpha*) exhibiting complex morphology (Butterfield, 2000), Archaeplastida are thought to have originated at least 1.2 billion years ago. This date has been pushed back even further with the recent description of putatively red algal fossils 1.6 billion years in age (Bengtson, Sallstedt, Belivanova, & Whitehouse, 2017). Some of the recent molecular clock data suggest that the Archaeplastida are ~1.5 billion years old (Parfrey, Lahr, Knoll, & Katz, 2011), which is only slightly younger than some estimates for the age of eukaryotes as a whole (cf. Eme, Sharpe, Brown, & Roger, 2014). These molecular clock-based data hence provide an impossibly narrow time frame for the obvious morphological complexity of these fossilized ancient red algae to have evolved. There are thus many uncertainties about the timing of the evolution of primary plastids, especially due to the rarity of fossils bearing on the early steps of eukaryote evolution (for a review, see Keeling et al., 2005). What we do know is that in addition to Archaeplastida, photosynthesis has spread across the eukaryotic tree by 'secondary' endosymbiosis (Fig. 1), which is the incorporation of a primary alga into another eukaryotic host (reviewed by Archibald, 2015a, 2015b; Keeling, 2013; Zimorski et al., 2014). Finally, there are numerous additional peculiar cases in plastid evolution, including plastid replacement in dinoflagellates (see, e.g. Dorrell & Howe, 2015) and

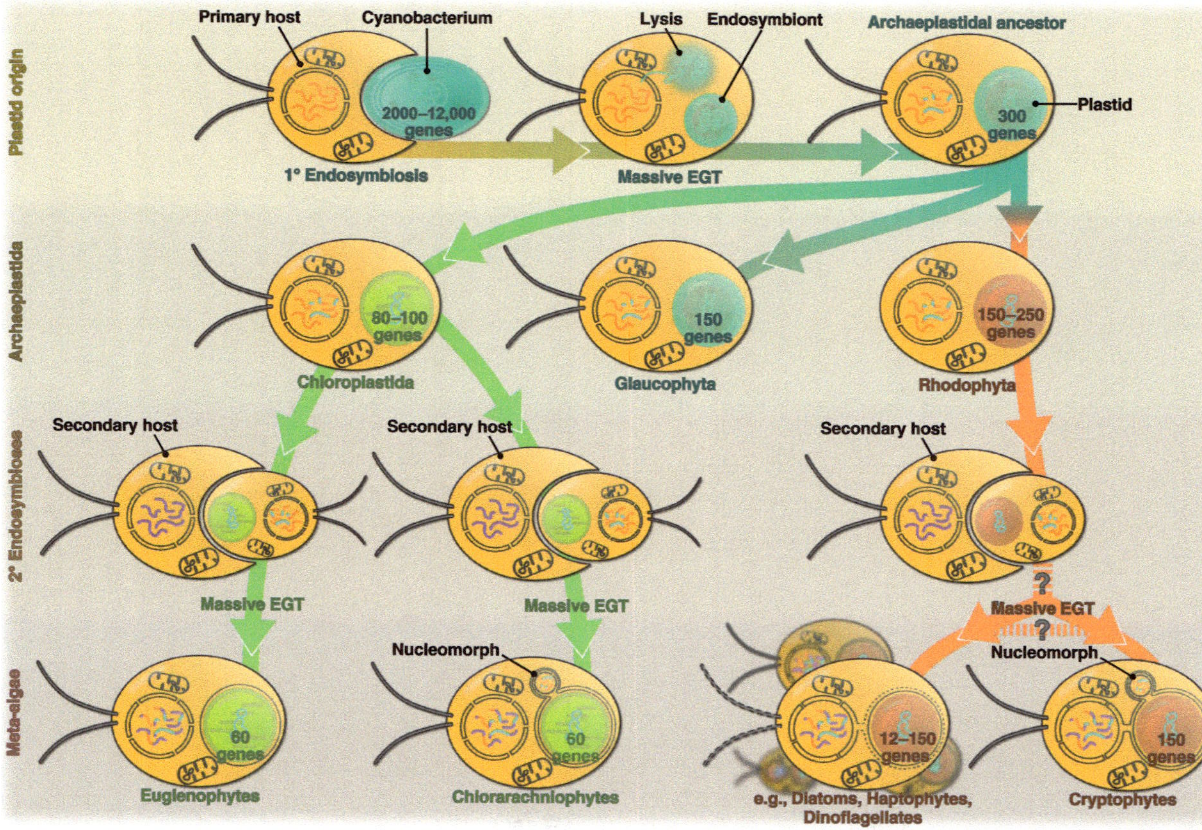

Fig. 1 See legend on next page.

plastid theft in various protists (reviewed by Dorrell & Howe, 2012; Johnson, 2011; Stoecker, Johnson, deVargas, & Not, 2009) and even in molluscs (reviewed by de Vries, Christa, & Gould, 2014; Rumpho, Pelletreau, Moustafa, & Bhattacharya, 2011; Serôdio, Cruz, Cartaxana, & Calado, 2014).

Primary endosymbiosis involves the engulfment of a whole prokaryotic cell. The cyanobacterial endosymbiont was likely covered with lipopolysaccharides and exuded oxygen and other biochemical waste; it also had a tough Gram-negative bacterial cell wall and a complex genome capable of synthesizing all of the proteins it needed for life. While the nature of the photosynthetic machinery senso lato (i.e. including thylakoids, the presence of photosystem II and photosystem I; see Hohmann-Marriott & Blankenship, 2011) speaks of a clear cyanobacterial ancestry, many other prokaryotic features have vanished during the course of plastid evolution. Yet, some features, such as the peptidoglycan (PG)-containing cell wall and a (reduced) genome, persist. Regarding the former new insight has recently emerged. It was long thought that the only remnant of the PG layer found in present-day plastids is that of the 'cyanelles' (muroplasts) of glaucophytes, an enigmatic group of exclusively freshwater algae (for an overview, see Jackson, Clayden, &

Fig. 1 Primary and secondary endosymbiosis and the origin of cyanobacterial genes in eukaryotes. Plastids originate from the endosymbiotic incorporation of a cyanobacterium by a heterotrophic protist (primary host; 1° endosymbiosis). The cyanobacterial plastid progenitor harboured a genome coding for a loosely defined set of proteins (2000–12,000, based on extant cyanobacteria; cyan DNA). Loss and endosymbiotic gene transfer (EGT), mediated by lysis and incorporation of genetic material by the host nucleus (note the orange host DNA), massively reduced this cyanobacterial genome. The lion's share of this initial reduction is thought to have occurred before the primary lineages of algae diverged (i.e. in the ancestor of all Archaeplastida), reducing the ancestral plastid genome to ~300 protein-coding genes (cf. Qiu, Lee, Yoon, & Bhattacharya, 2017). From this archaeplastidal ancestor, three lineages emerged: the Rhodophyta, Glaucophyta and Chloroplastida (from which land plants eventually emerged). In each lineage, further and independent EGT occurred. The Chloroplastida and Rhodophyta were involved in additional endosymbiotic events. In these cases secondary (2°; or even higher-order) eukaryote–eukaryote endosymbioses occurred. This happened at least two times independently involving Chloroplastida. The number of eukaryote–eukaryote endosymbiotic events involving Rhodophyta is still debated. In all of these higher-order endosymbioses, EGT mixed the genetic imprint of the (1) cyanobacterial plastid ancestor, (2) the archaeplastidal host and (3) the new host (purple DNA). Chlorarachniophyte (secondary green) and cryptophyte (secondary red) algae harbour nucleomorphs that, in both cases, contain a remnant genome of the primary host nucleus. *EGT*, endosymbiotic gene transfer.

Reyes-Prieto, 2015). However, the plastids of the moss *Physcomitrella patens* were recently found to contain a very thin PG layer (Hirano et al., 2016), which may also be true of other Archaeplastida within the green algal lineage. Indirect evidence comes from pharmacological inhibition of PG biosynthesis enzymes, which has been shown to result in plastid division defects in streptophyte algae and basal-branching land plants (Izumi, Ono, & Takano, 2003; Kasten & Reski, 1997; Matsumoto, Takechi, Sato, Takio, & Takano, 2012) up to, to a certain degree of division inhibition, ferns (Izumi, Kuroki, Nagafuji, Lin, & Takano, 2008); interestingly, the presence and antibiotic-responsiveness of PG components seems to cooccur with the presence of the *ftsZ3* gene (Grosche & Rensing, 2017). Nonetheless, only glaucophytes have a thick PG layer that equals that of cyanobacteria (Steiner & Löffelhardt, 2002; Steiner, Ma, Pfanzagl, & Löffelhardt, 2001), which might have had interesting implications regarding the evolution of host control over plastid division (for further discussion, see de Vries & Gould, 2017), e.g., through dynamin (cf. Miyagishima, Nakamura, Uzuka, & Era, 2014). Note that it has been hypothesized that the PG layer in glaucophytes might be essential for withstanding the turgor pressure that is created as a result of their carboxysomal-like carbon concentration mechanisms (Fathinejad et al., 2008). Next to the PG layer, one of the most demonstrably cyanobacterial features of modern-day plastids is the genome.

Genomic data show that most (~90% or more) of the genes present in the cyanobacterial endosymbiont were lost or transferred to the host nucleus early in archaeplastid evolution (Archibald, 2015a, 2015b; Martin & Herrmann, 1998; Qiu et al., 2017; Timmis, Ayliffe, Huang, & Martin, 2004; Fig. 1). This happened through a process known as endosymbiotic gene transfer (EGT; Martin, Brinkmann, Savonna, & Cerff, 1993; Martin et al., 1998). EGT is thought to work as follows: from a population of endosymbionts (i.e. the progenitors of the plastid), random lysis results in the release of DNA, which can occasionally (and randomly) be incorporated into the host nuclear genome by the activities of the host's DNA repair enzymes (Henze & Martin, 2001; Ricchetti, Fairhead, & Dujon, 1999; Timmis et al., 2004). This sets up the potential for genetic redundancy. The fortuitous expression of this gene, together with the evolution of a mechanism for targeting the gene product back to the endosymbiont/organelle (see below), lifts the constraints on retention of the organellar gene; the gene can then, by chance, be lost by mutational inactivation and/or outright deletion, making the organism reliant on the nuclear copy. The impact of EGT on the genome of plants and algae was huge; some estimates suggest that in the nuclear genome of the flowering plant *Arabidopsis* >4500 genes have a cyanobacterial ancestry (Martin et al., 2002), and EGT still occurs in plants and some algae (Huang, Ayliffe, & Timmis, 2003; Richly & Leister, 2004; Stegemann, Hartmann, Ruf, & Bock, 2003).

If EGT is/was rampant early in plastid evolution, why do these organelles retain a genome at all (e.g. Allen & Martin, 2016)? As we shall see, the answers to this question are both curious and instructive. There are in fact plastids that have completely lost their genomes (Molina et al., 2014; Smith & Lee, 2014) but, significantly, these organisms are all non-photosynthetic (Figueroa-Martinez, Nedelcu, Smith, & Reyes-Prieto, 2015). The reason(s) for genome retention presumably must hence revolve (at least to a certain degree) around photosynthesis. Below we explore hypotheses regarding the retention of plastid genes and genomes. We address the questions of (a) why the retention of plastid genes differs between the various algal and plant lineages and (b) how this impacts the biology of these organisms.

2. CYANOBACTERIAL GENES IN TWO GENETIC COMPARTMENTS

The cyanobacterial plastid progenitor left a substantial genetic imprint on the nuclear genome of the ancestral plastid-bearing eukaryote. We can assume that the endosymbiont had a 'typical' genome, containing all genes necessary for a photoautotrophic (free-living) lifestyle. Nevertheless, putting a number on the size and complexity of its genome is not straightforward. Extant cyanobacterial genomes exhibit significant differences in their coding capacity, with between fewer than 2000 and less than 12,000 protein-coding genes present (Dagan et al., 2013; Larsson, Nylander, & Bergman, 2011). What is more, we still do not know to which group of living cyanobacteria plastids are most closely related. Different analytical approaches have yielded different results, with some associating the plastid with (relatively) gene-poor early-branching cyanobacteria and others with gene-rich late-branching cyanobacteria (Dagan et al., 2013; de Alda, Esteban, Diago, & Houmard, 2014; Deusch et al., 2008; Ponce-Toledo et al., 2017). Protein gene presence/absence and amino acid sequence similarity data suggest a closer relationship between the more gene-rich cyanobacterial clades and the plastid progenitor (Dagan et al., 2013; Deusch et al., 2008); cyanobacterial genomes have nevertheless clearly undergone gene gains and losses (Larsson et al., 2011), complicating these inferences. However, such data have not been gathered for the recently discovered *Gloeomargarita* clade (cf. Couradeau et al., 2012; discussed in de Vries & Archibald, 2017a), which, on the basis of phylogenomics, shares specific ancestry with modern-day plastids (Ponce-Toledo et al., 2017).

To what degree we will ever be able to reconstruct the cyanobacterial genome of the plastid progenitor from nuclear genome sequences of Archaeplastida is unclear. Size estimates of the EGT 'footprint' across archaeplastidal nuclear genomes have varied from a few hundred genes to more than 4500 (Dagan et al., 2013; Martin et al., 2002; Price et al., 2012), the latter number being larger than the total number of genes found in many cyanobacterial genomes. Such estimates are complicated by the fact that after transfer, these cyanobacterial genes have experienced their own independent evolution, including independent losses, duplications and recombination (including domain shuffling) (Méheust, Zelzion, Bhattacharya, Lopez, & Bapteste, 2016). To the extent possible, the task of quantifying the genetic legacy of the cyanobacterial progenitor of the plastid hence calls for further comparative genomic analyses of a broader range of Archaeplastida. Regardless of the exact amount, the genetic material we know of (a) can ultimately be traced back to the cyanobacterial plastid progenitor (see also Fig. 1) and (b) speaks to the profound impact of endosymbiosis on Archaeplastida (see also Ku et al., 2015). For those genes still housed in the genome of the cyanobacterium-turned-plastid, the picture is much clearer.

3. PLASTID GENOMES: CODING CAPACITY AND FUNCTIONAL DIVERSITY

Plastid genomes harbour a tiny fraction of the genes present in their cyanobacterial ancestors. Why do these genes persist and what can we learn from consideration of their functions? Plastid genome sequences from diverse algae and plants show that they share (to a certain extent) a very similar set of genes (for comprehensive reviews, see, e.g. Allen, de Paula, Puthiyaveetil, & Nield, 2011; Green, 2011; Grzebyk, Schofield, Vetriani, & Falkowski, 2003). This residual gene set has been considered 'non-transferable' or at least transfer-resistant. Understanding why this is so is a fundamental question in evolutionary cell biology. It has long been known that organellar genomes are rich in genes associated with transcription and translation (summarized in Green, 2011), and some of the ribosomal genes in plastid genomes are even present in the same order as in cyanobacteria

(Stoebe & Kowallik, 1999). Furthermore, there is strong convergence between the set of ribosomal protein genes retained in plastid and mitochondrial genomes (Maier et al., 2013). Clearly the prokaryotic genetic information processing machinery is important for organelle biology and it is advantageous for its core components to be expressed 'on site'. Nevertheless, it is interesting to note that essentially all eukaryotes (including plants and algae) drive mitochondrial transcription by a single-subunit, nucleus-encoded, phage-type RNA polymerase (Cermakian, Ikeda, Cedergren, & Gray, 1996) (the only exception is an enigmatic group of protists called jakobids, whose mitochondria use a multisubunit bacterial-type polymerase, the genes for which reside in the organelle [Burger, Gray, Forget, & Lang, 2013; Lang et al., 1997]). The core transcription and translation machinery can thus be tinkered with.

The situation is somewhat similar for plastids. Modification of the plastids' prokaryotic system by host-derived components has occurred, which will later be discussed in the context of land plants. However, the only known cases in which the plastids' prokaryotic transcription apparatus has been completely functionally replaced by a eukaryotic version (which is, again, phage-derived; see Section 6) are found among nonphotosynthetic plants such as *Cuscuta* (Krause, Berg, & Krupinska, 2003). Further, considering the information processing genes in isolation fails to provide a satisfactory answer to the question of why is there a plastid genome in the first place: such genes are only necessary if there is genetic information to process. The reason(s) why in situ information processing is needed thus likely rests with the many other genes on the plastid genome. For these other genes, various hypotheses have been put forth to explain their persistence. Here we outline the most prominent two.

The first such hypothesis is the CoRR hypothesis of John Allen (Allen, 1993). CoRR stands for colocation for redox regulation; the hypothesis states that core genes associated with these processes must remain in the plastid genome because there is a direct feedback loop between the redox state of the light reaction or OXPHOS electron transport chain (cf. Allen, 2015). It is, hence, based on the fact that in both mitochondria and plastids (and here especially in the green lineage), a large fraction of organellar genes encode proteins associated with the electron transport chain of the oxidative phosphorylation pathway of mitochondria and the photosynthetic reaction centres of plastids (Allen, 1993, 2015; Allen & Martin, 2016). Indeed, many of the genes that are shared by all plastid genomes are associated with the major complexes in the photosynthesis light reaction: photosystem I and II, the cytochrome b_6f complex and the ATP synthase (Allen et al., 2011). Furthermore, experimental data show that redox regulation of these light reaction-associated genes occurs, including the gene coding for the high-turnover photosystem II protein psbA (Pfannschmidt, Nilsson, & Allen, 1999; Pfannschmidt, Nilsson, Tullberg, Link, & Allen, 1999). Sensing of the plastid redox status (via the plastoquinone pool connecting photosystem II and I) is carried out by the plastid-targeted protein CHLOROPLAST SENSOR KINASE (CSK), which was first characterized in *Arabidopsis thaliana* (Puthiyaveetil et al., 2008). This protein is found in all photosynthetic eukaryotes and regulates expression of photosystem components in a redox-dependent manner (Puthiyaveetil et al., 2008) by phosphorylating the plastid RNA polymerase (PEP) sigma factor-1 (SIG-1) (Puthiyaveetil et al., 2010). The existence of such feedback loops speaks to the importance of having core information processing machinery synthesized in the plastid (Allen, 2015). Indeed, there might be a codependency between endosymbiont-derived information processing and the feedback loops derived from ongoing photosynthesis. Nonphotosynthetic parasitic plants belonging to the Orobanchaceae are instructive in this context. They appear to lose the PEP genes from their plastid genome along with photosynthesis-associated genes (Wicke et al., 2016). Hence, once photosynthesis is lost, constraints on the retention of PEP genes in the plastid genome may be lifted.

Another hypothesis for the retention of organellar genomes revolves around the issue of hydrophobicity. In 1986, von Heijne proposed that (animal) mitochondria need to retain a genome because if the proteins were encoded by nuclear genes, their N-terminal hydrophobicity would result in them being misdirected (via the endoplasmatic reticulum [ER]) to the secretory pathway (von Heijne, 1986). Since genes retained in plastid genomes encode proteins that are eventually integrated into the thylakoid membrane, it is conceivable that the same problem applies to components of the light reaction chain of the plastid. However, plastids are known to import nucleus-encoded proteins that are very hydrophobic, and some proteins encoded by organellar genes are hydrophilic. Indeed, Cheung, Bogorad, van Montagu, and Schell (1988) showed that transformants expressing the gene *psbA* in the nucleus import its protein product (the D1 protein) into their thylakoid membrane. Intriguingly, Cheung et al. (1988) used a *psbA* gene that stemmed from an atrazine-resistant *Amaranthus hybridus* (*Ah*PsbA). The authors, hence, not only identified their successful transformants using a screen for antrazine tolerance but simultaneously showed that expressing *Ah*PsbA conferred atrazine resistance to tobacco. Thus, in this case, translocation of a hydrophobic, usually plastid-encoded, membrane protein from the cytosol into the chloroplast and, further, to the thylakoid membrane was feasible.

Further interest in the hydrophobicity hypothesis has come from the recent work of Björkholm, Ernst, Hagström, and Andersson (2017). Here the authors tested the retention of organellar genes for the 13 proteins encoded by genes in human mitochondrial DNA. The authors found that 12 out of 13 mitochondrial genes expressed in the nucleus (and equipped with coding sequences for potent mitochondrial targeting peptides) produce proteins that are (mis)directed to the ER, consistent with the idea that hydrophobicity can play a role in the 'transferability' of organellar genes, at least in some eukaryotes.

These models both seek to explain the retention of a common set of genes. What are we to make of the observed differences in plastid genome coding capacity from lineage to lineage? In the context of the CoRR hypothesis the question is why do some plastid genomes have more genes, i.e., a bigger or different set of genes. There seems to be a strong tendency to lose genes from the plastid genome, as Martin et al. (1998) showed that the very same genes have been transferred to the nucleus multiple times independently in different algal and plant lineages. Recently, it was proposed that in Rhodophyta, whose plastid genomes harbour 1.5 to 2 times as many genes as Chloroplastida, plastid gene retention served to counteract nuclear genome reduction early in red algal evolution (Qiu et al., 2017). Yet, given that EGT is such a strong force, it is not clear why this retention has remained so during the long period of time that red algae have been diverging from one another. This is especially intriguing given that red seaweeds are polyplastidic, a cell biological feature that correlates with an increased probability for EGT (for discussion of the relationship between polyplastidy and frequency of EGT, see Smith, Crosby, & Lee, 2011 and de Vries & Gould, 2017). We must therefore assume that there are strong evolutionary forces acting against gene transfer. Research suggests that environmental (abiotic) factors may play a role.

Given that photosynthetic eukaryotes dwell in a wide range of environments, they might have different requirements for the in situ regulation of plastid function (Simpson & Stern, 2002). The plastid genomes of diatoms of the genus *Thalassiosira* tell an interesting tale with regard to the location of the genes for plastid-localized proteins and environmental conditions. While many photosynthetic organisms in the ocean suffer from iron deficiency

(Browning et al., 2017), *Thalassiosira oceanica* can maintain high growth rates under low iron conditions (Strzepek & Harrison, 2004). Lommer et al. (2010) suggested that part of the tolerance for iron deficiency observed in *T. oceanica* is due to the very recent transfer of *petF* (coding for the iron-containing ferredoxin) to the nuclear genome—a gene that in *Thalassiosira pseudonana* is still in the plastid genome. The authors showed that differential regulation of the (now nuclear) *petF* occurs in an iron-dependent manner. Downregulation of *PETF* allows replacement of iron-dependent ferredoxin (i.e. PETF) with iron-free flavodoxin, which takes over PETF's role in the electron transport chain. It is only in the nucleus that integration into regulatory networks is possible (in this case those responding to the iron status of the cell). In turn, stoichiometry might also play out differently: Dorrell and Howe (2012) proposed that some of the genes that have the 'smoothest' passage to the nucleus are those that simply need to be expressed on a sufficiently high level to achieve full functionality (which might include peripheral proteins of the photosystems that ward off oxidative stress).

The aforementioned *Thalassiosira* example suggests that certain genes can indeed only successfully be transferred if there are strong evolutionary pressures to do so, in this case facilitated by environmental factors. In most cases the retention of certain genes in the plastid, where their expression is controlled in situ (as in the CoRR hypothesis), would seem to be 'preferred'. There is thus a balance to be struck between retention of genes in the plastid, where they act as part of a well-oiled machine, and transfer to the nucleus, where their protein products are one step removed from their site of action, but more sophisticated layers of control are possible. This includes integration of EGT-derived genes into preexisting eukaryotic gene regulatory networks, e.g., those involving phytohormone signalling, transcription factors, multistep signalling cascades and microRNAs. We will revisit this concept in the context of land plant evolution.

4. ROLE OF GENETIC AUTONOMY IN ENDOSYMBIOSIS

Together with protein import, EGT is a key force underlying the transition from endosymbiont to organelle. Once an essential set of genes has been transferred (and protein targeting to the endosymbiont has been established), the endosymbiont is bound to its host. Yet, by being bound more tightly to this host, the plastid also potentially becomes less able to be acquired by other hosts in the context of secondary (i.e. eukaryote–eukaryote) endosymbiosis (see bottom panels in Fig. 1). This is because, in the case of secondary endosymbiosis, plastid–nucleus communication has already been established in the primary alga prior to engulfment by the secondary host. With each secondary endosymbiotic event, the secondary host nucleus must take over the function of the primary nucleus. This complex situation underlies the so-called portable plastid hypothesis of Grzebyk et al. (2003), which was invoked primarily to explain the vast diversity of secondarily acquired red algal-type plastids in nature (cf. Grzebyk et al., 2003). In essence, the hypothesis states that, due to the presence of more genes in red algal plastids compared to those of green algae, the plastid of red algae was more 'transferable' into a new host.

More broadly interpreted, the portable plastid hypothesis suggests that "portable" plastids possess a bigger in situ functional core set: they bring with them more genes as well as a fine-tuned machinery to work with those genes to maintain proper organelle function. A bigger core plastid gene set requires less initial coordination with a new host and, in the case of a secondary plastid from the primary alga to the secondary host nuclear genome, also less EGT of essential plastid-targeted proteins (i.e. the nucleus-encoded core set) (cf. Grzebyk et al., 2003). For the permanent functional integration of a primary alga into its secondary host there is nevertheless the need for extensive transfer of genetic material. That such a process can take a long time or in some cases will never finish is apparent by the presence of nucleomorphs in cryptophyte and chlorarachniophyte algae.

Nucleomorphs are the reduced remnant nuclei of eukaryotic algal endosymbionts. We know of only two groups of algae with nucleomorphs: cryptophytes have a red algal-derived nucleomorph and plastid, while chlorarachniophytes have a green algal-derived nucleomorph and plastid (Douglas & Penny, 1999; Rogers, Gilson, Su, McFadden, & Keeling, 2007; Suzuki, Hirakawa, Kofuni, Sugita, & Ishida, 2016). Nucleomorphs have been reduced through the forces of EGT that act after acquisition of the primary alga into a secondary host (Curtis et al., 2012). While most secondary algae have completely lost the nucleomorph, there must have been a nucleomorph-bearing transition stage during the evolution of all secondary plastid-bearing organisms (see, e.g. Curtis et al., 2012; Gould, Maier, & Martin, 2015—although see later discussion about karyoklepty and kleptoplasty). Disentangling the genetic interactions between nucleus and plastid that were established during primary endosymbiosis is not easy. Nevertheless, we do observe examples of abrupt—but transient—disentanglement of the genetic interaction between host nucleus and plastid in nature. As discussed in the next section, this happens in the form of plastid theft.

5. KLEPTOPLASTY AND GENETIC SEMIAUTONOMY OF (STOLEN) PLASTIDS

Most plastids found in nature are heritable, with the requisite interaction between endosymbiont and nucleus established over the long term. This involved a balancing act between EGT and gene retention (and the associated coordination of gene expression that is intertwined with its physiological impact). We must assume that loss of genetic material and, potentially, traits that rendered the endosymbiont less beneficial to its host would have been selected against. Selection will have favoured—or at least not selected against—those descendants in which endosymbionts lost only obsolete properties (such as those associated with a free-living lifestyle). Yet, there are special cases where there is no long-term balancing act involved in plastid uptake. Kleptoplasty—plastid theft—is such a case.

Kleptoplasty is the acquisition of a preexisting plastid from an alga. It is known to involve a wide diversity of hosts (the thieves) and plastid donors. Most kleptoplastic organisms are protists belonging to the dinoflagellates, ciliates and foraminifera (see, e.g. Jauffrais et al., 2016; Pillet & Pawlowski, 2013; Stoecker et al., 2009). Yet, animals are also known to perform kleptoplasty, as highlighted by the various species of kleptoplastic sacoglossan sea slugs (de Vries, Christa, et al., 2014; de Vries, Rauch, Christa, & Gould, 2014; Rumpho et al., 2011; Serôdio et al., 2014). When it comes to the plastid donors (the prey), examples involving both primary and secondary plastids have been described. Most kleptoplastic slug species feed on primary green ulvophytes (sometimes also polyphagous, see, e.g. Christa, Wescott, Schäberle, König, & Wägele, 2013) but some species feed on secondary red xanthophytes (for an overview, see Christa, Händeler, Schäberle, König, & Wägele, 2014; de Vries, Christa, et al., 2014). Kleptoplastic foraminifera ingest secondary red diatom plastids (Pillet, de Vargas, & Pawlowski, 2011; Pillet & Pawlowski, 2013), while many freshwater ciliates obtain green algae (Stoecker et al., 2009). What all of these cases of kleptoplasty have in common is that they are nonpermanent but nevertheless involve plastids that are capable of continuing to function for a period of time in the absence of their original host's nucleus.

In such nonpermanent relationships, EGT bears in a different way upon the issue of suitability for symbiosis. As with the portable plastid hypothesis, a plastid thief might benefit from stealing a plastid with a higher degree of genetic autonomy. In a sense, kleptoplasty represents an extreme case of plastid portability. But while the portable plastid hypothesis seeks to explain the tempo and mode of secondary endosymbiosis (especially involving red algal-derived plastids), kleptoplasty is not a true symbiosis. In the case of kleptoplasty, organelle-to-thief nucleus EGT is not an option. The kleptoplast is already so reduced that the textbook trajectory from transient (endo)symbiont (stage one) to obligate endosymbiont (stage two) to organelle (stage three) cannot occur. (For discussion of the distinction between an endosymbiont and organelle, see Cavalier-Smith & Lee, 1985; briefly, an organelle's proteome is dependent on host nuclear gene products.) The kleptoplast already reached stage three in its 'previous owner' (i.e. the alga from which it was stolen). One can conceive of a simple solution to the problem: gene transfer from the prey nucleus (which itself bears the footprint of EGT) to the nucleus of the kleptoplastic host. However, in most kleptoplastidic systems that have been studied, no instances of lateral gene transfer from prey to thief have convincingly been detected (Bhattacharya, Pelletreau, Price, Sarver, & Rumpho, 2013; Pillet & Pawlowski, 2013; Rauch et al., 2015; Wägele et al., 2011).

If support from the nucleus does not occur in kleptoplastic associations, how do the stolen plastids maintain function? de Vries et al. (2013) proposed that the answer lies with the inherent properties of the plastid itself. Indeed, Green, Fox, and Rumpho (2005) observed that isolated plastids of *Vaucheria litorea* (the food source of the kleptoplastic sacoglossan sea slug *Elysia chlorotica*) stay physically intact for 2 weeks (and possibly longer); in contrast, using the same experimental setup, spinach plastids lost ~80% of their intactness within a single day. It was suggested (see de Vries, Christa, et al., 2014; de Vries et al., 2013) that the presence of certain genes in the plastid genome might confer the ability to utilize in situ (i.e. nucleus-independent) mechanisms for the maintenance of organelle biochemistry and integrity. In this case, the gene *ftsH*, which codes for a protease involved in the canonical photosystem II repair, was deemed significant (cf. de Vries et al., 2013; Janska, Kwasniak, & Szczepanowska, 2013; Lindahl et al., 2000; Nickelsen & Rengstl, 2013; Nixon, Michoux, Yu, Boehm, & Komenda, 2010); *ftsH*, along with other genes such as the chlorophyte version of *ycf1* or *tufA*, is found in the plastid genomes of many (green) algae but not higher branching streptophyte algae and land plants (Civáň, Foster, Embley, Séneca, & Cox, 2014; de Vries, Archibald, & Gould, 2017; de Vries et al., 2013; de Vries, Stanton, Archibald, & Gould, 2016). Given that the simple removal of damaged D1 (a key photosystem II subunit prone to damage; encoded by the *psbA* gene) can ward off the accumulation of reactive oxygen species (de Vries et al., 2013; Kato, Miura, Ido, Ifuku, & Sakamoto, 2009), encoding a few—but critical—additional proteins on the plastid genome could make a significant difference in the ability of a plastid to maintain functionality in the absence of algal nuclear support, rendering the plastids robust (de Vries, Christa, et al., 2014; de Vries et al., 2013). This suggests that these more robust plastids have an extended core set of genes. Such an extended set of plastid-encoded genes might be fine-tuned, especially relevant under high-stress conditions, by CoRR-based redox regulation. Such robust plastids are not autonomous (i.e. in a state that resembles the retention of the cyanobacterial plastid progenitor), far from it. They simply can stand their vigil longer when isolated from their accompanying nucleus. The reasons for this might trace back to the aforementioned high-stress environmental conditions the algae (from which the plastids are being acquired) usually dwell in. Therefore, 'robust' kleptoplasts are simply a consequence of the existence of an extended core gene set, which enables the curious phenomenon of kleptoplasty.

A peculiar exception might be the ciliate *Mesodinium rubrum*. This organism steals more than just the plastids: its theft includes the nucleus (kleptokaryon) of its cryptophyte prey (Johnson, Oldach, Delwiche, & Stoecker, 2007). These nuclei remain transcriptionally active and might thus through some means continue to support the kleptoplasts through, for example, supply of light-harvesting components (Johnson, 2011; Johnson et al., 2007). Indeed, based on this phenomenon Bodył (2017) recently proposed that kleptoplasty might offer a route (one might even say a short-cut) for acquiring permanent, heritable, plastids. Bodył's hypothesis is based upon the idea that these feeding behaviours are likely accompanied by very specific and sophisticated feeding mechanisms. These mechanisms, Bodył (2017) states, would facilitate plastid acquisition in a manner that is less 'messy' than the integration of an entire primary alga, which brings along its own eukaryotic cell biology including mitochondria and membrane systems. Bodył (2017) suggests that a kleptoplastidic protist would have be able to 'cherry-pick' the plastid while still being able to establish nutrient flow by, for example, modifying the phagosomal/symbiosomal membrane surrounding the newly acquired plastid. Bodył (2017) further hypothesizes that after the acquisition of the nucleus (kleptokaryon), the plastid thief gradually digests it and acquires the necessary nuclear genes for plastid-targeted proteins through the standard EGT process. In this context, it is noteworthy that in an RNAseq analysis of the kleptoplastic—but not kleptokaryotic—*Dinophysis acuminate*, Wisecaver and Hackett (2010) detected some transcripts harbouring targeting sequences that should direct them to the kleptoplasts. Hence, even in the absence of a kleptokaryon, there could be hitherto unexplored interactions between host nuclei and stolen plastids. Regardless, it is conceivable that a more robust plastid might be particularly amenable to kleptoplasty: the less a plastid requires constant nuclear support, the longer it will remain intact and make it, eventually, more portable.

6. PLASTID AUTONOMY AND THE EVOLUTION OF LAND PLANTS

Land plants have a high degree of nuclear control over plastid function. This is most apparent in the fact that the plastids of vascular land plants can differentiate into many different organelle types (reviewed by Jarvis & López-Juez, 2013). The diversity of these plastids is likely underpinned by the complexity of the plastid transcription machinery unique to land plants. Plastids still encode their own prokaryote-type (multisubunit) RNA polymerase (the PEP) that they utilize for transcription of plastid genes (recently reviewed by Börner, Aleynikova, Zubo, & Kusnetsov, 2015; Liebers et al., 2017; Pfannschmidt et al., 2015); all core subunits are encoded in the plastid (Ohyama et al., 1986; Sijben-Müller, Hallick, Alt, Westhoff, & Herrmann, 1986), although the sigma factors required for PEP activity are encoded in the nucleus (Hanaoka, Kanamaru, Takahashi, & Tanaka, 2003). Yet, angiosperms possess another plastid-targeted RNA polymerase, the nuclear-encoded RNA polymerase (NEP), and in dicots, there are even two of these (Azevedo et al., 2008; Hricová, Quesada, & Micol, 2006; Liere, Kaden, Maliga, & Börner, 2004 recently reviewed by Börner et al., 2015), which are possibly derived from a duplication event in basal-branching angiosperms (Liere, Weihe, & Börner, 2011; Yin, Richter, Börner, & Weihe, 2010). The core subunits of PEP most likely trace back to the RNA polymerase used for transcription by the cyanobacterial plastid progenitor (Hajdukiewicz, Allison, & Maliga, 1997; Kindgren & Strand, 2015; Martin et al., 1998; Wicke, Schneeweiss, dePamphilis, Müller, & Quandt, 2011). In contrast, the NEP is a phage-type polymerase (Lerbs-Mache, 1993) that likely evolved from the mitochondrial T3/T7 RNA polymerase (for discussion, see Liere et al., 2011). It most likely consists of only one subunit. Both NEP and PEP have their own promoters (Börner et al., 2015; Hajdukiewicz et al., 1997; Hricová et al., 2006; Liere et al., 2011). Intriguingly, activation of the PEP is controlled by the NEP (Börner et al., 2015). The NEP is hence downstream of the PEP in the chain that leads to transcriptional activity in the plastid. Not only is transcription of the genes that code for the PEP under control of the nucleus but also is the activity of the PEP itself.

The PEP is part of the 'transcription subdomain' of the plastid's transcriptionally active chromosome (pTAC; for an overview, see Pfalz & Pfannschmidt, 2013). In *Arabidopsis*, 12 additional key PEP-interacting proteins have been characterized (Arsova et al., 2010; Gao et al., 2011; Garcia et al., 2008; Myouga et al., 2008; Pfannschmidt et al., 2015; Pfalz, Liere, Kandlbinder, Dietz, & Oelmüller, 2006; Steiner, Schröter, Pfalz, & Pfannschmidt, 2011; Yagi, Ishizaki, Nakahira, Tozawa, & Shiina, 2012). These proteins have thus been coined PEP-associated proteins (PAPs). Disruption of any of these 12 PAPs results in lower activity of the PEP and *pap* mutant plants phenocopy the *rpoA*, *rpoB*, or *rpoC$_1$* mutants (Allison, Simon, & Maliga, 1996; De Santis-Maciossek et al., 1999; reviewed by Pfalz & Pfannschmidt, 2013). This highlights the fact that the plastid's core set discussed previously in the context of plastid portability and genetic autonomy has now been put under control of the nucleus. While the mechanisms behind plastid differentiation are still being dissected, it is obvious that they are linked to, e.g., transcriptional control (see discussion in Liebers et al., 2017).

Plastid transcriptional control sets land plants apart from all other photosynthetic eukaryotes known (de Vries et al., 2016). Could there be a reason that these specific alterations in the plastid occurred in the lineage that eventually gave rise to land plants? We recently proposed that it is linked to an increased need for coordination of plastid–nucleus communication (de Vries et al., 2016). We observed a change in plastid genome coding capacity involving specific proteins such as *ycf1* (de Vries et al., 2017, 2016), pointing to a change in the plastid biology of streptophytes. We suggested that among the factors that made the streptophyte algal ancestor to land plants successful was the proper integration of plastid-derived signals into a stress response (cf. de Vries et al., 2016).

Dry land is a high-stress environment for any photosynthetic organism. Terrestrial algae come from various lineages and have equally various strategies for dealing with these stressors (for an overview, see Holzinger & Karsten, 2013; Holzinger & Pichrtová, 2016; Raven & Edwards, 2014). If we now revisit the CoRR hypothesis, we can ask the question of whether there might be limitations to what a plastid can do with regard to its response to severe environmental stress. It might be that CoRR-based regulation is sufficient for most environments, but that very harsh conditions, such as terrestrial environments, pose a challenge that the CoRR alone cannot meet. These challenges are dealt with using various photoprotection mechanisms (see, e.g. Holzinger & Pichrtová, 2016; Karsten & Holzinger, 2014). Yet, the ancestors that gave rise to the land plant lineage (and land plants themselves) were obviously highly successful in dealing with these abiotic factors; only they 'rose above their substrate' (see full discussions in Becker & Marin, 2009 and de Vries & Archibald, 2017b). The reason(s) for this might be tied to the evolution of sophisticated ways of regulating plastid function beyond those associated with the CoRR hypothesis. As mentioned above, the nuclear-encoded proteins involved in regulation of transcriptional activity of the plastid are necessary for proper differentiation of the plastid in land plants (cf. Börner et al., 2015; Liebers et al., 2017; Pfannschmidt et al., 2015). Further, one of the few proteins that is only transcribed by the NEP is the *rpoB* gene that codes for the β subunit of the PEP (Börner et al., 2015; Zhelyazkova et al., 2012). What this means is that without the activity of these nuclear-encoded proteins, none of the CoRR-defined proteins can act. This represents an absolute level of nuclear control over plastid function, as it overrides the actions that could be carried out by in situ plastid regulatory processes such as those outlined by the CoRR hypothesis.

One possible future direction for research on nuclear control over plastid function is to address to what degree operational signalling (acting during ongoing photochemistry) and biogenic signalling (during differentiation, most famously the proplastid-to-chloroplast transition) are intertwined with their evolutionary history (for recent review and more information on operational and biogenic signalling, see Chan, Phua, Crisp, McQuinn, & Pogson, 2016). It is conceivable that factors that are now involved in biogenic signalling were initially involved in the fine-tuning of the plastid during stress—hence operational signalling. Regardless of the implications for plastid biogenic signalling, it is easily conceivable that a fine-tuned operational plastid signalling was pivotal for dealing with terrestrial stressors and, hence, plant terrestrialization. Dealing with these stressors must have been the earliest, most critical, steps in land plant evolution. The degree to which tissue-dependent differentiation might have played a role in this depends on the complexity of the thallus of the first plants. Yet, all current data suggest that it was rather simple (recently galvanized in Delwiche & Cooper, 2015). Hence, differentiation of plastid types based on tissue types likely arose later. Indeed, from the relatively early-diverging mosses we know that they bear chloroplasts even in their spores (Reski, 1998).

Research on photosynthetic organisms is focused to a very large degree on land plants. Therefore, much of what we know about plastid–nucleus communication might be biased, leading us to think that land plants are exceptional. The reality is that no matter which photosynthetic lineage we consider, the genes for the vast majority of proteins required for plastid function are found in the nucleus. Therefore, it is likely that most lineages can coordinate nuclear gene expression of plastid-targeted proteins based on the needs of the organelle. Indeed, research on *Chlamydomonas* has shown that some of the retrograde signalling factors known from land plants also function in green algae (von Gromoff, Alawady, Meinecke, Grimm, & Beck, 2008). At the present time, little is known about retrograde signalling in red algae and glaucophytes, especially in light of key differences in the composition of classical photosynthesis-associated nuclear gene expression, which is to a large degree made up of components of the light-harvesting complex (LHC) (for review, see Chan et al., 2016; Nott, Jung, Koussevitzky, & Chory, 2006). Since glaucophytes and rhodophytes have a phycobilisome-based antenna system (instead of LHC-based), many of the proteins for which are encoded on their plastid genome (Allen et al., 2011; Ohta et al., 2003; Price et al., 2012; Stirewalt, Michalowski, Löffelhardt, Bohnert, & Bryant, 1995), one might expect key differences in their plastid–nucleus communication as it revolves around antenna protein biosynthesis.

7. OUTLOOK

Most research on the biology of plastids has been focused on the detailed characterization of a very limited set of green plastids and, in particular, those of land plants. From such a green-centric point of view, one might get the impression that plastids are all more or less equally reduced. They are not. There can be more than twofold differences in the protein-coding gene content of plastid genomes between the green (Chloroplastida) and red (Rhodophyta) lineages, including the red- and green-derived secondary plastids. In this chapter, we have discussed some of the implications these differences might have with regard to organelle genetic semiautonomy and the concomitant influence on in situ control over plastid function. There is also much to be learned from the study of genes found in some plastid genomes but not others; combined with knowledge of an organism's environment, such studies can help us better understand the evolutionary forces that influence a genes subcellular location. By learning what a plastid can do on its own, we can open up new avenues of biotechnological research: combining an extended genetic repertoire for in situ responses with sophisticated nuclear control mechanisms might be a fresh approach to plastid engineering for novel photophysiological properties.

ACKNOWLEDGEMENTS

J.d.V. gratefully acknowledges support through a German Research Foundation (DFG) Research Fellowship (VR 132/1-1). J.M.A. is a Senior Fellow of the Canadian Institute for Advanced Research and thanks the Natural Sciences and Engineering Research Council of Canada for support in the form of a Discovery grant. We thank two reviewers for the helpful comments.

REFERENCES

Adl, S. M., Simpson, A. G. B., Lane, C. E., Lukeš, J., Bass, D., Bowser, S. S., et al. (2012). The revised classification of eukaryotes. *Journal of Eukaryotic Microbiology, 59*, 429–514.

Allen, J. F. (1993). Control of gene expression by redox potential and the requirement for chloroplast and mitochondrial genomes. *Journal of Theoretical Biology, 165*, 609–631.

Allen, J. F. (2015). Why chloroplasts and mitochondria retain their own genomes and genetic systems: Colocation for redox regulation of gene expression. *Proceedings of the National Academy of Sciences of the United States of America, 112*, 10231–10238.

Allen, J. F., de Paula, W. B. M., Puthiyaveetil, S., & Nield, J. (2011). A structural phylogenetic map for chloroplast photosynthesis. *Trends in Plant Science, 16*, 645–655.

Allen, J. F., & Martin, W. F. (2016). Why have organelles retained genomes? *Cell Systems, 2*, 70–72.

Allison, L. A., Simon, L. D., & Maliga, P. (1996). Deletion of *rpoB* reveals a second distinct transcription system in plastids of higher plants. *The EMBO Journal, 15*, 2802–2809.

Archibald, J. M. (2015a). Endosymbiosis and eukaryotic cell evolution. *Current Biology, 25*, R911–R921.

Archibald, J. M. (2015b). Genomic perspectives on the birth and spread of plastids. *Proceedings of the National Academy of Sciences of the United States of America, 112*, 10147–10153.

Arsova, B., Hoja, U., Wimmelbacher, M., Greiner, E., Üstün, S., Melzer, M., et al. (2010). Plastidial thioredoxin *z* interacts with two fructokinase-like proteins in a thiol-dependent manner: Evidence for an essential role in chloroplast development in *Arabidopsis* and *Nicotiana benthamiana*. *The Plant Cell, 22*, 1498–1515.

Azevedo, J., Courtois, F., Hakimi, M.-A., Demarsy, E., Lagrange, T., Alcaraz, J.-P., et al. (2008). Intraplastidial trafficking of a phage-type RNA polymerase is mediated by a thylakoid RING-H2 protein. *Proceedings of the National Academy of Sciences of the United States of America, 105*, 9123–9128.

Becker, B., & Marin, B. (2009). Streptophyte algae and the origin of embryophytes. *Annals of Botany, 103*, 999–1004.

Bengtson, S., Sallstedt, T., Belivanova, V., & Whitehouse, M. (2017). Three-dimensional preservation of cellular and subcellular structures suggests 1.6 billion-year-old crown-group red algae. *PLoS Biology, 15*, e2000735.

Bhattacharya, D., Pelletreau, K. N., Price, D. C., Sarver, K. E., & Rumpho, M. E. (2013). Genome analysis of *Elysia chlorotica* egg DNA provides no evidence for horizontal gene transfer into the germ line of this kleptoplastic mollusc. *Molecular Biology and Evolution, 30*, 1843–1852.

Björkholm, P., Ernst, A. M., Hagström, E., & Andersson, S. G. E. (2017). Why mitochondria need a genome revisited. *FEBS Letters*, *591*, 65–75.

Bodył, A. (2017). Did some red alga-derived plastids evolve via kleptoplastidy? A hypothesis. *Biological Reviews*, in press [Epub ahead of print], https://doi.org/10.1111/brv.12340.

Börner, T., Aleynikova, A. Y., Zubo, Y. O., & Kusnetsov, V. V. (2015). Chloroplast RNA polymerases: Role in chloroplast biogenesis. *Biochimica et Biophysica Acta—Bioenergetics*, *1847*, 761–769.

Browning, T. J., Achterberg, E. P., Yong, J. C., Rapp, I., Utermann, C., Engel, A., et al. (2017). Iron limitation of microbial phosphorus acquisition in the tropical North Atlantic. *Nature Communications*, *8*, 15465.

Bryant, D. A., & Frigaard, N.-U. (2006). Prokaryotic photosynthesis and phototrophy illuminated. *Trends in Microbiology*, *14*, 488–496.

Burger, G., Gray, M. W., Forget, L., & Lang, F. B. (2013). Strikingly bacteria-like and gene-rich mitochondrial genomes throughout jakobid protists. *Genome Biology and Evolution*, *5*, 418–438.

Butterfield, N. J. (2000). *Bangiomorpha pubescens* n. gen., n. sp.: Implications for the evolution of sex, multicellularity, and the Mesoproterozoic/Neoproterozoic radiation of eukaryotes. *Paleobiology*, *26*, 386–404.

Cavalier-Smith, T. (1982). The origins of plastids. *Biological Journal of the Linnean Society*, *17*, 289–306.

Cavalier-Smith, T., & Lee, J. J. (1985). Protozoa as hosts for endosymbioses and the conversion of symbionts into organelles. *Journal of Protozoology*, *32*, 376–379.

Cermakian, N., Ikeda, T. M., Cedergren, R., & Gray, M. W. (1996). Sequences homologous to yeast mitochondrial and bacteriophage T3 and T7 RNA polymerases are widespread throughout the eukaryotic lineage. *Nucleic Acids Research*, *24*, 648–654.

Chan, K. X., Phua, S. Y., Crisp, P., McQuinn, R., & Pogson, B. J. (2016). Learning the languages of the chloroplast: Retrograde signaling and beyond. *Annual Review of Plant Biology*, *67*, 25–53.

Cheung, A. Y., Bogorad, L., van Montagu, M., & Schell, J. (1988). Relocating a gene for herbicide tolerance: A chloroplast gene is converted into a nuclear gene. *Proceedings of the National Academy of Sciences of the United States of America*, *85*, 391–395.

Christa, G., Händeler, K. H., Schäberle, T. F. S., König, G. M., & Wägele, H. (2014). Identification of sequestered chloroplasts in photosynthetic and non-photosynthetic sacoglossan sea slugs (Mollusca, Gastropoda). *Frontiers in Zoology*, *11*, 15.

Christa, G., Wescott, L., Schäberle, T. F., König, G. M., & Wägele, H. (2013). What remains after 2 months of starvation? Analysis of sequestered algae in a photosynthetic slug, *Plakobranchus ocellatus* (Sacoglossa, Opisthobranchia), by barcoding. *Planta*, *237*, 559–572.

Civáň, P., Foster, P. G., Embley, M. T., Séneca, A., & Cox, C. J. (2014). Analyses of charophyte chloroplast genomes help characterize the ancestral chloroplast genome of land plants. *Genome Biology and Evolution*, *6*, 897–911.

Couradeau, E., Benzerara, K., Gérard, E., Moreira, D., Bernard, S., Brown, G. E., et al. (2012). An early-branching microbialite cyanobacterium forms intracellular carbonates. *Science*, *336*, 459–462.

Curtis, B. A., Tanifuji, G., Burki, F., Gruber, A., Irimia, M., Maruyama, S., et al. (2012). Algal genomes reveal evolutionary mosaicism and the fate of nucleomorphs. *Nature*, *492*, 59–65.

Dagan, T., Roettger, M., Stucken, K., Landan, G., Koch, R., Major, P., et al. (2013). Genomes of stigonematalean cyanobacteria (subsection V) and the evolution of oxygenic photosynthesis from prokaryotes to plastids. *Genome Biology and Evolution*, *5*, 31–44.

de Alda, J. A. G. O., Esteban, R., Diago, M. L., & Houmard, J. (2014). The plastid ancestor originated among one of the major cyanobacterial lineages. *Nature Communications*, *5*, 4937.

Delwiche, C. F., & Cooper, E. D. (2015). The evolutionary origin of a terrestrial flora. *Current Biology*, *25*, R899–R910.
De Santis-Maciossek, G., Kofer, W., Bock, A., Schoch, S., Maier, R. M., Wanner, G., et al. (1999). Targeted disruption of the plastid RNA polymerase genes *rpoA*, *B* and *C1*: Molecular biology, biochemistry and ultrastructure. *The Plant Journal*, *18*, 477–489.
Deusch, O., Landan, G., Roettger, M., Gruenheit, N., Kowallik, K. V., Allen, J. F., et al. (2008). Genes of cyanobacterial origin in plant nuclear genomes point to a heterocyst-forming plastid ancestor. *Molecular Biology and Evolution*, *25*, 748–761.
de Vries, J., & Archibald, J. M. (2017a). Endosymbiosis: Did plastids evolve from a freshwater cyanobacterium? *Current Biology*, *27*, R103–R105.
de Vries, J., & Archibald, J. M. (2017b). Plant evolution: Landmarks on the path to terrestrial life. *New Phytologist*, in press.
de Vries, J., Archibald, J. M., & Gould, S. B. (2017). The carboxy terminus of YCF1 contains a motif conserved throughout >500 million years of streptophyte evolution. *Genome Biology and Evolution*, *9*, 473–479.
de Vries, J., Christa, G., & Gould, S. B. (2014). Plastid survival in the cytosol of animal cells. *Trends in Plant Science*, *19*, 347–350.
de Vries, J., & Gould, S. B. (2017). The monoplastidic bottleneck in algae and plant evolution. *Journal of Cell Science*, in press [Epub ahead of print], https://doi.org/10.1242/jcs.203414.
de Vries, J., Habicht, J., Woehle, C., Huang, C., Christa, G., Wägele, H., et al. (2013). Is *ftsH* the key to plastid longevity in sacoglossan slugs? *Genome Biology and Evolution*, *5*, 2540–2548.
de Vries, J., Rauch, C., Christa, G., & Gould, S. B. (2014). A sea slug's guide to plastid symbiosis. *Acta Societatis Botanicorum Poloniae*, *83*, 415–421.
de Vries, J., Stanton, A., Archibald, J. M., & Gould, S. B. (2016). Streptophyte terrestrialization in light of plastid evolution. *Trends in Plant Science*, *21*, 467–476.
Dorrell, R. G., & Howe, C. J. (2012). What makes a chloroplast? Reconstructing the establishment of photosynthetic symbioses. *Journal of Cell Science*, *125*, 1865–1875.
Dorrell, R. G., & Howe, C. J. (2015). Integration of plastids with their hosts: Lessons learned from dinoflagellates. *Proceedings of the National Academy of Sciences of the United States of America*, *112*, 10247–10254.
Douglas, S. E., & Penny, S. L. (1999). The plastid genome of the cryptophyte alga, *Guillardia theta*: Complete sequence and conserved synteny groups confirm its common ancestry with red algae. *Journal of Molecular Evolution*, *48*, 236–244.
Eme, L., Sharpe, S. C., Brown, M. W., & Roger, A. J. (2014). On the age of eukaryotes: Evaluating evidence from fossils and molecular clocks. *Cold Spring Harbor Perspectives in Biology*, *6*, a016139.
Fathinejad, S., Steiner, J. M., Reipert, S., Marchetti, M., Allmaier, G., Burey, S. C., et al. (2008). A carboxysomal carbon-concentrating mechanism in the cyanelles of the 'coelacanth' of the algal world, *Cyanophora paradoxa*? *Physiologia Plantarum*, *133*, 27–32.
Figueroa-Martinez, F., Nedelcu, A. M., Smith, D. R., & Reyes-Prieto, A. (2015). When the lights go out: The evolutionary fate of free-living colorless green algae. *New Phytologist*, *206*, 972–982.
Gao, Z. P., Yu, Q. B., Zhao, T. T., Ma, Q., Chen, G. X., & Yang, Z. N. (2011). A functional component of the transcriptionally active chromosome complex, Arabidopsis pTAC14, interacts with pTAC12/HEMERA and regulates plastid gene expression. *Plant Physiology*, *157*, 1733–1745.
Garcia, M., Myouga, F., Takechi, K., Sato, H., Nabeshima, K., Nagata, N., et al. (2008). An *Arabidopsis* homolog of the bacterial peptidoglycan synthesis enzyme MurE has an essential role in chloroplast development. *The Plant Journal*, *53*, 924–934.

Gould, S. B., Maier, U. G., & Martin, W. F. (2015). Protein import and the origin of red complex plastids. *Current Biology*, *25*, R515–R521.
Green, B. R. (2011). Chloroplast genomes of photosynthetic eukaryotes. *The Plant Journal*, *66*, 34–44.
Green, B. J., Fox, T. C., & Rumpho, M. E. (2005). Stability of isolated algal chloroplasts that participate in a unique mollusc/kleptoplast association. *Symbiosis*, *40*, 31–40.
Grosche, C., & Rensing, S. A. (2017). Three rings for the evolution of plastid shape: A tale of land plant FtsZ. *Protoplasma*, *254*, 1879–1885.
Grzebyk, D., Schofield, O., Vetriani, C., & Falkowski, P. G. (2003). The mesozoic radiation of eukaryotic algae: The portable plastid hypothesis. *Journal of Phycology*, *39*, 259–267.
Hajdukiewicz, P. T., Allison, L. A., & Maliga, P. (1997). The two RNA polymerases encoded by the nuclear and the plastid compartments transcribe distinct groups of genes in tobacco plastids. *The EMBO Journal*, *16*, 4041–4048.
Hanaoka, M., Kanamaru, K., Takahashi, H., & Tanaka, K. (2003). Molecular genetic analysis of chloroplast gene promoters dependent on SIG2, a nucleus-encoded sigma factor for the plastid-encoded RNA polymerase, in *Arabidopsis thaliana*. *Nucleic Acids Research*, *31*, 7090–7098.
Henze, K., & Martin, W. (2001). How do mitochondrial genes get into the nucleus? *Trends in Genetics*, *17*, 383–387.
Hirano, T., Tanidokoro, K., Shimizu, Y., Kawarabayasi, Y., Ohshima, T., Sato, M., et al. (2016). Moss chloroplasts are surrounded by a peptidoglycan wall containing D-amino acids. *The Plant Cell*, *28*, 1521–1532.
Hohmann-Marriott, M. F., & Blankenship, R. E. (2011). Evolution of photosynthesis. *Annual Review of Plant Biology*, *62*, 515–548.
Holzinger, A., & Karsten, U. (2013). Desiccation stress and tolerance in green algae: Consequences for ultrastructure, physiological, and molecular mechanisms. *Frontiers in Plant Science*, **4**, 327.
Holzinger, A., & Pichrtová, M. (2016). Abiotic stress tolerance of charophyte green algae: New challenges for omics techniques. *Frontiers in Plant Science*, 7, 678.
Hricová, A., Quesada, V., & Micol, J. L. (2006). The *SCABRA3* nuclear gene encodes the plastid RpoTp RNA polymerase, which is required for chloroplast biogenesis and mesophyll cell proliferation in Arabidopsis. *Plant Physiology*, *141*, 942–956.
Huang, C. Y., Ayliffe, M. A., & Timmis, J. N. (2003). Direct measurement of the transfer rate of chloroplast DNA into the nucleus. *Nature*, *422*, 72–76.
Izumi, Y., Kuroki, J., Nagafuji, H., Lin, X., & Takano, H. (2008). Effects of antibiotics that inhibit bacterial peptidoglycan synthesis on plastid division in pteridophytes. *Cytologia*, *73*, 393–400.
Izumi, Y., Ono, K., & Takano, H. (2003). Inhibition of plastid division by ampicillin in the pteridophyte *Selaginella nipponica* Fr. et Sav. *Plant and Cell Physiology*, *44*, 183–189.
Jackson, C., Clayden, S., & Reyes-Prieto, A. (2015). The Glaucophyta: The blue-green plants in a nutshell. *Acta Societatis Botanicorum Poloniae*, *84*, 149–165.
Jackson, C., & Reyes-Prieto, A. (2014). The mitochondrial genomes of the glaucophytes *Gloeochaete wittrockiana* and *Cyanoptyche gloeocystis*: Multilocus phylogenetics suggests a monophyletic archaeplastida. *Genome Biology and Evolution*, *6*, 2774–2785.
Janska, H., Kwasniak, M., & Szczepanowska, J. (2013). Protein quality control in organelles—AAA/FtsH story. *Biochimica et Biophysica Acta—Molecular Cell Research*, *1833*, 381–387.
Jarvis, P., & López-Juez, E. (2013). Biogenesis and homeostasis of chloroplasts and other plastids. *Nature Reviews Molecular Cell Biology*, *14*, 787–802.
Jauffrais, T., Jesus, B., Metzger, E., Mouget, J.-L., Jorissen, F., & Geslin, E. (2016). Effect of light on photosynthetic efficiency of sequestered chloroplasts in intertidal benthic foraminifera (*Haynesina germanica* and *Ammonia tepida*). *Biogeosciences*, *13*, 2715–2726.

Johnson, M. D. (2011). Acquired phototrophy in ciliates: A review of cellular interactions and structural adaptations. *Journal of Eukaryotic Microbiology*, *58*, 185–195.
Johnson, M. D., Oldach, D., Delwiche, C. F., & Stoecker, D. K. (2007). Retention of transcriptionally active cryptophyte nuclei by the ciliate *Myrionecta rubra*. *Nature*, *445*, 426–428.
Karsten, U., & Holzinger, A. (2014). Green algae in alpine biological soil crust communities: Acclimation strategies against ultraviolet radiation and dehydration. *Biodiversity and Conservation*, *23*, 1845–1858.
Kasten, B., & Reski, R. (1997). β-Lactam antibiotics inhibit chloroplast division in a moss (*Physcomitrella patens*) but not in tomato (*Lycopersicon esculentum*). *Journal of Plant Physiology*, *150*, 137–140.
Kato, Y., Miura, E., Ido, K., Ifuku, K., & Sakamoto, W. (2009). The variegated mutants lacking chloroplastic FtsHs are defective in D1 degradation and accumulate reactive oxygen species. *Plant Physiology*, *151*, 1790–1801.
Keeling, P. J. (2013). The number, speed, and impact of plastid endosymbioses in eukaryotic evolution. *Annual Review of Plant Biology*, *64*, 583–607.
Keeling, P. J., Burger, G., Durnford, D. G., Lang, B. F., Lee, R. W., Pearlman, R. E., et al. (2005). The tree of eukaryotes. *Trends in Ecology & Evolution*, *20*, 670–676.
Kindgren, P., & Strand, Å. (2015). Chloroplast transcription, untangling the Gordian Knot. *New Phytologist*, *206*, 889–891.
Krause, K., Berg, S., & Krupinska, K. (2003). Plastid transcription in the holoparasitic plant genus *Cuscuta*: Parallel loss of the *rrn16* PEP-promoter and of the *rpoA* and *rpoB* genes coding for the plastid-encoded RNA polymerase. *Planta*, *216*, 815–823.
Ku, C., Nelson-Sathi, S., Roettger, M., Sousa, F. L., Lockhart, P. J., Bryant, D., et al. (2015). Endosymbiotic origin and differential loss of eukaryotic genes. *Nature*, *524*, 427–432.
Lang, F. B., Burger, G., O'Kelly, C. J., Cedergren, R., Golding, G. B., Lemieux, C., et al. (1997). An ancestral mitochondrial DNA resembling a eubacterial genome in miniature. *Nature*, *387*, 493–497.
Larkum, A. W. D., Lockhart, P. J., & Howe, C. J. (2007). Shopping for plastids. *Trends in Plant Science*, *12*, 189–195.
Larsson, J., Nylander, J. A., & Bergman, B. (2011). Genome fluctuations in cyanobacteria reflect evolutionary, developmental and adaptive traits. *BMC Evolutionary Biology*, *11*, 187.
Lerbs-Mache, S. (1993). The 110-kDa polypeptide of spinach plastid DNA-dependent RNA polymerase: Single-subunit enzyme or catalytic core of multimeric enzyme complexes? *Proceedings of the National Academy of Sciences of the United States of America*, *90*, 5509–5513.
Liebers, M., Grübler, B. R., Chevalier, F., Lerbs-Mache, S., Merendino, L., Blanvillain, R., et al. (2017). Regulatory shifts in plastid transcription play a key role in morphological conversions of plastids during plant development. *Frontiers in Plant Science*, *8*, 23.
Liere, K., Kaden, D., Maliga, P., & Börner, T. (2004). Overexpression of phage-type RNA polymerase RpoTp in tobacco demonstrates its role in chloroplast transcription by recognizing a distinct promoter type. *Nucleic Acids Research*, *32*, 1159–1165.
Liere, K., Weihe, A., & Börner, T. (2011). The transcription machineries of plant mitochondria and chloroplasts: Composition, function, and regulation. *Journal of Plant Physiology*, *168*, 1345–1360.
Lindahl, M., Spetea, C., Hundal, T., Oppenheim, A. B., Adam, Z., & Andersson, B. (2000). The thylakoid FtsH protease plays a role in the light-induced turnover of the photosystem II D1 protein. *The Plant Cell*, *12*, 419–431.
Lommer, M., Roy, A.-S., Schilhabel, M., Schreiber, S., Rosenstiel, P., & LaRoche, J. (2010). Recent transfer of an iron-regulated gene from the plastid to the nuclear genome in an oceanic diatom adapted to chronic iron limitation. *BMC Genomics*, *11*, 718.
Mackiewicz, P., & Gagat, P. (2014). Monophyly of Archaeplastida supergroup and relationships among its lineages in the light of phylogenetic and phylogenomic studies. Are we close to a consensus? *Acta Societatis Botanicorum Poloniae*, *83*, 263–280.

Maier, U. G., Zauner, S., Woehle, C., Bolte, K., Hempel, F., Allen, J. F., et al. (2013). Massively convergent evolution for ribosomal protein gene content in plastid and mitochondrial genomes. *Genome Biology and Evolution, 5*, 2318–2329.

Martin, W., Brinkmann, H., Savonna, C., & Cerff, R. (1993). Evidence for a chimeric nature of nuclear genomes: Eubacterial origin of eukaryotic glyceraldehyde-3-phosphate dehydrogenase genes. *Proceedings of the National Academy of Sciences of the United States of America, 90*, 8692–8696.

Martin, W., & Herrmann, R. G. (1998). Gene transfer from organelles to the nucleus: How much, what happens, and why? *Plant Physiology, 118*, 9–17.

Martin, W., Rujan, T., Richly, E., Hansen, A., Cornelsen, S., Lins, T., et al. (2002). Evolutionary analysis of *Arabidopsis*, cyanobacterial, and chloroplast genomes reveals plastid phylogeny and thousands of cyanobacterial genes in the nucleus. *Proceedings of the National Academy of Sciences of the United States of America, 99*, 12246–12251.

Martin, W., Stoebe, B., Goremykin, V., Hansmann, S., Hasegawa, M., & Kowallik, K. V. (1998). Gene transfer to the nucleus and the evolution of chloroplasts. *Nature, 393*, 162–165.

Matsumoto, H., Takechi, K., Sato, H., Takio, S., & Takano, H. (2012). Treatment with antibiotics that interfere with peptidoglycan biosynthesis inhibits chloroplast division in the desmid *Closterium. PLoS One, 7*, e40734.

Méheust, R., Zelzion, E., Bhattacharya, D., Lopez, P., & Bapteste, E. (2016). Protein networks identify novel symbiogenetic genes resulting from plastid endosymbiosis. *Proceedings of the National Academy of Sciences of the United States of America, 113*, 3579–3584.

Mereschkowsky, C. (1905). Über Natur und Ursprung der Chromatophoren im Pflanzenreiche. *Biologisches Centralblatt, 25*, 593–604.

Miyagishima, S.-Y., Nakamura, M., Uzuka, A., & Era, A. (2014). FtsZ-less prokaryotic cell division as well as FtsZ- and dynamin-less chloroplast and non-photosynthetic plastid division. *Frontiers in Plant Science, 5*, 459.

Molina, J., Hazzouri, K. M., Nickrent, D., Geisler, M., Meyer, R. S., Pentony, M. M., et al. (2014). Possible loss of the chloroplast genome in the parasitic flowering plant *Rafflesia lagascae* (Rafflesiaceae). *Molecular Biology and Evolution, 31*, 793–803.

Myouga, F., Hosoda, C., Umezawa, T., Iizumi, H., Kuromori, T., Motohashi, R., et al. (2008). A heterocomplex of iron superoxide dismutases defends chloroplast nucleoids against oxidative stress and is essential for chloroplast development in *Arabidopsis. The Plant Cell, 20*, 3148–3162.

Nickelsen, J., & Rengstl, B. (2013). Photosystem II assembly: From cyanobacteria to plants. *Annual Review of Plant Biology, 64*, 609–635.

Nixon, P. J., Michoux, F., Yu, J., Boehm, M., & Komenda, J. (2010). Recent advances in understanding the assembly and repair of photosystem II. *Annals of Botany, 106*, 1–16.

Nott, A., Jung, H.-S., Koussevitzky, S., & Chory, J. (2006). Plastid-to-nucleus retrograde signaling. *Annual Review of Plant Biology, 57*, 739–759.

Ohta, N., Matsuzaki, M., Misumi, O., Miyagishima, S.-Y., Nozaki, H., Tanaka, K., et al. (2003). Complete sequence and analysis of the plastid genome of the unicellular red alga *Cyanidioschyzon merolae. DNA Research, 10*, 67–77.

Ohyama, K., Fukuzawa, H., Kohchi, T., Shirai, H., Sano, T., Sano, S., et al. (1986). Chloroplast gene organization deduced from complete sequence of liverwort *Marchantia polymorpha* chloroplast DNA. *Nature, 322*, 572–574.

Parfrey, L. W., Lahr, D. J. G., Knoll, A. H., & Katz, L. A. (2011). Estimating the timing of early eukaryotic diversification with multigene molecular clocks. *Proceedings of the National Academy of Sciences of the United States of America, 108*, 13629.

Pfalz, J., Liere, K., Kandlbinder, A., Dietz, K.-J., & Oelmüller, R. (2006). pTAC2, -6, and -12 are components of the transcriptionally active plastid chromosome that are required for plastid gene expression. *The Plant Cell, 18*, 176–197.

Pfalz, J., & Pfannschmidt, T. (2013). Essential nucleoid proteins in early chloroplast development. *Trends in Plant Science*, *18*, 186–194.

Pfannschmidt, T., Blanvillain, R., Merendino, L., Courtois, F., Chevalier, F., Liebers, M., et al. (2015). Plastid RNA polymerases: Orchestration of enzymes with different evolutionary origins controls chloroplast biogenesis during the plant life cycle. *Journal of Experimental Botany*, *66*, 6957–6973.

Pfannschmidt, T., Nilsson, A., & Allen, J. F. (1999). Photosynthetic control of chloroplast gene expression. *Nature*, *397*, 625–628.

Pfannschmidt, T., Nilsson, A., Tullberg, A., Link, G., & Allen, J. F. (1999). Direct transcriptional control of the chloroplast genes *psbA* and *psaAB* adjusts photosynthesis to light energy distribution in plants. *IUBMB Life*, *48*, 271–276.

Pillet, L., de Vargas, C., & Pawlowski, J. (2011). Molecular identification of sequestered diatom chloroplasts and kleptoplastidy in foraminifera. *Protist*, *162*, 394–404.

Pillet, L., & Pawlowski, J. (2013). Transcriptome analysis of foraminiferan *Elphidium margaritaceum* questions the role of gene transfer in kleptoplastidy. *Molecular Biology and Evolution*, *30*, 66–69.

Ponce-Toledo, R. I., Deschamps, P., Lopez-Garcia, P., Zivanovic, Y., Benzerara, K., & Moreira, D. (2017). An early-branching freshwater cyanobacterium at the origin of plastids. *Current Biology*, *27*, 386–391.

Price, D. C., Chan, C. X., Yoon, H. S., Yang, E. C., Qiu, H., Weber, A. P. M., et al. (2012). *Cyanophora paradoxa* genome elucidates origin of photosynthesis in algae and plants. *Science*, *335*, 843–847.

Puthiyaveetil, S., Ibrahim, I. M., Jeličić, B., Tomašic, A., Fulgosi, H., & Allen, J. F. (2010). Transcriptional control of photosynthesis genes: The evolutionarily conserved regulatory mechanism in plastid genome function. *Genome Biology and Evolution*, *2*, 888–896.

Puthiyaveetil, S., Kavanagh, T. A., Cain, P., Sullivan, J. A., Newell, C. A., Gray, J. C., et al. (2008). The ancestral symbiont sensor kinase CSK links photosynthesis with gene expression in chloroplasts. *Proceedings of the National Academy of Sciences of the United States of America*, *105*, 10061–10066.

Qiu, H., Lee, J. M., Yoon, H. S., & Bhattacharya, D. (2017). Hypothesis: Gene-rich plastid genomes in red algae may be an outcome of nuclear genome reduction. *Journal of Phycology*, *53*, 715–719.

Rauch, C., de Vries, J., Rommel, S., Rose, L. E., Woehle, C., Christa, G., et al. (2015). Why it is time to look beyond algal genes in photosynthetic slugs. *Genome Biology and Evolution*, *7*, 2602–2607.

Raven, J. A., & Edwards, D. (2014). Photosynthesis in early land plants: Adapting to the terrestrial environment. In D. Hanson & S. Rice (Eds.), *Advances in photosynthesis and respiration (including bioenergy and related processes), Vol 37 Photosynthesis in bryophytes and early land plants*. Dordrecht: Springer.

Reski, R. (1998). Development, genetics and molecular biology of mosses. *Botanica Acta*, *111*, 1–15.

Ricchetti, M., Fairhead, C., & Dujon, B. (1999). Mitochondrial DNA repairs double-strand breaks in yeast chromosomes. *Nature*, *402*, 96–100.

Richly, E., & Leister, D. (2004). NUPTs in sequenced eukaryotes and their genomic organization in relation to NUMTs. *Molecular Biology and Evolution*, *21*, 1972–1980.

Rodríguez-Ezpeleta, N., Brinkmann, H., Burey, S. C., Roure, B., Burger, G., Löffelhardt, W., et al. (2005). Monophyly of primary photosynthetic eukaryotes: Green plants, red algae, and glaucophytes. *Current Biology*, *15*, 1325–1330.

Rogers, M. B., Gilson, P. R., Su, V., McFadden, G. I., & Keeling, P. J. (2007). The complete chloroplast genome of the chlorarachniophyte *Bigelowiella natans*: Evidence for independent origins of chlorarachniophyte and euglenid secondary endosymbionts. *Molecular Biology and Evolution*, *24*, 54–62.

Rumpho, M. E., Pelletreau, K. N., Moustafa, A., & Bhattacharya, D. (2011). The making of a photosynthetic animal. *Journal of Experimental Biology*, *214*, 303–311.
Serôdio, J., Cruz, S., Cartaxana, P., & Calado, R. (2014). Photophysiology of kleptoplasts: Photosynthetic use of light by chloroplasts living in animal cells. *Philosophical Transactions of the Royal Society B: Biological Sciences*, *369*, 20130242.
Sijben-Müller, G., Hallick, R. B., Alt, J., Westhoff, P., & Herrmann, R. G. (1986). Spinach plastid genes coding for initiation factor IF-1, ribosomal protein S11 and RNA polymerase α-subunit. *Nucleic Acids Research*, *14*, 1029–1044.
Simpson, C. L., & Stern, D. B. (2002). The treasure trove of algal chloroplast genomes. Surprises in architecture and gene content, and their functional implications. *Plant Physiology*, *129*, 957–966.
Smith, D. R., Crosby, K., & Lee, R. W. (2011). Correlation between nuclear plastid DNA abundance and plastid number supports the limited transfer window hypothesis. *Genome Biology and Evolution*, *3*, 365–371.
Smith, D. R., & Lee, R. W. (2014). A plastid without a genome: Evidence from the nonphotosynthetic green algal genus *Polytomella*. *Plant Physiology*, *164*, 1812–1819.
Stegemann, S., Hartmann, S., Ruf, S., & Bock, R. (2003). High-frequency gene transfer from the chloroplast genome to the nucleus. *Proceedings of the National Academy of Sciences of the United States of America*, *100*, 8828–8833.
Steiner, J. M., & Löffelhardt, W. (2002). Protein import into cyanelles. *Trends in Plant Science*, *7*, 72–77.
Steiner, J., Ma, Y., Pfanzagl, B., & Löffelhardt, W. (2001). *Cyanophora paradoxa* as a model organism for plastid evolution. *Endocytobiosis and Cell Research*, *14*, 127–128.
Steiner, S., Schröter, Y., Pfalz, J., & Pfannschmidt, T. (2011). Identification of essential subunits in the plastid-encoded RNA polymerase complex reveals building blocks for proper plastid development. *Plant Physiology*, *157*, 1043–1055.
Stirewalt, V. L., Michalowski, C. B., Löffelhardt, W., Bohnert, H. J., & Bryant, D. A. (1995). Nucleotide sequence of the cyanelle genome from *Cyanophora paradoxa*. *Plant Molecular Biology Reporter*, *13*, 327–332.
Stoebe, B., & Kowallik, K. V. (1999). Gene-cluster analysis in chloroplast genomics. *Trends in Genetics*, *15*, 344–347.
Stoecker, D. K., Johnson, M. D., deVargas, C., & Not, F. (2009). Acquired phototrophy in aquatic protists. *Aquatic Microbial Ecology*, *57*, 279–310.
Strzepek, R. F., & Harrison, P. J. (2004). Photosynthetic architecture differs in coastal and oceanic diatoms. *Nature*, *431*, 689–692.
Suzuki, S., Hirakawa, Y., Kofuni, R., Sugita, M., & Ishida, K.-I. (2016). Plastid genome sequences of *Gymnochlora stellata*, *Lotharella vacuolata*, and *Partenskyella glossopodia* reveal remarkable structural conservation among chlorarachniophyte species. *Journal of Plant Research*, *129*, 581–590.
Timmis, J. N., Ayliffe, M. A., Huang, C. Y., & Martin, W. (2004). Endosymbiotic gene transfer: Organelle genomes forge eukaryotic chromosomes. *Nature Reviews Genetics*, *5*, 123–135.
von Gromoff, E. D., Alawady, A., Meinecke, L., Grimm, B., & Beck, C. F. (2008). Heme, a plastid-derived regulator of nuclear gene expression in *Chlamydomonas*. *The Plant Cell*, *20*, 552–567.
von Heijne, G. (1986). Why mitochondria need a genome. *FEBS Letters*, *198*, 1–4.
Wägele, H., Deusch, O., Händeler, K., Martin, R., Schmitt, V., Christa, G., et al. (2011). Transcriptomic evidence that longevity of acquired plastids in the photosynthetic slugs *Elysia timida* and *Plakobranchus ocellatus* does not entail lateral transfer of algal nuclear genes. *Molecular Biology and Evolution*, *28*, 699–706.
Wicke, S., Müller, K. F., dePamphilis, C. W., Quandt, D., Bellote, S., & Schneeweiss, G. M. (2016). Mechanistic model of evolutionary rate variation en route to a nonphotosynthetic lifestyle in plants. *Proceedings of the National Academy of Sciences of the United States of America*, *113*, 9045–9050.

Wicke, S., Schneeweiss, G. M., dePamphilis, C. W., Müller, K. F., & Quandt, D. (2011). The evolution of the plastid chromosome in land plants: Gene content, gene order, gene function. *Plant Molecular Biology*, *76*, 273–297.

Wisecaver, J. H., & Hackett, J. D. (2010). Transcriptome analysis reveals nuclear-encoded proteins for the maintenance of temporary plastids in the dinoflagellate *Dinophysis acuminata*. *BMC Genomics*, *11*, 366.

Yagi, Y., Ishizaki, Y., Nakahira, Y., Tozawa, Y., & Shiina, T. (2012). Eukaryotic-type plastid nucleoid protein pTAC3 is essential for transcription by the bacterial-type plastid RNA polymerase. *Proceedings of the National Academy of Sciences of the United States of America*, *109*, 7541–7546.

Yin, C., Richter, U., Börner, T., & Weihe, A. (2010). Evolution of plant phage-type RNA polymerases: The genome of the basal angiosperm *Nuphar advena* encodes two mitochondrial and one plastid phage-type RNA polymerases. *BMC Evolutionary Biology*, *10*, 379.

Zhelyazkova, P., Sharma, C. M., Förstner, K. U., Liere, K., Vogel, J., & Börner, T. (2012). The primary transcriptome of barley chloroplasts: Numerous noncoding RNAs and the dominating role of the plastid-encoded RNA polymerase. *The Plant Cell*, *24*, 123–136.

Zimorski, V., Ku, C., Martin, W. F., & Gould, S. B. (2014). Endosymbiotic theory for organelle origins. *Current Opinion in Microbiology*, *22*, 38–48.

CHAPTER TWO

Lost in the Light: Plastid Genome Evolution in Nonphotosynthetic Algae

David R. Smith[1]
University of Western Ontario, London, ON, Canada
[1]Corresponding author: e-mail address: dsmit242@uwo.ca

Contents

Abstract

Photosynthesis is an awe-inspiring process. It has shaped, coloured, and diversified the biological world in innumerable ways and supplies us with the air we breathe. Photosynthetic organisms are literally our lifelines on Earth. Without them we perish. Perhaps this is why many of us are uncomfortable with and confused by the concept of a photosynthetic organism forfeiting its ability to convert sunlight into chemical energy, giving up its life-sustaining powers. Indeed, the evolutionary loss of photosynthesis, which has occurred countless times throughout evolution, remains a poorly understood and underappreciated topic, both among researchers and the general public. This is unfortunate because nonphotosynthetic plants and algae represent some of the most diverse and interesting (and even deadly) species on the planet, and they can teach us a lot about photosynthesis and biology as a whole. Here, I review the origins and evolution of nonphotosynthetic eukaryotic algae. I portray these biologically "broken light bulbs" in a contemporary framework, paying particular attention to their plastid genomes, which are much more complex and architecturally varied than one might expect. If you are anything of a rebel and prefer misfits over conformists, trouble makers over the straight-laced, and mysteries over simple plotlines, then you will not be disappointed by the eclectic assemblage of algae that have relinquished their hold on the sun.

Advances in Botanical Research, Volume 85
ISSN 0065-2296
https://doi.org/10.1016/bs.abr.2017.10.001

1. INTRODUCTION

Your absence has gone through me
Like thread through a needle.
Everything I do is stitched with its color.

W.S. Merwin

When I was an undergraduate student in Biology, I detested courses on plants and algae. Human genetics, animal behaviour, disease-causing bacteria—bring it on! Anything but botany. I still cringe at the thought of having to memorize the life cycle of a fern, and the only time I perked up in an entire semester of plant physiology was when the instructor talked about *Psilocybe* species (magic mushrooms). Who would have guessed at the time that I would go on to have a career in the plant sciences, studying the genes and genomes of eukaryotic algae? Certainly not my plant phys prof, who graciously gave me a passing grade.

I was a late bloomer. It would take another 2 years and strong persuading from my eventual PhD supervisor before I finally saw the proverbial photosynthetic light and made the scientific leap to the realm of chloroplast-containing organisms. My gateway drug into this verdant domain was not what you might expect. It wasn't some beautiful, mellifluous flower or a magnificent 200-ft. redwood. It wasn't even the bright kaleidoscopic colours of chlorophyll that first swayed me. It was something more drab and faded, and went by the name *Polytomella*.

At the first meeting with my prospective PhD supervisor, Robert Lee, he led me into the hallway outside of his cluttered office and pointed enthusiastically to a four-by-four-foot poster on the wall, which described an obscure green alga called *Polytomella*. "Have you ever heard of this critter?" asked Bob, tapping his hand against the poster. I hadn't. "That's a shame, because it is one awesome little unicell," he exclaimed. "It's free living has four flagella and a plastid, but lacks chlorophyll and can no longer derive energy from sunlight. In other words, it's a photosynthetic burnout, a green alga that isn't even green." That was my introduction to the world of non-photosynthetic algae. Being a bit of a burnout myself, I was immediately hooked and itching with curiosity.

How did achromatic algae evolve and how do they survive? Why do they lug around a plastid (the epicentre of photosynthesis) if they're non-photosynthetic? Are there different types of colourless algae, or is *Polytomella* the only one? Have certain land plants also lost photosynthetic capabilities?

Why, in Darwin's name, did I not hear anything about this in my undergraduate biology courses? And where do I sign up to start researching these organisms? Soon, I would have even more questions as I trudged through a 5-year PhD on the organelle genetics of *Polytomella* and its close relatives. I would quickly come to realize that nonphotosynthetic algae and land plants are surprisingly diverse and among the most intriguing and enigmatic species on the tree of life.

The forfeiting of photosynthesis has occurred numerous times and in disparate lineages throughout eukaryotic evolution (Blouin & Lane, 2012; Figueroa-Martinez, Nedelcu, Smith, & Reyes-Prieto, 2015; Keeling, 2013; Krause, 2008). Wherever you find photosynthesis, you will also find examples of its loss (Keeling, 2013). Nonphotosynthetic plastid-bearing species can be found in almost every kind of environment and ecosystem. They can be mind bogglingly beautiful or downright ugly, abundant or scarce, benign or deadly. Some are prolific predators, others are peaceful osmotrophs, and many are terrifying parasites with global health and economic implications (Figueroa-Martinez et al., 2015; Janouškovec et al., 2015). Most are incredibly tiny, often going unnoticed by even the keenest observers, and a few are gargantuan, by any standard of the word.

Indeed, the infamous nonphotosynthetic parasitic land plant *Rafflesia* has the largest known flower of any angiosperm, measuring, in some species, over 3ft. in diameter and weighing over 20lb (Meijer, 1984). But woe betide to anyone who goes looking for this floral behemoth, for if they are lucky enough to find it, they may get an unfortunate surprise:

> *Much has been made of the smell produced by* Rafflesia *flowers: an early traveler once described it as 'a penetrating odour more repulsive than any buffalo carcass in an advanced stage of decomposition' ... Given their rarity and unpredictability, it is remarkable that anyone ever sees a* Rafflesia *flower in all its glory. But of course, they do. Two localities in Sabah [Borneo] offer a reasonable chance of success. ... If one should bloom a sign immediately appears on the main road that a* Rafflesia *is flowering, and they charge passerby a fee to see their prized flower. Make no mistake, on a local scale this is big business, as several hundred tourists have been known to see a single flower over the course of a five- to six-day blooming period.*
>
> **Garbutt and Prudente (2006)**

Rafflesia aside, most species that have lost photosynthesis are not tourist attractions, but they are the focal point for cutting-edge research. Studies of colourless algae have improved our understanding of endosymbiosis (Janouškovec et al., 2015), cell biology (McFadden & Yeh, 2017), genome evolution (Smith & Lee, 2014), and the diversification of life (Burki et al., 2016). They have also redefined how we view plastids (Fichera & Roos, 1997) and raised questions about what defines an alga or plant (Janouškovec et al., 2017). Some colourless lineages retain many of the features and machineries of their close photosynthetic relatives and are reliant on their plastid and plastid genome (plastome), others have completely done away with plastid DNA (ptDNA) and its associated gene expression system (Smith & Asmail, 2014), and some have gone a step further abandoning the plastid entirely (Gornik et al., 2015). If that weren't enough, there are organisms that have lost and regained plastids (Janouškovec et al., 2015).

As I tell my students whenever they get bored of my proselytizing about plastid evolution, research on colourless algae is not limited to basic science and, in fact, might hold the secrets for curing deadly diseases. For example, the malaria parasite (*Plasmodium falciparum*) and the causative agents of toxoplasmosis (*Toxoplasma gondii*) each have a nonphotosynthetic plastid called an apicoplast, and ever since it was first discovered in the mid 1990s scientists have been proclaiming its potential for therapeutic intervention (Fichera & Roos, 1997). The cyanobacterial-derived pathways within the apicoplast "are all very distant from human host metabolism and cellular processes, leaving room to design or discover specific inhibitors that would perturb the apicoplast but have no side effects" (McFadden & Yeh, 2017). Scientists are desperately trying, and have had some moderate success, in designing drugs blocking key apicoplast pathways, including those connected to the replication, transcription, and translation of ptDNA (Goodman, Pasaje, Kennedy, McFadden, & Ralph, 2016). It's not just humans who are at the mercy of parasitic nonphotosynthetic algae: the apicoplast-containing genera *Babesia*, *Eimeria*, and *Theileria* can cause serious diseases in domesticated (and undomesticated) animals, such as cattle, chickens, and other livestock (Foth & McFadden, 2003). But don't let these parasites bias you against nonphotosynthetic algae. Many, like *Polytomella*, are benign, do more good than harm, and are poised to become model research species.

Below, I explore the good and the bad sides of nonphotosynthetic algae, focusing on recent major discoveries in plastid genomics. I highlight the remarkable diversity in ptDNA architecture among colourless protists and how these data have advanced the fields of organelle genetics and plastid biology. But before we can discuss the nitty-gritty of nonphotosynthetic plastids, we first need to examine how photosynthetic plastids and their genomes came to be.

Perhaps, the most amazing thing about plastids is that they exist at all. Their labyrinthine journey from free-living bacteria to integral and inalienable components of algae and land plants involved countless winding, diverging, and colliding roads, and a lot of luck. The story of plastids has many plots, many characters, is replete with whimsy and mystery, and is still ongoing. Certain aspects of plastid evolution remain unresolved and are mired in debate, confusion, and controversy, but thankfully we now have a clear understanding of the key players and main events that first gave rise to eukaryotic phototrophy.

2. AND THEN THERE WAS LIGHT

When you think about the complexity of our natural world—plants using quantum mechanics for photosynthesis, for example—a smartphone begins to look like a pretty dumb object.

Jeff VanderMeer

Today, eukaryotic life is teeming with photosynthesis; it occurs in at least half of the currently defined supergroups (Burki, 2014). But it wasn't always like this. For the first few hundred million years of eukaryotic evolution there were no plastids. Eukaryotes owe their existence to a 1.8-billion-year-old cellular merger between two obligate heterotrophs: a bacterial endosymbiont (which resembled present-day alphaproteobacteria) and an archaeal host (which is thought to be linked to the Lokiarchaeota) (Gray, 2012; Spang et al., 2015). Early eukaryotes and the initial lineages that they gave rise to were entirely devoid of photosynthesis. Things would have remained that way until relatively recently[a] (Nowack, 2014) if it weren't for a fortuitous primary endosymbiotic event between a photosynthetic bacterium (the endosymbiont) and a unicellular nonphotosynthetic eukaryote (the host) about one and a half billion years ago (Archibald, 2015; Smith, 2017).

It makes intuitive sense why a heterotroph would want to hijack a cyanobacterium—for the sweet rewards of photosynthesis, of course—but precisely how this enslavement occurred is not so straightforward.

[a]The unicellular eukaryote *Paulinella chromatophora* (Rhizaria, Cercozoa) has a recently acquired cyanobacterial endosymbiont. Between 60 and 200 million years ago, the ancestor of this little-known amoeboid alga transitioned from a heterotrophic bacterivorous existence, sustained in part by feeding on cyanobacteria, to a phototrophic one, dependent on a cyanobacterial endosymbiont called a chromatophore (Nowack, 2014). *P. chromatophora* is the only known example of primary acquisition of a photosynthetic organelle outside of that which generated the Archaeplastida.

Undergraduate textbooks like to depict it as a single step: an image of a Pac-Man-esque eukaryote gobbling up an unsuspecting green dot. And then, voilà, a fully integrated chloroplast, with all the bells and whistles, within a modern-day plant or alga. Don't be fooled, this fast-tracked version of primary endosymbiosis is an oversimplification. Plastid organellogenesis was undoubtedly more complex, drawn out, and multifaceted than many textbooks would have us believe, occurring at a population level and an evolutionary timescale, and likely involving multiple contributing partners. Some of these complexities are described in the "shopping bag model" (Larkum, Lockhart, & Howe, 2007) of primary plastid evolution:

> *It seems unlikely that the stable [cyanobacterial] symbiont ultimately acquired by the host cell would be the first one it had ever acquired. The acquisition would almost certainly have been preceded by the uptake of other photosynthetic organisms. … [E]arly rounds of failed endosymbiosis, with some would-be endosymbionts eventually lysing and liberating DNA into the cytosol, would result in integration of endosymbiont DNA into the nuclear genome. This DNA would have persisted in the nucleus for a period of time, even if there were no longer functional symbionts in the host cytoplasm. If, finally, a symbiont [was] able to establish a balanced relationship with the host, the reservoir of sequences in the host nucleus that were derived from previous photosynthetic organisms would have provided a pool of sequences to encode proteins to be imported into the newly established plastid.*
>
> ***Howe, Barbrook, Nisbet, Lockhart, and Larkum (2008)***

As provocative as the shopping bag scenario may be, it remains to be determined how many, if any, failed endosymbioses preceded the successful cyanobacterial endosymbiont—and should be stressed that early plastid evolution is an ongoing area of debate (Dagan et al., 2013; reference therein). However, there is strong evidence that the ultimate progenitor of all plastids was a fan of freshwater and hails from a newly uncovered clade called *Gloeomargarita* (Ponce-Toledo et al., 2017). Using a comprehensive phylogenomic dataset, Ponce-Toledo et al. (2017) showed that *Gloeomargarita lithophora*—a deep-branching, biofilm-forming cyanobacterium—is the closest known prokaryotic relative of plastids. What's more, the entire *Gloeomargarita* group appears to be restricted to freshwater environments, suggesting that eukaryotic photosynthesis first emerged in a terrestrial freshwater setting.

So, after a long, fortuitous start and some help from *Gloeomargarita* et al., photosynthesis became firmly established within the eukaryotic domain, eventually giving rise to the supergroup Archaeplastida (Adl et al., 2012), which is made up of red algae, green algae, land plants, and glaucophytes.

Each of these archaeplastidal lineages can trace their photosynthetic properties directly back to the *Gloeomargarita*-like endosymbiont and as such are said to have primary plastids, which contain two membranes (Keeling, 2013; Reyes-Prieto, Weber, & Bhattacharya, 2007). Not surprisingly, the first lineage to diverge within the Archaeplastida (the Glaucophyta) (Ponce-Toledo et al., 2017) is completely restricted to freshwater environments, thus, following in the footsteps of its cyanobacterial progenitor (Delwiche & Cooper, 2015). But the other archaeplastidal lineages, in addition to being found on land and in freshwater, have successfully colonized saltwater ecosystems (Keeling, 2013; Reyes-Prieto et al., 2007).

If life was simple and evolution was a straight road the story of eukaryotic photosynthesis would stop here. But as any card-carrying biologist will tell you, evolution can be a crooked and winding process, and is not opposed to taking the odd sidestep. Accordingly, plastids and photosynthesis have jumped horizontally from the Archaeplastida to other supergroups via eukaryote–eukaryote endosymbioses (Archibald, 2015; Burki, 2017; Keeling, 2013). It is a dog-eat-dog world and many heterotrophic protists make their living by devouring eukaryotic algae. Factor in a little evolutionary indigestion and some of the ideas from the shopping bag model and before you know it the photosynthetic food has become a photosynthetic endosymbiont, and then fast-forward a few more million years and it's now a bona fide photosynthetic organelle. Red algae are no stranger to this narrative, having weaved their photosynthetic powers and plastids into some pretty remote phylogenetic corners. For example, haptophyte algae (e.g. *Emiliania*), diatom algae (e.g. *Phaeodactylum*), golden algae (e.g. *Ochromonas*), and brown algae (kelp) all have red-algal-derived plastids, as do apicomplexan parasites, such as *P. falciparum*, and most dinoflagellates (e.g. *Symbiodinium*) (Archibald, 2015; Burki, 2017; Keeling, 2013). The number of eukaryote-to-eukaryote endosymbiotic events that occurred to give rise to the complex red-algal-derived plastids is hotly debated (Burki, 2017). Green algae are in on the action as well, transferring their plastids to euglenophytes (e.g. *Euglena*) and the dinoflagellate lineage *Lepidodinium* (Kamikawa, Tanifuji, Kawachi, et al., 2015) in separate secondary endosymbiotic events.

One of the major goals and outcomes of evolutionary genomics has been disentangling the convoluted history of plastids derived from one eukaryote merging with another (commonly referred to as complex plastids). As it currently stands, plastids have moved laterally from one eukaryotic lineage to another no fewer than five times (Archibald, 2015). Tracking these movements can literally be a game of "keep your eyes on the plastid." On at least three separate occasions, a heterotrophic eukaryote has snatched a plastid (via tertiary endosymbiosis) from an alga that itself acquired its plastid secondarily from a red alga (Burki, 2017). Equally as convoluted are serial endosymbioses, whereby a secondary plastid is replaced by another plastid (Kamikawa, Tanifuji, Kawachi, et al., 2015).

In most cases, all that remains of these secondary, tertiary, or serial endosymbiotic events is the final product: an integrated, functional plastid with one or more extra membranes—a consequence of all that jumping around and the reason behind the name "complex" plastid. But sometimes the crime scene has not been entirely cleared. For cryptophytes and chlorarachniophytes, the nuclei and nuclear genomes of the engulfed primary algae—a red alga and green alga, respectively—persist in the host cell (alongside the plastid) as highly reduced organelles called nucleomorphs (Moore & Archibald, 2009).

Algae with complex plastids may seem a bit like endosymbiotic circus acts, but keep in mind that they carry out a significant proportion of the photosynthesis that occurs on Earth, and thus play an important role in reducing global atmospheric carbon dioxide levels. But, as described later, it is not always bright and sunny in the world of complex or primary plastids. Both of these kinds of plastid have discarded their photosynthetic abilities on many occasions.

3. BURNING OUT: THE EVOLUTIONARY LOSS OF PHOTOSYNTHESIS

The world breaks everyone and afterward many are strong at the broken places. But those that will not break it kills. It kills the very good and the very gentle and the very brave impartially. If you are none of these you can be sure it will kill you too but there will be no special hurry.

Ernest Hemingway—A Farewell to Arms

As counter intuitive as it may seem, a large number of algae and plants can no longer convert carbon dioxide and water into sugar and oxygen (Blouin & Lane, 2012; Figueroa-Martinez et al., 2015; Keeling, 2013; Krause, 2008). Most colourless algae are not easy to observe with the naked eye, and it is really only those who study them in the lab under a microscope that have seen one up close and personal. Nonphotosynthetic land plants, on the other hand, are hard to miss, even if they are not all as massive as *Rafflesia*; they can even be quite beautiful and ghostly, given their lack of chlorophylls, as anyone who has gazed upon the porcelain-like petals of fringed pinesap, Indian pipe, hillside broomrape, or the flatglobe dodder can attest. Beautiful or not, why would any self-respecting and sound-minded alga or plant forsake photosynthesis, especially after all the trouble and time to acquire a plastid? The answer to this question is not as mysterious or baffling as you might expect, and has its roots in a feeding strategy called mixotrophy.

As the name implies, mixotrophic algae and plants can make use of both inorganic and organic carbon sources via photoautotrophy and chemoheterotrophy, respectively. The latter is achieved by phagocytosing entire cells (i.e. predation) or through the endocytosis or osmosis of organic compounds—or simply put: engulfing or absorbing things from the environment. Sounds like a great strategy, right? Make sugar while the sun is shining and the gettin' is good, and keep filling the coffers even if things go dark and you're stuck, for instance, under Arctic sea ice for 6 months. Being mixotrophic also means that a random mutation knocking out photosynthetic (or heterotrophic) capabilities would not necessarily be lethal, which it would be in an obligate photoautotroph.

Despite its obvious benefits, mixotrophy is a mixed blessing because it is metabolically expensive to sustain both trophic strategies, so much so that mixotrophic algae are thought to expel five times more energy and nutrients on preserving photosynthesis than on the upkeep of heterotrophy (Raven, 1997). Therefore, given the right conditions, such as when the metabolic costs of maintaining the photosynthetic machinery exceed the benefits, doing away with photoautotrophy can arguably be advantageous, even when light conditions are favourable (de Castro, Gaedke, & Boenigk, 2009). Such a view is supported by the fact that the loss of photosynthesis is not uncommon among mixotrophic species:

> *Extant colorless algal lineages have either phagotrophic or osmotrophic lifestyles, and this is generally a reflection of the heterotrophic strategy employed by their mixotrophic relatives. For example, phagotrophic colorless algae can be found among dinoflagellates, stramenopiles and cryptophytes; this lifestyle is consistent with the presence of phagotrophism in their close mixotrophic relatives. Other colorless algae, such as the chlorophyte green algae* Helicosporidium, Prototheca, Polytoma, *and* Polytomella, *are closely related to osmo-mixotrophic chlorophytes and adopted an osmotrophic strategy where the source of dissolved organic matter can be either a host (in the case of pathogenic/parasitic species) or the environment (in free-living species).*
>
> ***Figueroa-Martinez et al. (2015)***

Thus, for colourless algae, mixotrophy appears to be a prerequisite for losing photoautotrophic functions. The same theme also emerges from work on nonphotosynthetic land plants (Julou et al., 2005; Selosse, Charpin, & Not, 2017). One could debate whether photosynthetic loss is adaptive (e.g. shedding the burden of photosynthesis) or nonadaptive (e.g. random genetic drift), but there is no denying that a single mutation in the right place to the right gene is sometimes all it takes to bring down the entire photosynthetic apparatus and dramatically change phenotype and lifestyle.

Work on the model green alga *Chlamydomonas reinhardtii* has shown that point mutations to photopigment genes can shut down photosynthesis (McCarthy, Kobayashi, & Niyogi, 2004; Meinecke et al., 2010). A nonphotosynthetic mutant of *C. reinhardtii* defective for phytoene synthase—one of the first enzymes in carotenoid biosynthesis—bears a remarkable resemblance to naturally occurring colourless algae, exhibiting starch accumulation, a disorganized eyespot, and no pyrenoid (Inwood, Yoshihara, Zalpuri, Kim, & Kustu, 2008). Moreover, the lack of carotenoids leads to plastids with no stacked thylakoidal membranes, paralleling the situation in other nonphotosynthetic chlamydomonadaleans (Inwood et al., 2008). This mutant can also grow in the dark with acetate as a carbon source implying "that mutations of this type would be nearly neutral in environments where photosynthesis is not critical for carbon assimilation and offers an ecological scenario and a plausible explanation for the origin of free-living heterotrophic colourless algae" (Inwood et al., 2008).

Although colourless algae have often taken a similar route to arriving at heterotrophy, the outcome following the loss of photosynthesis can vary within and among lineages. It can result in obligate parasitism (e.g. *P. falciparum*) or an opportunistic pathogenic existence (e.g. the green alga *Prototheca wickerhamii*), a voracious predatory lifestyle (e.g. the colpodellid *Voromonas pontica*), or a harmless osmotrophic one (e.g. the green alga *Polytomella*). With respect to the Apicomplexa, the evolutionary loss of photosynthesis spawned an entire phylum of dangerous obligate animal parasites. Conversely, for green algae, nonphotosynthetic parasites, infecting everything from plants to insects to humans, have evolved multiple times independently within closely related lineages interspersed with photosynthetic taxa, and the same is true for free-living colourless green algae (Figueroa-Martinez et al., 2015). Similar trends are observed in red algae, which are estimated to have the largest number of recently photosynthetic parasites of any major group, including nearly half of all recognized floridiophytes (Blouin & Lane, 2012). And don't get me started on the various flavours of parasitic nonphotosynthetic land plants, described in detail in Wicke and Naumann (2018).

It might be easy to do away with photosynthesis, but it is not so easy to dump a plastid—all known nonphotosynthetic members of the Archaeplastida, for example, retain one (Archibald, 2015; Keeling, 2013). This is because as plastid endosymbiosis took hold, the host became dependent upon its cyanobacterial (or plastid-donating) partner for much more than photosynthesis. In plants and algae, many vital biochemical pathways unrelated to photosynthesis are outsourced entirely or partly to the plastid, such as the biosynthesis of aromatic and hydrophobic side-chain amino acids, tetrapyrroles, and terpenoids (Gould, Waller, & McFadden, 2008). Although nearly all the enzymes involved in these pathways are nuclear encoded, most nonphotosynthetic plastids still retain a genome, albeit one that is typically highly reduced with a much smaller gene content than that in photosynthetic taxa (Figueroa-Martinez et al., 2015; Graham, Lam, & Merckx, 2017; Krause, 2008). As described in the following sections, the plastomes of nonphotosynthetic species are architecturally diverse and can tell us a lot about the processes involved with and the consequences of forgoing photosynthesis.

4. GENETIC BALL AND CHAIN: PLASTOMES IN COLOURLESS ALGAE

Any half-awake materialist well knows—that which you hold holds you.

Tom Robbins

Unless you are in the field of plastid genetics, your idea of a plastome probably looks something like this: an intact, AT-rich circular molecule of approximately 150 kilobases (kb) encoding a few dozen proteins mostly involved in photosynthesis. Yes, this image fits the classic plastid genome map of *Arabidopsis* or corn or rice, but it is not representative of most ptDNAs. For both photosynthetic and colourless species, plastomes span the gamut of size, structure, and content (Green, 2011; Smith & Keeling, 2015).

The plastomes of photosynthetic algae, for instance, can be enormous, exceeding a million base pairs and 90% noncoding DNA in the red alga

Corynoplastis japonica (Muñoz-Gómez et al., 2017) and the green alga *Acetabularia acetabulum* (based on partial on ptDNA sequence; de Vries et al., 2013), or small and compact, like the 66-kb ptDNA of the dinoflagellate *Lepidodinium chlorophorum* (Kamikawa, Tanifuji, Kawachi, et al., 2015). They can be contained in long linear chromosomes with monomeric, concatenated, or branched structures (Bendich, 2004; Smith & Keeling, 2015), or fragmented into dozens of small circular molecules, as exemplified by the *Symbiodinium* ptDNA (Barbrook, Voolstra, & Howe, 2014). They can be biased in adenine and thymine or guanine and cytosine (Smith, 2012), and can contain fewer than 25 genes or as many as 250 (Janouškovec et al., 2013). And the expression of these genomes can involve nonstandard codes, the removal of dozens of introns (even introns within introns), and complicated forms of posttranscriptional processing—dinoflagellate ptDNAs are an amusement park for substitutional RNA editing (Knoop, 2011; Smith & Keeling, 2016). Thus, plastomes are much more multifarious and bizarre than most scientists might think.

The standard narrative for what happens to ptDNA after the forfeiture of photosynthesis is one of gene loss and an overall reduction in complexity. Take the 56-kb plastome of the nonphotosynthetic green alga and opportunistic animal pathogen *P. wickerhamii*. When compared to its close free-living photosynthetic relative *Auxenochlorella protothecoides*, it looks like someone came along and surgically removed nearly all of the genes connected to photosynthesis from the *P. wickerhamii* ptDNA, leaving behind 27 tRNAs, a few rRNAs, and 40 protein-coding genes (Yan et al., 2015). Nearly all of these remaining genes are involved in plastid gene expression—a complicated process involving both plastid- and nuclear-encoded machinery (Gould et al., 2008). What makes this gene loss all the more striking is that the *P. wickerhamii* and *A. protothecoides* ptDNAs are completely syntenic, photosynthetic genes notwithstanding (Yan et al., 2015).

The *P. wickerhamii* ptDNA, however, still bears the marks of its photosynthetic past, harbouring a nearly full complement of chloroplast ATP synthase subunit genes, which are typically associated with the electron transport chain of photosynthesis. These same genes have also been found in the plastomes from two other nonphotosynthetic unicellular algae (the cryptophyte *Cryptomonas paramecium* and the diatom *Nitzschia* sp.) and several parasitic plants (Donaher et al., 2009; Kamikawa, Tanifuji, Ishikawa, et al., 2015). This, alongside the absence of other photosynthesis-related genes from these genomes, has left researchers scratching their heads as to why ATP synthase subunits are retained in some colourless plastids. A Japanese group working on *Nitzschia* has an interesting hypothesis:

> *It is possible that these ATP synthase complexes might be retained for ATP synthesis using a proton gradient generated through an as yet unknown, photosynthesis-independent mechanism. Here, we suggest an alternative function: ... that following loss of photosynthesis, the ATP synthase complex in the nonphotosynthetic diatom plastids has functioned to hydrolyze ATP to maintain a proton gradient between the thylakoid lumen and stroma, required for the Tat-dependent protein translocation system. ... we suggest that the Tat system also functions (or has worked) in [other] nonphotosynthetic plastids, and could again be the main reason for the retention of ATP synthase genes ....*
>
> ***Kamikawa, Tanifuji, Ishikawa, et al. (2015)***

Supporting this hypothesis is the presence of a gene for Tat in the plastome of *Nitzschia* sp., but such a gene is lacking from the ptDNAs of *P. wickerhamii* and *C. paramecium* (Donaher et al., 2009; Yan et al., 2015). And by no means do the ptDNAs of all nonphotosynthetic algae contain ATP synthase genes (Figueroa-Martinez, Nedelcu, Smith, & Reyes-Prieto, 2017). In fact, at least one is a pseudogene in *C. paramecium* (Donaher et al., 2009), and they have been entirely lost from the ultracompact 37-kb ptDNA of *Helicosporidium* sp., a nonphotosynthetic pathogen and very close relative of *P. wickerhamii* (de Koning & Keeling, 2006).

Like *Helicosporidium*, the plastomes of apicomplexan parasites are paragons of compactness, ranging from about 30 to 40 kb, having as little as 5% intergenic DNA, and encoding around 30 proteins, mostly for transcribing and translating ptDNA, and none representing subunits of ATP synthase (Foth & McFadden, 2003; Janouškovec et al., 2015). For the longest time, the Apicomplexa held the record for the smallest ptDNAs ever observed. But in recent years more extreme examples of plastid genomic reduction have come from heterotrophic land plants, such as the orchid *Epipogium roseum* (19 kb) and the holoparasite *Pilostyles aethiopica* (11.4 kb) (Bellot & Renner, 2015; Schelkunov et al., 2015).

Whether you are talking about the ptDNA of colourless algae or heterotrophic plants, some common themes arise, including a small genome size, a reduced coding repertoire, a paucity of intergenic and intronic DNA, genomic rearrangements, a particularly high AT content, and elevated rates of sequence evolution (de Koning & Keeling, 2006; Figueroa-Martinez, Nedelcu, Smith, et al., 2017; Garg et al., 2014; Wicke, Müller, Quandt, Bellot, & Schneeweiss, 2016). But as biologists explore more and more ptDNAs, they are finding that these trends do not always hold. The plastome of the free-living colourless alga *Euglena longa* is far from intron-poor, boasting 61 introns (Gockel & Hachtel, 2000)—although keep in mind that the ptDNA of its close photosynthetic relative *Euglena gracilis* has an unprecedentedly large number of introns (160) (Hallick et al., 1993). Another strange thing about *E. longa* is that its ptDNA encodes the large subunit of the enzyme RuBisCO (RBCL), and the small subunit of this enzyme (RBCS) is encoded in the nuclear genome as a precursor polyprotein comprising multiple RBCS repeats (Chan, Keller, Canaday, Weil, & Imbault, 1990). What on Earth is a nonphotosynthetic species doing with RuBisCO? A team of Czech researchers think the answer may be "absolutely nothing."

> *Both the RBCL and RBCS proteins are synthesized in* E. longa, *but their abundance is very low compared to* E. gracilis. *No RBCS monomers could be detected in* E. longa, *suggesting that processing of the precursor polyprotein is inefficient in this species. The abundance of RBCS is regulated post-transcriptionally. Indeed, blocking the cytoplasmic translation by cycloheximide has no immediate effect on the RBCS stability in photosynthetically grown* E. gracilis, *but in* E. longa, *the protein is rapidly degraded. Altogether, our results revealed signatures of evolutionary degradation (becoming defunct) of RuBisCO in* E. longa *and suggest that its biological role in this species may be rather unorthodox, if any.*
>
> ***Záhonová, Füssy, Oborník, Eliáš, and Yurchenko (2016)***

The *E. longa* RuBisCO enigma exemplifies another common thread running through the field of nonphotosynthetic plastid genomics—that there are usually one or more genes kicking around in the ptDNA whose function in a nonphotosynthetic context is not easily explained. Other protein-coding genes that meet this criterion include *clpP*, *ftsH*, and *ycf1*, which have been independently conserved in the ptDNAs of diverse colourless algae (Figueroa-Martinez, Nedelcu, Smith, et al., 2017), but arguably do not have clearly defined roles in heterotrophic taxa. The *clpP* gene product (a subunit of a ClpP peptidase) is thought to be involved in protein homeostasis (Ramundo et al., 2014), that of *ftsH* is believed to be an essential protease (de Vries et al., 2013; Maul et al., 2002), and the precise function of *ycf1* is unknown (de Vries, Sousa, Bölter, Soll, & Gould, 2015; Nakai, 2015) but might be related to membrane anchorage and/or nucleic acid binding (Boudreau et al., 1997; Drescher, Ruf, Calsa, Carrer, & Bock, 2000; Ozawa et al., 2009).

The idea that nonphotosynthetic ptDNAs can harbour genes for essential pathways apart from photosynthesis is one of the main arguments for why most colourless species still sustain a plastid genome and all that entails

(Janouškovec et al., 2015). Plastid genome replication and gene expression require a complex infrastructure spanning two genetic compartments and involving hundreds of proteins. It might seem wasteful and inefficient for such an exhaustive system to persist so that only a few (or less) key metabolic genes from the ptDNA can be expressed. But if the gene or genes in question are essential and haven't successfully moved to another compartment, then the ptDNA is indispensable and the genomic bureaucracy must endure. Consequently, it was long believed that nonphotosynthetic plastids were irreversibly tied to their genomes (Barbrook, Howe, & Purton, 2006; Nair & Striepen, 2011), but now it is known that at least some species have broken free of this genetic "ball and chain."

5. ADIÓS PTDNA: THE OUTRIGHT LOSS OF A PLASTOME

Perfection is achieved, not when there is nothing more to add, but when there is nothing left to take away.

Antoine de Saint-Exupery

When I was a PhD student, my supervisor Bob (who made a cameo in the beginning of this chapter) would always march into the lab with grandiose ideas and flamboyant hypotheses. "Here's what we're going to do, Smitty," he'd say. "We're going to merge *Chlamydomonas* with *Polytomella*! What do you think—shall we call it *Chlamydomella* or *Polytomonas*?" Most of his proclamations, like *Chlamydomella*, were merely meant to produce a smile or a laugh, but sometimes he'd come up with intriguing ideas formed from years of careful observation and hours of critical thought. Shortly after I arrived in the lab, Bob became adamant that *Polytomella* (a colourless chlamydomonadalean green alga, in case you forgot) was missing a plastid genome, something the other lab members, including myself, were sceptical about. Bob's assertion was based in part on the inability to detect plastid rRNA in *Polytomella* using Northern blot or PCR experiments (Nedelcu, 2001; Nedelcu, Spencer, Denovan-Wright, & Lee, 1996). But as every scientist knows, it is much harder to prove that something doesn't exist than prove that it does exist. After a number of inconclusive experiments on the presence/absence of *Polytomella* ptDNA, next-generation sequencing technologies arrived to the rescue.

High-throughput sequencing of total cellular DNA or RNA from an alga or plant, including nonphotosynthetic ones, typically yields a large number of plastid-derived reads, which can be used to assemble complete or nearly complete plastid genomes or transcriptomes (Shi et al., 2016; Smith, 2013). However, extensive Illumina sequencing of four different *Polytomella* species uncovered not a single identifiable ptDNA or RNA sequence (Smith & Lee, 2014). Although encouraging, this observation by itself was not enough to confidently conclude that *Polytomella* algae have no plastid genome. The real smoking gun came from an exhaustive bioinformatics search and characterization of nuclear-encoded, plastid-targeted proteins from *Polytomella*. This search uncovered a diversity of biochemical pathways occurring in the *Polytomella* plastid, such as isoprenoid biosynthesis and amino acid metabolism, but not one associated with replicating, repairing, transcribing, or translating a plastome (Asmail & Smith, 2016; Smith & Lee, 2014). So, after nearly a decade of working on the organelle genetics of *Polytomella*, Bob and I were finally able to provide sufficient data to support outright plastid genome loss in this colourless genus. On the day that the paper was accepted, we had champagne on ice ready to celebrate the first example of a plastid-bearing lineage with no ptDNA only to discover that another team had beaten us to the summit by only a few weeks. Like *Polytomella*, the nonphotosynthetic and parasitic angiosperm *Rafflesia lagascae* appears to have entirely shed its ptDNA (Molina et al., 2014).

The authors of the *Rafflesia* paper sequenced and assembled vast amounts of whole genomic DNA isolated from an *R. lagascae* floral bud and then scanned the resulting reads and contigs for plastid-derived sequences. Although they easily identified a large number of high-coverage contigs corresponding to the mitochondrial genome, they found very few with similarity to genic or intergenic sequences normally found in land plant plastomes. Moreover, not one of the plastid-like contigs contained a complete gene or an intact open reading frame, nor were they phylogenetically associated with close relatives of *Rafflesia*, but instead affiliated with species closely related to *Tetrastigma* (the plant that *R. lagascae* parasitizes). Based on these findings, Molina et al. (2014) argued that the plastid sequences recovered from the Illumina sequencing came from the nuclear (and in a few cases mitochondrial) genome and were horizontally transferred to *R. lagascae* from the plastome of *Tetrastigma*. Unfortunately, there were no accompanying data on nuclear-encoded, plastid-targeted proteins in *R. lagascae* to support the hypothesis of plastid genome loss—but see (Lee et al., 2016). Another concern with the interpretation of the data from *R. lagascae*, as pointed out by Krause (2015), is the current lack of physical evidence for the existence of a plastid compartment at all. [Note: a plastid clearly exists in

Polytomella (Moore, Cantor, Sheeler, & Kahn, 1970).] Krause goes on to suggest that something sneaky may be going on in *Rafflesia*:

> *It is feasible that the intimate association between* Rafflesia *and its host has led to parasite cells being populated with host plastids. The sequestration of host plastids could have relieved the parasite of the selective pressure to keep its own plastid genome. Thus, the phylogenetic loss of the plastid genome may be tolerable for the parasite because it can ontogenetically 'hijack' host organelles.*
>
> ***Krause (2015)***

A fascinating hypothesis, and not without precedent. The appropriation of plastids by nonphotosynthetic organisms (kleptoplasty) is a well-documented phenomenon, performed by some dinoflagellates (Gast, Moran, Dennett, & Caron, 2007) and even animals, such as the sea slug *Elysia chlorotica*, which steals plastids from the heterokont alga *Vaucheria litorea* (Pelletreau et al., 2011). However, there are currently no confirmed examples of kleptoplasty being performed by any land plant, or archaeplastid for that matter.

To some, it may come as a surprise that the first convincing cases for ptDNA loss (*Polytomella* and *Rafflesia*) came from lineages whose plastids descend directly from a primary endosymbiosis of a cyanobacterium and not from those whose plastids derive from eukaryote–eukaryote endosymbioses (i.e. complex algae). However, there is mounting evidence that nonphotosynthetic plastids from certain complex algae have ditched their genomes. Genomic and/or transcriptomic analyses of the colpodellids *Alphamonas edax*, *V. pontica*, and *Colpodella angusta* (free-living heterotrophic relatives of apicomplexans), the dinoflagellates *Dinophysis acuminate*, *Noctiluca scintillans*, *Oxyrrhis marina*, as well as the perkinsid *Perkinsus marinus* (a close colourless relative of dinoflagellates) are consistent with these species harbouring a plastid but lacking ptDNA (Janouškovec et al., 2017, 2015). As scientists explore evermore remote and esoteric regions of the eukaryotic tree of life, they will likely discover many more species that have rid themselves of the burden and bureaucracy of ptDNA. I predict that not only will researchers expose many different reasons for hanging on to a plastome long after dropping photosynthesis, but they will discover a diversity of ways to discard of one.

What about scrapping the plastid completely? To the best of my knowledge, there are only two clear cases of plastid loss from the entire eukaryotic domain: the apicomplexan *Cryptosporidium parvum* (one of several species that cause cryptosporidiosis) and the basal dinoflagellate *Hematodinium* sp.

(a parasite of crustaceans) (Abrahamsen et al., 2004; Gornik et al., 2015). The fact that both of these parasites salvage metabolites from their host could have alleviated their metabolic dependence on a plastid. Outright plastid loss has never been observed in free-living heterotrophs, perhaps because they are dependent on plastid-derived metabolites that they cannot glean from their food or the environment (Janouškovec et al., 2017). But one particular free-living heterotroph has a claim to fame that no parasite has yet matched: plastid genomic inflation.

6. NONPHOTOSYNTHETIC PTDNA: NOT SO SMALL AFTER ALL

Improvement makes straight roads, but the crooked roads without improvement, are roads of genius.

William Blake

Closely related to *Polytomella* is another nonphotosynthetic lineage represented by *Polytoma uvella*, a free-living unicellular osmotroph. Despite the similar sounding names and modes of existence, the *P. uvella* and *Polytomella* lineages lost photosynthesis independently of one another, and unlike the latter, the former has a plastid genome (Figueroa-Martinez et al., 2015; Nedelcu, 2001). However, it wasn't until very recently that researchers learnt about the size and coding content of this genome (Figueroa-Martinez, Nedelcu, Smith, et al., 2017). Given the close phylogenetic proximity of *P. uvella* and *Polytomella* species, one might have expected *P. uvella* to have a very small ptDNA, but the opposite was true.

P. uvella currently has the largest plastome ever found in a nonphotosynthetic species: ~230 kb and 75% noncoding (Figueroa-Martinez, Nedelcu, Smith, et al., 2017). Even more impressive, the genome is tens of thousands of nucleotides larger than those of its closest known photosynthetic relatives, *Chlamydomonas leiostraca* (167 kb) and *C. applanata* (~203 kb), a trend not previously observed in any other close photosynthetic–nonphotosynthetic duo (Figueroa-Martinez, Nedelcu, Smith, et al., 2017). Regardless of its large size, the *P. uvella* plastome has, like other nonphotosynthetic ptDNAs, undergone significant gene loss, shedding all coding regions for photosynthetic pathways. But unlike other nonphotosynthetic ptDNAs that of *P. uvella* has highly expanded intergenic regions.

Maybe the tightening of intergenic regions in heterotrophic ptDNAs has less to do with the loss of photosynthesis and more to do with another life-history feature common among many nonphotosynthetic lineages: parasitism. With some exceptions, the transition from a free living to a parasitic existence (particularly an obligate one) is associated with widespread genomic compaction (McCutcheon & Moran, 2012; Poulin & Randhawa, 2015). *P. uvella*, however, is free living and there is no reason to believe that it had a recent parasitic ancestor. Thus, the lack of genomic compaction in this colourless alga might partly be a consequence of it not being a parasite. One should also stress that the absence of parasitism certainly does not preclude a plastome from being compact, be it in a nonphotosynthetic or a photosynthetic species, and there are a number of nonparasitic colourless plants and algae with very little noncoding DNA in their plastomes (Donaher et al., 2009). But a parasitic lifestyle, in many cases, probably contributes to the extreme genomic compaction found in some ptDNAs (Figueroa-Martinez, Nedelcu, Reyes-Prieto, & Smith, 2017).

At first glance, the ptDNAs of *P. uvella* and *Polytomella* appear to have taken opposite paths following the loss of photosynthesis: genomic inflation vs complete genome loss. But, as noted by the authors of the *P. uvella* ptDNA sequence, such a claim might be misleading:

> *The evolutionary processes leading to these different events are not mutually exclusive and can occur in parallel. The loss of a plastid genome centers on coding DNA and involves the deletion of genes and the outsourcing of ptDNA-dependent pathways to other genetic compartments (Barbrook et al., 2006; Smith & Lee, 2014). Conversely, the expansion of a plastid genome acts on noncoding DNA, whereby error-prone DNA maintenance processes or selfish elements, for example, result in insertions in intergenic DNA. Therefore, the increase in noncoding DNA in a plastid genome does not preclude that genome from ultimately being lost. In fact, as noted above, repeat-rich noncoding DNA may even promote gene loss. In other words, there is no reason to assume that the nonphotosynthetic ancestor of* Polytomella *did not have a large, repeat-rich ptDNA or that* P. uvella *will not eventually lose its plastid genome. What is clear is that some chlamydomonadalean algae, whether they are photosynthetic or nonphotosynthetic, have a remarkable tendency toward extremes in organelle genome size.*
>
> ***Figueroa-Martinez, Nedelcu, Smith, et al. (2017)***

In fact, the order to which both *Polytomella* spp. and *P. uvella* belong—the Chlamydomonadales—has a propensity for plastid genomic inflation, with at least six members known to have ptDNAs in excess of 250kb (Featherston, Arakaki, Nozaki, Durand, & Smith, 2016).

There has been much debate about the forces driving organelle genomic expansion, with some arguing that it might be a consequence of random genetic drift, mutation rate, and/or inefficient and finicky DNA maintenance processes (Smith & Keeling, 2015). The identification of an inflated ptDNA in a heterotrophic alga only adds a further layer of complexity to the already complicated conundrum of genome size evolution. If anything, the *P. uvella* plastome reinforces the idea that no type of chromosome is immune to genomic expansion, even those that exist in the dark.

7. CONCLUDING THOUGHTS

Well, now,
if little by little you stop loving me
I shall stop loving you little by little.
If suddenly
you forget me
do not look for me,
for I shall already have forgotten you.
If you think it long and mad,
the wind of banners
that passes through my life,
and you decide
to leave me at the shore
of the heart where I have roots,
remember
that on that day,
at that hour,
I shall lift my arms
and my roots will set off
to seek another land.

Pablo Neruda

Nonphotosynthetic algae remind us of the fallacy that evolution is progressive. No, evolution does not produce organisms perfectly suited to their environments. It leads to the survival of species with a diversity of traits—species that are "good enough" to get by, and colourless algae, despite the lack of photoautotrophy, certainly do get by. Plastid-bearing heterotrophs also reinforce the idea that evolution is not always adaptive. Through mutation and random genetic drift, a population can evolve in ways that are not necessarily catered to the environment in which it exists. Indeed, holding on to a resource heavy plastid and plastid genome long after relinquishing photosynthetic capabilities may not always be the best strategy, but it persists nevertheless. To fully appreciate the cellular and genomic architecture of nonphotosynthetic algae, we need to assess them in a range of evolutionary lights. I hope that when you think of these eclectic organisms and their genomes, you do not just see broken light bulbs and a lack of chlorophyll, but also see them for all the dark and light shades of life that they encompass.

ACKNOWLEDGEMENTS

D.R.S. is supported by a Discovery Grant from the Natural Sciences and Engineering Research Council (NSERC) of Canada. He can be found online at www.arrogantgenome.com.

REFERENCES

Abrahamsen, M. S., Templeton, T. J., Enomoto, S., Abrahante, J. E., Zhu, G., Lancto, C. A., et al. (2004). Complete genome sequence of the apicomplexan, *Cryptosporidium parvum*. *Science*, *304*(5669), 441–445.

Adl, S. M., Simpson, A. G., Lane, C. E., Lukeš, J., Bass, D., Bowser, S. S., et al. (2012). The revised classification of eukaryotes. *Journal of Eukaryotic Microbiology*, *59*, 429–514.

Archibald, J. M. (2015). Endosymbiosis and eukaryotic cell evolution. *Current Biology*, *25*, R911–R921.

Asmail, S. R., & Smith, D. R. (2016). Retention, erosion, and loss of the carotenoid biosynthetic pathway in the nonphotosynthetic green algal genus *Polytomella*. *New Phytologist*, *209*, 899–903.

Barbrook, A. C., Howe, C. J., & Purton, S. (2006). Why are plastid genomes retained in non-photosynthetic organisms? *Trends in Plant Science*, *11*, 101–108.

Barbrook, A. C., Voolstra, C. R., & Howe, C. J. (2014). The chloroplast genome of a *Symbiodinium* sp. clade C3 isolate. *Protist*, *165*, 1–13.

Bellot, S., & Renner, S. S. (2015). The plastomes of two species in the endoparasite genus *Pilostyles* (Apodanthaceae) each retain just five or six possibly functional genes. *Genome Biology and Evolution*, *8*, 189–201.

Bendich, A. J. (2004). Circular chloroplast chromosomes: The grand illusion. *Plant Cell*, *16*, 1661–1666.

Blouin, N. A., & Lane, C. E. (2012). Red algal parasites: Models for a life history evolution that leaves photosynthesis behind again and again. *BioEssays*, *34*, 226–235.

Boudreau, E., Turmel, M., Goldschmidt-Clermont, M., Rochaix, J. D., Sivan, S., Michaels, A., et al. (1997). A large open reading frame (*orf1995*) in the chloroplast DNA of *Chlamydomonas reinhardtii* encodes an essential protein. *Molecular and General Genetics*, *253*, 649–653.

Burki, F. (2014). The eukaryotic tree of life from a global phylogenomic perspective. *Cold Spring Harbor Perspectives in Biology*, *6*, a016147.

Burki, F. (2017). The convoluted evolution of eukaryotes with complex plastids. *Advances in Botanical Research*, *84*, 1–30.

Burki, F., Kaplan, M., Tikhonenkov, D. V., Zlatogursky, V., Minh, B. Q., Radaykina, L. V., et al. (2016). Untangling the early diversification of eukaryotes: A phylogenomic study of the evolutionary origins of Centrohelida, Haptophyta and Cryptista. *Proceedings of the Royal Society B: Biological Sciences*, *283*, 20152802.

Chan, R. L., Keller, M., Canaday, J., Weil, J. H., & Imbault, P. (1990). Eight small subunits of *Euglena* ribulose 1-5 bisphosphate carboxylase/oxygenase are translated from a large mRNA as a polyprotein. *EMBO Journal*, *9*, 333.

Dagan, T., Roettger, M., Stucken, K., Landan, G., Koch, R., Major, P., et al. (2013). Genomes of Stigonematalean cyanobacteria (subsection V) and the evolution of oxygenic photosynthesis from prokaryotes to plastids. *Genome Biology and Evolution*, *5*, 31–44.

de Castro, F., Gaedke, U., & Boenigk, J. (2009). Reverse evolution: Driving forces behind the loss of acquired photosynthetic traits. *PLoS One*, *4*, e8465.
de Koning, A. P., & Keeling, P. J. (2006). The complete plastid genome sequence of the parasitic green alga *Helicosporidium* sp. is highly reduced and structured. *BMC Biology*, *4*, 12.
Delwiche, C. F., & Cooper, E. D. (2015). The evolutionary origin of a terrestrial flora. *Current Biology*, *25*, R899–R910.
de Vries, J., Habicht, J., Woehle, C., Huang, C., Christa, G., Wägele, H., et al. (2013). Is *ftsH* the key to plastid longevity in sacoglossan slugs? *Genome Biology and Evolution*, *5*, 2540–2548.
de Vries, J., Sousa, F. L., Bölter, B., Soll, J., & Gould, S. B. (2015). YCF1: A green TIC? *Plant Cell*, *27*, 1827–1833.
Donaher, N., Tanifuji, G., Onodera, N. T., Malfatti, S. A., Chain, P. S., Hara, Y., et al. (2009). The complete plastid genome sequence of the secondarily nonphotosynthetic alga *Cryptomonas paramecium*: Reduction, compaction, and accelerated evolutionary rate. *Genome Biology and Evolution*, *1*, 439–448.
Drescher, A., Ruf, S., Calsa, T., Carrer, H., & Bock, R. (2000). The two largest chloroplast genome-encoded open reading frames of higher plants are essential genes. *Plant Journal*, *22*, 97–104.
Featherston, J., Arakaki, Y., Nozaki, H., Durand, P. M., & Smith, D. R. (2016). Inflated organelle genomes and a circular-mapping mtDNA probably existed at the origin of coloniality in volvocine green algae. *European Journal of Phycology*, *51*, 369–377.
Fichera, M. E., & Roos, D. S. (1997). A plastid organelle as a drug target in apicomplexan parasites. *Nature*, *390*, 407–409.
Figueroa-Martinez, F., Nedelcu, A. M., Reyes-Prieto, A., & Smith, D. R. (2017). The plastid genomes of nonphotosynthetic algae are not so small after all. *Communicative & Integrative Biology*, *10*, e1283080.
Figueroa-Martinez, F., Nedelcu, A. M., Smith, D. R., & Reyes-Prieto, A. (2015). When the lights go out: The evolutionary fate of free-living colorless green algae. *New Phytologist*, *206*, 972–982.
Figueroa-Martinez, F., Nedelcu, A. M., Smith, D. R., & Reyes-Prieto, A. (2017). The plastid genome of *Polytoma uvella* is the largest known among colorless algae and plants and reflects contrasting evolutionary paths to nonphotosynthetic lifestyles. *Plant Physiology*, *173*(2), 932–943.
Foth, B. J., & McFadden, G. I. (2003). The apicoplast: A plastid in *Plasmodium falciparum* and other apicomplexan parasites. *International Review of Cytology*, *224*, 57–110.
Garbutt, N., & Prudente, J. C. (2006). *Wild Borneo: The wildlife and scenery of Sabah, Sarawak, Brunei and Kalimantan*. United Kingdom: New Holland.
Garg, A., Stein, A., Zhao, W., Dwivedi, A., Frutos, R., Cornillot, E., et al. (2014). Sequence and annotation of the apicoplast genome of the human pathogen *Babesia microti*. *PLoS One*, *9*, e107939.
Gast, R. J., Moran, D. M., Dennett, M. R., & Caron, D. A. (2007). Kleptoplasty in an Antarctic dinoflagellate: Caught in evolutionary transition? *Environmental Microbiology*, *9*, 39–45.
Gockel, G., & Hachtel, W. (2000). Complete gene map of the plastid genome of the nonphotosynthetic euglenoid flagellate *Astasia longa*. *Protist*, *151*, 347–351.
Goodman, C. D., Pasaje, C. F. A., Kennedy, K., McFadden, G. I., & Ralph, S. A. (2016). Targeting protein translation in organelles of the Apicomplexa. *Trends in Parasitology*, *32*, 953–965.
Gornik, S. G., Cassin, A. M., MacRae, J. I., Ramaprasad, A., Rchiad, Z., McConville, M. J., et al. (2015). Endosymbiosis undone by stepwise elimination of the plastid in a parasitic dinoflagellate. *Proceedings of the National Academy of Sciences of the United States of America*, *112*, 5767–5772.

Gould, S. B., Waller, R. F., & McFadden, G. I. (2008). Plastid evolution. *Annual Review of Plant Biology, 59*, 491–517.
Graham, S. W., Lam, V. K., & Merckx, V. S. (2017). Plastomes on the edge: The evolutionary breakdown of mycoheterotroph plastid genomes. *New Phytologist, 214*, 48–55.
Gray, M. W. (2012). Mitochondrial evolution. *Cold Spring Harbor Perspectives in Biology, 4*, a011403.
Green, B. R. (2011). Chloroplast genomes of photosynthetic eukaryotes. *The Plant Journal, 66*, 34–44.
Hallick, R. B., Hong, L., Drager, R. G., Favreau, M. R., Monfort, A., Orsat, B., et al. (1993). Complete sequence of *Euglena gracilis* chloroplast DNA. *Nucleic Acids Research, 21*, 3537–3544.
Howe, C. J., Barbrook, A. C., Nisbet, R. E. R., Lockhart, P. J., & Larkum, A. W. D. (2008). The origin of plastids. *Philosophical Transactions of the Royal Society B: Biological Sciences, 363*, 2675–2685.
Inwood, W., Yoshihara, C., Zalpuri, R., Kim, K. S., & Kustu, S. (2008). The ultrastructure of a *Chlamydomonas reinhardtii* mutant strain lacking phytoene synthase resembles that of a colorless alga. *Molecular Plant, 1*, 925–937.
Janouškovec, J., Gavelis, G. S., Burki, F., Dinh, D., Bachvaroff, T. R., Gornik, S. G., et al. (2017). Major transitions in dinoflagellate evolution unveiled by phylotranscriptomics. *Proceedings of the National Academy of Sciences of the United States of America, 114*, E171–E180.
Janouškovec, J., Liu, S.-L., Martone, P. T., Carré, W., Leblanc, C., Collén, J., et al. (2013). Evolution of red algal plastid genomes: Ancient architectures, introns, horizontal gene transfer, and taxonomic utility of plastid markers. *PLoS One, 8*, e59001.
Janouškovec, J., Tikhonenkov, D. V., Burki, F., Howe, A. T., Kolisko, M., Mylnikov, A. P., et al. (2015). Factors mediating plastid dependency and the origins of parasitism in apicomplexans and their close relatives. *Proceedings of the National Academy of Sciences of the United States of America, 112*(33), 10200–10207.
Julou, T., Burghardt, B., Gebauer, G., Berveiller, D., Damesin, C., & Selosse, M. A. (2005). Mixotrophy in orchids: Insights from a comparative study of green individuals and nonphotosynthetic individuals of *Cephalanthera damasonium*. *New Phytologist, 166*, 639–653.
Kamikawa, R., Tanifuji, G., Ishikawa, S. A., Ishii, K. I., Matsuno, Y., Onodera, N. T., et al. (2015). Proposal of a twin arginine translocator system-mediated constraint against loss of ATP synthase genes from nonphotosynthetic plastid genomes. *Molecular Biology and Evolution, 32*, 2598–2604.
Kamikawa, R., Tanifuji, G., Kawachi, M., Miyashita, H., Hashimoto, T., & Inagaki, Y. (2015). Plastid genome-based phylogeny pinpointed the origin of the green-colored plastid in the dinoflagellate *Lepidodinium chlorophorum*. *Genome Biology and Evolution, 7*, 1133–1140.
Keeling, P. J. (2013). The number, speed, and impact of plastid endosymbioses in eukaryotic evolution. *Annual Review of Plant Biology, 64*, 583–607.
Knoop, V. (2011). When you can't trust the DNA: RNA editing changes transcript sequences. *Cellular and Molecular Life Sciences, 68*, 567–586.
Krause, K. (2008). From chloroplasts to "cryptic" plastids: Evolution of plastid genomes in parasitic plants. *Current Genetics, 54*, 111–121.
Krause, K. (2015). Grand-scale theft: Kleptoplasty in parasitic plants? *Trends in Plant Science, 20*, 196–198.
Larkum, A. W. D., Lockhart, P. J., & Howe, C. J. (2007). Shopping for plastids. *Trends in Plant Science, 12*, 189–195.
Lee, X. W., Mat-Isa, M. N., Mohd-Elias, N. A., Aizat-Juhari, M. A., Goh, H. H., Dear, P. H., et al. (2016). Perigone lobe transcriptome analysis provides insights into *Rafflesia cantleyi* flower development. *PLoS One, 11*, e0167958.

Maul, J. E., Lilly, J. W., Cui, L., Miller, W., Harris, E. H., & Stern, D. B. (2002). The *Chlamydomonas reinhardtii* plastid chromosome islands of genes in a sea of repeats. *Plant Cell, 14*, 2659–2679.

McCarthy, S. S., Kobayashi, M. C., & Niyogi, K. K. (2004). White mutants of *Chlamydomonas reinhardtii* are defective in phytoene synthase. *Genetics, 168*, 1249–1257.

McCutcheon, J. P., & Moran, N. A. (2012). Extreme genome reduction in symbiotic bacteria. *Nature Reviews Microbiology, 10*, 13–26.

McFadden, G. I., & Yeh, E. (2017). The apicoplast: Now you see it, now you don't. *International Journal for Parasitology, 47*, 137–144.

Meijer, W. (1984). New species of *Rafflesia* (Rafflesiaceae). *Blumea, 30*, 209–215.

Meinecke, L., Alawady, A., Schroda, M., Willows, R., Kobayashi, M. C., Niyogi, K. K., et al. (2010). Chlorophyll-deficient mutants of *Chlamydomonas reinhardtii* that accumulate magnesium protoporphyrin IX. *Plant Molecular Biology, 72*, 643–658.

Molina, J., Hazzouri, K. M., Nickrent, D., Geisler, M., Meyer, R. S., Pentony, M. M., et al. (2014). Possible loss of the chloroplast genome in the parasitic flowering plant *Rafflesia lagascae* (Rafflesiaceae). *Molecular Biology and Evolution, 31*, 793–803.

Moore, C. E., & Archibald, J. M. (2009). Nucleomorph genomes. *Annual Review of Genetics, 43*, 251–264.

Moore, J., Cantor, M. H., Sheeler, P., & Kahn, W. (1970). The ultrastructure of *Polytomella agilis*. *Journal of Eukaryotic Microbiology, 17*, 671–676.

Muñoz-Gómez, S. A., Mejía-Franco, F. G., Durnin, K., Colp, M., Grisdale, C. J., Archibald, J. M., et al. (2017). The new red algal subphylum Proteorhodophytina comprises the largest and most divergent plastid genomes known. *Current Biology, 27*(11), 1677–1684.

Nair, S. C., & Striepen, B. (2011). What do human parasites do with a chloroplast anyway? *PLoS Biology, 9*, e1001137.

Nakai, M. (2015). The TIC complex uncovered: The alternative view on the molecular mechanism of protein translocation across the inner envelope membrane of chloroplasts. *Biochimica et Biophysica Acta (BBA), 1847*, 957–967.

Nedelcu, A. M. (2001). Complex patterns of plastid 16S rRNA gene evolution in nonphotosynthetic green algae. *Journal of Molecular Evolution, 53*, 670–679.

Nedelcu, A. M., Spencer, D. F., Denovan-Wright, E. M., & Lee, R. W. (1996). Discontinuous mitochondrial and chloroplast large subunit ribosomal RNAs among green algae: Phylogenetic implications. *Journal of Phycology, 32*, 103–111.

Nowack, E. C. (2014). *Paulinella chromatophora*-rethinking the transition from endosymbiont to organelle. *Acta Societatis Botanicorum Poloniae, 83*, 387–397.

Ozawa, S. I., Nield, J., Terao, A., Stauber, E. J., Hippler, M., Koike, H., et al. (2009). Biochemical and structural studies of the large Ycf4-photosystem I assembly complex of the green alga *Chlamydomonas reinhardtii*. *Plant Cell, 21*, 2424–2442.

Pelletreau, K. N., Bhattacharya, D., Price, D. C., Worful, J. M., Moustafa, A., & Rumpho, M. E. (2011). Sea slug kleptoplasty and plastid maintenance in a metazoan. *Plant Physiology, 155*, 1561–1565.

Ponce-Toledo, R. I., Deschamps, P., López-García, P., Zivanovic, Y., Benzerara, K., & Moreira, D. (2017). An early-branching freshwater cyanobacterium at the origin of plastids. *Current Biology, 27*, 386–391.

Poulin, R., & Randhawa, H. S. (2015). Evolution of parasitism along convergent lines: From ecology to genomics. *Parasitology, 142*(S1), S6–S15.

Ramundo, S., Casero, D., Mühlhaus, T., Hemme, D., Sommer, F., Crèvecoeur, M., et al. (2014). Conditional depletion of the *Chlamydomonas* chloroplast ClpP protease activates nuclear genes involved in autophagy and plastid protein quality control. *Plant Cell, 26*, 2201–2222.

Raven, J. A. (1997). Phagotrophy in phototrophs. *Limnology and Oceanography, 42*, 198–205.

Reyes-Prieto, A., Weber, A. P. M., & Bhattacharya, D. (2007). The origin and establishment of the plastid in algae and plants. *Annual Review of Genetics, 41*, 147–168.
Schelkunov, M. I., Shtratnikova, V. Y., Nuraliev, M. S., Selosse, M. A., Penin, A. A., & Logacheva, M. D. (2015). Exploring the limits for reduction of plastid genomes: A case study of the mycoheterotrophic orchids *Epipogium aphyllum* and *Epipogium roseum*. *Genome Biology and Evolution*, 7, 1179–1191.
Selosse, M. A., Charpin, M., & Not, F. (2017). Mixotrophy everywhere on land and in water: The grand écart hypothesis. *Ecology Letters, 20*, 246–263.
Shi, C., Wang, S., Xia, E. H., Jiang, J. J., Zeng, F. C., & Gao, L. Z. (2016). Full transcription of the chloroplast genome in photosynthetic eukaryotes. *Scientific Reports, 6*, 30135.
Smith, D. R. (2012). Updating our view of organelle genome nucleotide landscape. *Frontiers in Genetics, 3*, 175.
Smith, D. R. (2013). RNA-Seq data: A goldmine for organelle research. *Briefings in Functional Genomics, 12*, 454–456.
Smith, D. R. (2017). Let there be light: A contemporary primer on primary plastid endosymbiosis. *Advances in Botanical Research, 84*, 31–56.
Smith, D. R., & Asmail, S. R. (2014). Next-generation sequencing data suggest that certain nonphotosynthetic green plants have lost their plastid genomes. *New Phytologist, 204*, 7–11.
Smith, D. R., & Keeling, P. J. (2015). Mitochondrial and plastid genome architecture: Reoccurring themes, but significant differences at the extremes. *Proceedings of the National Academy of Sciences of Sciences of the United States of America, 112*, 10177–10184.
Smith, D. R., & Keeling, P. J. (2016). Protists and the wild, wild west of gene expression: New frontiers, lawlessness, and misfits. *Annual Review of Microbiology, 70*, 161–178.
Smith, D. R., & Lee, R. W. (2014). A plastid without a genome: Evidence from the nonphotosynthetic green algal genus *Polytomella*. *Plant Physiology, 164*, 1812–1819.
Spang, A., Saw, J. H., Jørgensen, S. L., Zaremba-Niedzwiedzka, K., Martijn, J., Lind, A. E., et al. (2015). Complex archaea that bridge the gap between prokaryotes and eukaryotes. *Nature, 521*, 173–179.
Wicke, S., & Naumann, J. (2018). Molecular evolution of plastid genomes in parasitic flowering plants. *Advances in Botanical Research, 85*, 315–347.
Wicke, S., Müller, K. F., Quandt, D., Bellot, S., & Schneeweiss, G. M. (2016). Mechanistic model of evolutionary rate variation en route to a nonphotosynthetic lifestyle in plants. *Proceedings of the National Academy of Sciences of the United States of America, 113*, 9045–9050.
Yan, D., Wang, Y., Murakami, T., Shen, Y., Gong, J., Jiang, H., et al. (2015). *Auxenochlorella protothecoides* and *Prototheca wickerhamii* plastid genome sequences give insight into the origins of non-photosynthetic algae. *Scientific Reports, 5*, 14465.
Záhonová, K., Füssy, Z., Oborník, M., Eliáš, M., & Yurchenko, V. (2016). RuBisCO in non-photosynthetic alga *Euglena longa*: Divergent features, transcriptomic analysis and regulation of complex formation. *PLoS One, 11*, e0158790.

FURTHER READING

Hemingway, E. (1957). *A farewell to arms*. New York: Scribner.

CHAPTER THREE

Plastid Genomes in the Myzozoa

Sergio A. Muñoz-Gómez, Claudio H. Slamovits[1]
Centre for Comparative Genomics and Evolutionary Bioinformatics, Dalhousie University, Halifax, NS, Canada
[1]Corresponding author: e-mail address: claudio.slamovits@dal.ca

Contents

Abstract

The myzozoa encompasses quite disparate protists, like the infamous apicomplexan parasites, or the famous dinoflagellate phytoplankton. Collectively, myzozoans display a wide diversity of plastids; they all most likely descended from a common myzozoan plastid ancestor. Some myzozoan plastids are photosynthetic whereas others are not; some have plastid genomes (plastomes) but others have lost them. The only two eukaryotes known to have lost plastids altogether are myzozoans. In this chapter, we explore the diversity and evolution of myzozoan plastids and plastomes, and compare them to those of other

Advances in Botanical Research, Volume 85
ISSN 0065-2296
https://doi.org/10.1016/bs.abr.2017.11.015

photosynthetic eukaryotes. Myzozoan plastomes are remarkable for encompassing the smallest photosynthesis-supporting plastomes known (in peridinin dinophytes) and for having the lowest GC content of all plastomes (in sporozoans). Myzozoan plastomes also have the smallest gene repertoires among red lineage plastomes, and such a state seems to have been reached through at least four episodic events of plastome reduction; two of these episodes appear to be associated with symbiogeneses. Myzozoans have played an important role in our understanding of plastid and plastome reduction among eukaryotes. Future discoveries of 'environmental' plastomes will allow us to increase the diversity and better reconstruct the diversification of myzozoan plastomes.

1. INTRODUCTION

Myzozoans comprise a group of protists that is remarkable for displaying a great diversity of plastids. The reason for this is that their evolutionary diversification has produced parasites, mutualistic endosymbionts, predators, algae (strict photosynthesizers), and mixotrophs (cells capable of predatory heterotrophy but also photosynthesis). Most myzozoans are heterotrophic (sporozoans, colpodellids, perkinsozoans, and half of dinoflagellates), but they are ancestrally plastid-bearing mixotrophs. Conveniently for us, the Myzozoa also turns out to be one of the best sampled groups in terms of plastid diversity. Myzozoans are named after their inferred ancestral capacity to feed by myzocytosis (Cavalier-Smith & Chao, 2004). Myzocytosis is a feeding mode in which the cytoplasmic contents of the prey cell are sucked leaving the plasmalemma outside—this contrasts with phagocytosis in which the whole prey cell is ingested (Schnepf & Deichgräber, 1984). The most commonly known myzozoan protists are apicomplexans and dinoflagellates. The formers are known to be deadly parasites of animals (e.g. malaria), whereas the latter are known as important primary producers or to cause harmful algal blooms (e.g. red tides) in waters. But apicomplexans are not the direct sisters to dinoflagellates; each lineage has closer but less diverse myzozoan relatives. Recent discoveries of algae on the apicomplexan side of the Myzozoa tree have given us more confidence in reconstructing the early steps in the evolution of plastids in this group. In this chapter, we explore the diversity and evolution of myzozoan plastid genomes or plastomes.

2. THE MYZOZOA

2.1 What Are Apicomplexans?

Apicomplexans are eukaryotic unicells (protists) that, in the broad sense, comprise both intracellular and extracellular endosymbionts (or individuals living inside another that can be commensals, parasites, or even mutualists; apicomplexans in the strict sense), as well as their closest relatives (free-living predators as well as free-living and endosymbiotic photosynthesizers). Classical apicomplexans (parvphylum Sporozoa, see below for a reference taxonomic scheme) are infamous parasites of animals. Some examples are *Plasmodium falciparum*, the cause of malaria in humans, *Cryptosporidium* and *Cyclospora*, causes of gastrointestinal diseases with diarrhoea in humans, and *Babesia* and *Theileria* that infect domestic animals like cattle. Gregarines are parasites, perhaps commensals, of invertebrates. It is believed that every animal species serves as a host for a corresponding coevolved sporozoan parasite. In tropical forests, apicomplexan parasites are the most abundant and diverse protists, at least matching the diversity of vertebrates and invertebrates (Mahé et al., 2017). This makes of apicomplexan parasites perhaps the most diversified and successful group of parasites on Earth.

Apicomplexans are a phylogenetically cohesive group. This has been conclusively shown by several single-gene and multigene phylogenies (e.g. Fast, Xue, Bingham & Keeling, 2002; Harper, Waanders & Keeling, 2005). The group is also united ultrastructurally by possessing a rostrum made of cystoskeletal structures (e.g. a pseudoconoid) and endomembranes (e.g. micronemes) at their cell apex, the so-called more developed apical complex in sporozoans, that is used for attachment and invasion (in gregarines), penetration of host cells (in haematozoans, piroplasms, and some coccidians), or feeding through myzocytosis (in colpodellids). They have also retained ancestral features shared with other myzozoans or alveolates such as cortical alveoli and micropores. Apicomplexans are inferred to have evolved from a plastid-bearing photosynthetic and flagellated myzozoan ancestor that had a precursor apical rostrum, was able myzocytose, made cysts, and reproduced by schyzogony (Cavalier-Smith & Chao, 2004).

2.2 Taxonomy of the Apicomplexa

The latest evolutionary taxonomic scheme for the group has the infraphylum Apicomplexa subdivided into two parvphyla: Apicomonada and Sporozoa (Fig. 1; Cavalier-Smith, 2017; see also Votýpka et al., 2016 for a compatible non-Linnean scheme). The Sporozoa comprises gregarines, the probably mutualistic *Nephromyces*, and classical apicomplexans such as coccidians, piroplasms, and haemosporidians; whereas the Apicomonada comprises the free-living and heterotrophic colopodellid predators, and the chromerid algae which are intracellular photosynthetic endosymbionts of, or free-living and associated to, corals (Fig. 1). A great diversity of

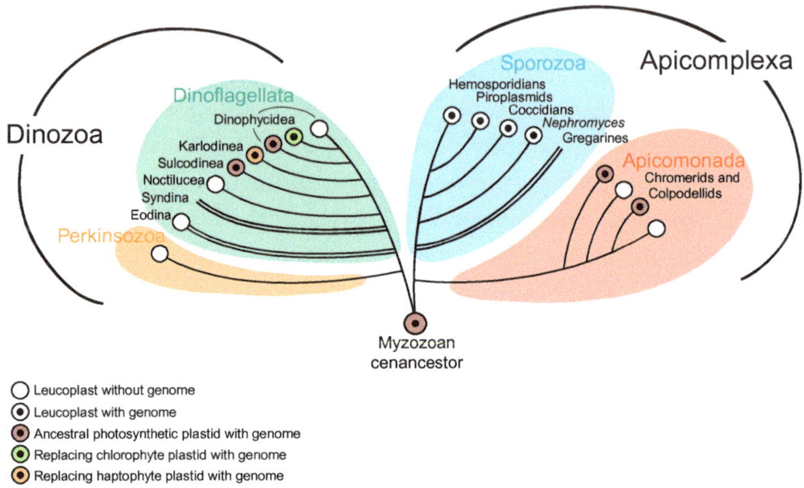

Fig. 1 A schematic phylogeny of the Myzozoa. The diagram summarizes and synthesizes the phylogenetic relationships among myzozoans based on Adl et al. (2012), Votýpka, Modrý, Oborník, Šlapeta, and Lukeš (2016), Janouškovec et al. (2015, 2017), and Cavalier-Smith (2017). For dinozoans, the evolutionary taxonomic scheme and taxon names of Cavalier-Smith (2017) are adopted. For apicomplexans, informal names are used for the particular major lineages discussed within the text, but Cavalier-Smith (2017) is followed for taxa above the parvphylum level. The distribution of plastids and their genomes is shown by different combinations of *coloured* and *inside circles*. Dinophytes with barely reduced ochrophyte endosymbionts (dinotoms), as well as cryptophyte-derived kleptoplastids in the dinophyte *Dinophysis* are not shown. The aplastidic myzozoans *Cryptosporidium* and *Haematodinium* are phylogenetically contained within gregarines and Syndina, respectively. *Double branches* denote paraphyly.

apicomplexans is known only from environmental surveys, apicomplexan-related lineages I–VIII (ARLs), or environmental clades I–XI (Janouškovec, Horák, Barott, Rohwer, & Keeling, 2013; Janouškovec et al., 2015). But these novel species have yet to be cultured and further studied so to be incorporated into formal classification schemes. Another more phylogenetic and cladistic scheme restricts the Apicomplexa clade to the classical endosymbiotic apicomplexans (*sensu stricto*), excluding their free-living and photosynthetic relatives (Adl et al., 2012). We will here refer to apicomplexans in their broadest sense (*sensu lato*) which also includes their free-living and photosynthetic relatives.

2.3 What Are Dinozoans?

Dinozoans encompass a great diversity of protists. About half of them are heterotrophic (either predatory or parasitic), whereas the other half are photosynthetic (either obligate or mixotrophic). They are all aquatic, and live in both marine and freshwaters. Some dinozoans are popularly known as bioluminescent plankton in seas (*Noctiluca*), algal endosymbionts of corals (*Symbiodinium*), and makers of red tides (*Alexandrium*). Most dinoflagellates, and specially the dinokaryotes, have their cell bodies divided into two parts, the episome and the hyposome (see Janouškovec et al., 2017 for character evolution mapped onto an updated phylogeny). Some 'basal' dinoflagellates, like *Perkinsus* and *Psamossa* (see below for their taxonomy), have apical rostra that are homologous to the specialized apical complex of sporozoans (e.g. see Okamoto & Keeling, 2014). Many also move by means of two flagella, a longitudinal flagellum that sticks out of the cell and propels it, and a ribbon-like transversal flagellum that wraps around the cell and makes it rotate as it swims forward. A subgroup of dinozoans (within the Dinophycidae) evolved heavily armoured cells by building thick cellulose thecal plates within their alveoli (the so-called thecate dinoflagellates). Some dinoflagellates evolved extraordinary structures for predatory feeding, like the peduncle (a flexible tube for sucking up on prey cells) and the pallium (a cytoplasmic veil that entirely covers prey cells). Most photosynthetic dinoflagellates (here informally called dinophytes) are also active predators (i.e. mixotrophs) and this ultimately allowed some dinophyte groups to replace their ancestral peridinin plastids for others of chlorophyte or haptophyte origin. Another very unusual feature found among dinoflagellates (in the Dinokaryota, which covers the Noctilucea, Sulcodinea, and Peridinea) is the 'dinokaryon', a nucleus which has permanently condensed chromosomes, phycodnavirus-like and bacterial histone-like basic proteins instead of proper histones to package bulk DNA, and massive amounts of DNA (Gornik et al., 2012; Janouškovec et al., 2017). Dinozoans are inferred to have descended from a plastid-bearing mixotrophic ancestor quite like that from which apicomplexans (sensu lato) are thought to have evolved. Further aspects of dinozoan biology can be found in Saldarriaga and Taylor (2017).

2.4 Taxonomy of the Dinozoa

Here we follow the updated scheme of Cavalier-Smith (2017) for Dinozoa taxonomy, which is largely in agreement with the latest phylogeny of Janouškovec et al. (2017). The dinozoans comprise both the Perkinsozoa and the Dinoflagellata (Fig. 1). Within the Perkinsozoa, we so far only find intracellular parasites of animals (*Perkinsus*), dinoflagellates (*Parvilucifera*), or cryptophyte algae (*Rastrimonas*) (Reñé, Alacid, Ferrera, & Garcés, 2017).

The Dinoflagellata comprises the rest of dinozoan diversity (see above for some examples). The 'basal groups' of dinoflagellates include predatory flagellates (like *Oxyrrhis* and *Psamossa*; the Eodina) and diverse intracellular parasites (the Syndina). Both groups are probably paraphyletic, and the Syndina includes the marine alveolate groups (MAGs) I and II. The Dinokaryota, informally known as the 'core dinoflagellates', contains the bioluminescent and the giant predatory *Noctiluca* (Noctilucea), and all other dinoflagellates (Sulcodinea, Karlodinea, and Dinophycidea in Fig. 1) among which we first find examples of the ancestral photosynthetic peridinin plastid. All nondinokaryote dinozoans are heterotrophic and some, like *Oxyrrhis* and even the perkinsozoan *Perkinsus*, have relicts of the peridinin plastid found among dinophytes (see Fig. 1). Sister to the Noctilucea is a large group of dinoflagellates that comprises such diverse unicells as all photosynthetic dinozoans (i.e. dinophytes), armoured (thecate) dinoflagellates, ocelloid-guided predators, and kleptoplastidic mixotrophs and among others. Within the Dinophycidea are the common orders of the Gymnodiniales, Gonyaulacales, Peridiniales, Prorocentrales, Dinophysiales, and Suessiales ($\approx$Symbiodiniaceae). For a conservative and more informal scheme, see Saldarriaga and Taylor (2017). For a morphology-based scheme, see Hoppenrath (2017).

3. THE ORIGIN OF MYZOZOAN PLASTIDS

There are currently two main competing groups of ideas about how myzozoans came to have plastids. The first group views the origin of myzozoan plastids as direct vertical descendants from a plastid common ancestor shared with all other red meta-algae. This view is epitomized by the chromalveolate hypothesis that states that the plastids of red meta-algae (i.e. cryptophytes, haptophytes, ochrophytes, dinophytes, chromerid algae, and their nonphotosynthetic apicomplexan descendants) were inherited vertically from a single and ancestral secondary endosymbiosis between a protozoan and a red algal unicell (Cavalier-Smith, 1999). The second group views the origin of plastids in red meta-algae by a succession of higher-order endosymbiosis (lateral spreading), usually starting with a secondary endosymbiosis with a red alga to give rise to the plastids of cryptophytes.

Different hypotheses exist on how secondary red plastids were transferred among red meta-algae. In regard to myzozoans, Sanchez-Puerta and Delwiche (2008) first suggested that myzozoans acquired their plastids from a single (tertiary) endosymbiosis with either a haptophyte or a hacrobian ancestor (of both cryptophytes and haptophytes). Bodyl, Stiller, and Mackiewicz (2009) suggested a haptophyte origin of the dinophyte peridinin-containing plastid, but remained vague about the precise origin of the apicomplexan plastid. Dorrell and Smith (2011) more generally suggested a haptophyte origin of the myzozoan plastid. Petersen et al. (2014) postulated independent origins for the apicomplexan and dinozoan plastids without specifying donors. More recently, and based on new plastid phylogenies, Ševčíková et al. (2015) suggested that apicomplexan plastids evolved from an ochrophyte most closely related to a limnistan (eustigamtophycean or chrysophycean) alga. However, the support for this phylogenetic association was equivocal and might stem from artefacts in tree reconstruction due to the high divergences (long stems in trees resulting in long-branch attraction artefacts) of apicoplast and eustigmatophycean plastid genomes. Based on the findings of Ševčíková et al. (2015), Füssy and Oborník (2017) argued that it is possible that, early in their evolution, apicomplexans replaced an ancestral myzozoan plastid with one of ochrophyte origin. Bodył (2017) now postulates that myzozoan plastids evolved from a quaternary endosymbiosis with an ochrophyte, but dinophytes later replaced this ancestral plastid with another one of haptophyte origin to give rise to the typical peridinin plastid.

In summary, four possibilities have been imagined (almost every possibility) for the origin and evolution of myzozoan plastids: (1) myzozoans ancestrally had a plastid that has been inherited vertically from a distant ancestor (i.e. a single ancestral secondary endosymbiosis, the chromalveolate hypothesis); (2) myzozoans ancestrally had a plastid, but it was acquired through a higher-order endosymbiosis (from a haptophyte or an ochrophyte) before their diversification; (3) myzozoans ancestrally had a plastid (by either 1 or 2), but dinozoans (or apicomplexans; Füssy & Oborník, 2017) replaced this ancestral plastid to give rise to their divergent perdinin plastid; or (4) the taxa Apicomplexa and Dinozoa acquired their plastids independently from each other after their divergence from a common nonphotosynthetic myzozoan ancestor (Waller & Kořený, 2017). The most parsimonious views, in our opinion, assume a single ancestral myzozoan plastid that was inherited vertically by both dinozoans and apicomplexans (compatible with 1 or 2).

4. DIVERSITY OF PLASTIDS IN THE MYZOZOA

Many plastid types arose from the diversification of the ancestral myzozoan plastid. Photosynthetic plastids (sometimes referred to as chloroplasts) are found on both sides of the Myzozoa tree: in many dinokaryotes (Dinozoa) and in some apicomonads (chromerid algae; Apicomplexa). All other plastids found in the Myzozoa are nonphotosynthetic; these are called leucoplasts. Some leucoplasts have plastomes (like in sporozoanas) but others have lost them (like colpodellids and perkinsozoans). We now know that leucoplasts were lost at least twice in the Myzozoa; once in the Dinozoa (*Hematodinium*) and once in the Apicomplexa (*Cryptosporidium*). See Fig. 1 for a distribution of different plastid types across the major myzozoan lineages.

4.1 Sporozoan Leucoplasts Still Retain Plastomes

Sporozoans have small genomes (plastomes) in the stroma of their biosynthetic relict plastids. Actually, sporozoan plastids are the only myzozoan leucoplasts with plastomes (see Fig. 1; but see Gavelis et al., 2015 and Fawcett & Parrow, 2014 for the description of two understudied dinoflagellates that might have also retained plastomes in their leucoplasts, *Nematodinium* sp., and one strain of *Esoptrodinium* sp, respectively). The plastid DNA (ptDNA; a 35-Kb circular DNA molecule) was first identified in 1975, but it was first thought to be mitochondrial DNA (Kilejian, 1975). Only later was the true mitochondrial DNA identified (a 6-Kb linear DNA molecule; Suplick, Akella, Saul, & Vaidya, 1988; Vaidya, Akella, & Suplick, 1989), and the real ptDNA localized to spherical bodies (Köhler et al., 1997; McFadden, Reith, Munholland, & Lang-Unnasch, 1996). The plastidic nature of the ptDNA was confirmed by restriction mapping and sequencing of some of its genes (Gardner, Feagin, et al., 1991; Gardner, Williamson, & Wilson, 1991). Spherical bodies were then renamed 'apicoplasts' for *api*complexan *plast*id (Köhler et al., 1997). Apicoplasts turned out to be surrounded by four membranes and because of their nonphotosynthetic nature they lack all pigments and thylakoids. The presence of derived plastids within apicomplexan parasites immediately pointed to their algal ancestry.

4.2 Apicoplasts Have a Red Algal Ancestry

The first attempts to decipher the phylogeny of apicoplasts debated the origin of apicoplasts from either a red or a green alga (Funes, Reyes-Prieto, Pérez-Martínez, & González-Halphen, 2004). Conflicting evidence fuelled this controversy (Arisue & Hashimoto, 2015). Support for a green algal origin of apicoplasts came from some single-gene (Funes et al., 2002; Köhler et al., 1997) and multigene phylogenies (Cai, Fuller, McDougald, & Zhu, 2003; Lau, McElwain, Brayton, Knowles, & Roalson, 2009), but also from a rare split of the mitochondrial *cox2* gene that is shared between apicomplexans and green algae (Funes et al., 2002). In contrast, support for a red algal origin of the apicoplast came from phylogenies of the plastid 16S rRNA gene (Zhang, Green, & Cavalier-Smith, 2000), the plastid but nucleus-encoded GAPDH gene (Fast, Kissinger, Roos, & Keeling, 2001; Harper & Keeling, 2003), and similarities in the organization of apicoplast and red algal plastomes (Blanchard & Hicks, 1999). Today, it is well accepted that apicoplasts ultimately descended from a red alga. The phylogenetic affiliation of apicoplast genes to those of green plastids was shown to be artefactual, and the rare split in the *cox2* gene was found to be convergent (Waller & Keeling, 2006; Waller, Keeling, van Dooren, & McFadden, 2003). The evidence also seems to be strong enough to view apicoplasts as sisters to the peridinin-containing plastids of dinoflagellates, and chromerid plastids as links between the two; all of them having descended vertically from a common myzozoan plastid ancestor (Janouškovec, Horák, Oborník, Lukes, & Keeling, 2010).

4.3 Dinozoans Exhibit a Great Diversity of Plastids

Only half of the known species of dinozoans have photosynthetic plastids (Fig. 1; Saldarriaga, Taylor, Keeling, & Cavalier-Smith, 2001). Of these, most have a type of plastid that is thought to be ancestral to dinozoans, the peridinin plastid, and which is likely to be a divergent descendant of the ancestral myzozoan plastid (see discussion on the origins of myzozoan plastids above). This peridinin plastid was early on shown to be of red algal origin and to be related to those of other red meta-algae (Zhang et al., 2000). But the peridinin plastid has some unique features that distinguish it from those of all other red meta-algal plastids. Besides the accessory carotenoid pigment peridinin, the archetypical dinophyte plastid also has chlorophyll *a* and c_2, a three-membraned envelope and a greatly divergent plastome. Another bizarre feature of dinophyte peridinin plastids is their RuBisCO type II (to fix CO_2) of proteobacterial rather than cyanobacterial origin. This ancestral replacement by lateral gene transfer was first thought to be a unique and defining feature of peridinin plastids, but is now also known to be shared with apicomonad algae—a laterally acquired RuBisCO was present in the ancestral myzozoan (Janouškovec et al., 2010).

Some groups of dinophytes have replaced their ancestral peridinin plastid (Fig. 1). In some, the newly acquired plastid or endosymbiont could be alongside a no longer photosynthetic peridinin plastid (this is clearly the case in the 'dinotoms' and *Dinophysis*). There are two clear examples of replacing plastids among dinophytes. The first involves some members of the Gymnodiniaceae (*Lepidodinium chlorophorum* and *Lepidodinium viridae*) which have (secondary) green plastids with chlorophyll *a*, *b* but no peridinin. These green plastids have a pigment composition typical of green algae, and they are also surrounded by four membranes; they also have a 'nucleomorph'. Moreover, their green algal affinity has been confirmed by ultrastructure, biochemistry, and phylogeny (Matsumoto et al., 2011; Matsumoto, Kawachi, Miyashita, & Inagaki, 2012). The specific green algal donor of the green plastid of *L. chlorophorum* was shown to be a pedinophyte (a chlorophyte) based on plastome phylogenies (Kamikawa et al., 2015). The second example is that of *Karenia*, *Karlodinium*, and *Takayama* (Kareniacea) which now have a so-called (tertiary) fucoxanthin plastid. This plastid has the typical pigment composition of a haptophyte plastid (e.g. chlorophyll c_1, c_2, and fucoxanthin but no peridinin) and is also surrounded by four membranes (but no nucleomorph). The origin of the fucoxanthin plastid in the Kareniaceae has also been strongly demonstrated based on phylogenies of plastid- and nucleus-encoded genes for plastid proteins (Gabrielsen et al., 2011; Tengs et al., 2000; Yoon et al., 2005).

A subgroup in the Peridinales, the so-called 'dinotoms', has recently acquired tertiary diatom (Ochrophyta) endosymbionts, which have plastids of red algal origin themselves. These endosymbionts are barely reduced (only the diatom outer shell or frustule seems to have been lost) and thus are not properly called organelles yet. Indeed, the dinotoms *Kryptoperidinium foliaceum* and *Durinskia baltica* derive photosynthate from their endosymbionts (Hehenberger, Burki, Kolisko, & Keeling, 2016). Even though 'dinotoms' are a monophyletic group within the Peridiniales, their diatom endosymbionts have been acquired multiple times independently. Indeed a remarkable example of endosymbiotic convergence likely facilitated by some sort of a constraint (Yamada, Sym, & Horiguchi, 2017). Many other diverse dinoflagellates are also known for engaging in kleptoplastidy, or the stealing of prey's plastids to temporarily tap on them (Waller & Kořený, 2017). A classic example of a kleptoplastidic dinoflagellate is *Dinophysis* which harbours kleptoplastids of cryptophyte origin that are acquired indirectly through the ciliate *Mesodinium rubrum*.

4.4 Apicomonads Have Ancestral-Type Plastids

The closest photosynthetic relatives of sporozoans are the chromerids or apicomonad algae *Chromera velia* and *Vitrella brassicaformis* (Moore et al., 2008; Oborník et al., 2012). *Chromera* and *Vitrella* are not each other's closest relatives but are more closely related to free-living heterotrophic myzozoan predators called colpodellids (Fig. 1; Janouškovec et al., 2015). Apicomonad photosynthetic plastids constitute 'missing links' between sporozoan and dinophyte plastids by possessing features that are present in either one or the other. For example, apicomonad plastids are photosynthetic like dinophyte plastids, but are surrounded by four membranes like the leucoplasts of sporozoans and the dinozoan *Perkinsus*. Despite being more closely related to sporozoan apicoplasts, chromerid plastids share several features with dinophyte perdinin plastids like thylakoids stacked in triplets, a type II RuBisCO, and polyuridinylated plastome transcripts. In terms of major photosynthetic pigments, chromerids have chlorophyll *a* but no chlorophyll *c*, unlike peridinin dinophytes that have both (chlorophyll *c* which is the hallmark pigment of red meta-algae; Janouškovec et al., 2010; Moore et al., 2008). The plastomes of apicomonad photosynthetic plastids have gene contents that encompass the nonoverlapping sets found in both sporozoan and dinophyte plastids. Phylogenies of plastomes have also confirmed that chromerid plastids are more closely related to apicoplasts and perdinin plastids than to other red meta-algae (Janouškovec et al., 2010).

4.5 Why Do Myzozoans Retain Leucoplasts?

The reason why leucoplasts, like apicoplasts, are retained by many myzozoans (and some other ancestrally but no longer photosynthetic groups) is that plastids have become highly integrated with the overall cytosolic metabolism of their host cells. Host cells came to rely on plastids not only for photosynthesis, which is dispensable depending on life style, but also for the biosynthesis of fatty acids, isoprenoids, haeme, and iron–sulfur (Fe–S) clusters. For example, apicomplexans plastids export fatty acids, isoprenoids, and haeme to the cytosol (or mitochondrion), whereas iron–sulfur clusters are required for the biogenesis of plastid enzymes involved in the biosynthesis of fatty acids and isoprenoids (van Dooren & Hapuarachchi, 2017).

The ultimate evolutionary answer to the issue of leucoplast retention, though, might be a combination of historical constraints and efficiency through compartmentalization (selective constraints). Even if the leucoplast plastome is lost by transferring its remaining genes to the nucleus, leucoplasts remain a place for important metabolic pathways (e.g. isoprenoid biosynthesis) on which the cytosolic metabolism relies—some myzozoan have plastome-less leucoplasts. And whole pathways might not be easy to relocate to the cytosol. For this to happen, all plastid-targeted enzymes should lose their plastid localization simultaneously. So there has been strong phylogenetic inertia for the location of this plastid enzymes, i.e., their relocation to the cytosol would require multiple improbable changes whose intermediate states would be detrimental. On the other hand, it is also possible, but less plausible, that there is an adaptive value in compartmentalizing plastid biosynthetic pathways in a small compartment like the sporozoan apicoplast. Metabolic compartmentalization improves efficiency (by increasing concentrations of metabolites and enzymes) and might contain potential toxic metabolic intermediates.

The first myzozoan ancestor (or an earlier ancestor) was a chimeric cell with redundant metabolism as a result of both plastid and cytosolic pathways for the synthesis of haeme (tetrapyrroles), isoprenoids, and fatty acids. But during myzozoan diversification, metabolic redundancy allowed for the chancy loss of cytosolic pathways, leaving the cell dependent on plastid pathways. Isoprenoid biosynthesis appears to be the most indispensable plastid pathway because it is conserved by every myzozoan that has retained a plastid organelle (Janouškovec et al., 2015; Waller, Gornik, Koreny, & Pain, 2016).

4.6 Why Do Some Sporozoan Leucoplasts Retain Plastomes?

Numerous hypotheses have been formulated to explain why endosymbiotic organelles retain genomes. However, only few of them apply to non-photosynthetic plastids, as they have dispensed with an electron transport (photosynthetic) chain and their plastomes do not encode particularly hydrophobic proteins (Barbrook, Howe, & Purton, 2006). Why do some sporozoans keep their apicoplast plastomes? Apicoplasts are the only non-photosynthetic plastids among myzozoans that are known to retain plastomes. Most of the genes encoded by the apicoplast plastome are transcription and translation genes such as ribosomal proteins, tRNAs, and a RNA polymerase. The only apicoplast plastome-encoded genes that fall outside these categories are *sufB*, *clpC*, and *ycf93*. Therefore, all other apicoplast plastome-encoded genes are there to support the expression of *sufB*, *clpC*, and *ycf93*. The 'limited transfer window' hypothesis best explains the persistence of a plastome among myzozoan leucoplasts (Barbrook, Howe, et al., 2006). The 'limited window transfer' hypothesis states that species with few or one plastid per cell have extremely low rates of gene transfer (or endosymbiotic gene transfer, EGT) from the plastome to the nuclear genome (Barbrook, Howe, et al., 2006). EGT is primarily driven by the release of ptDNA from lysed organelles that get incorporated into nuclear genomes. If the single apicoplast of a sporozoan cell lyses there is no way to regenerate this organelle and the cell would die. This in turn suggests that the reason why plastomes remain in apicoplasts is simply because some genes like *sufB*, *clpC*, or *ycf93* have not had a chance to be successfully transferred to the nuclear genome. Because examples of successful transfers of *sufB* and *clpC* to the nucleus of some myzozoans are known (see Janouškovec et al., 2015), the retention of plastomes in sporozoans is best seen as a simple historical accident. The adaptationistic alternative, the 'essential tRNA' hypothesis, runs into important counterexamples among sporozoans (see Janouškovec et al., 2015 for a discussion).

4.7 Plastome Loss in Some Myzozoans

The most extreme cases of plastome reduction would be exemplified by the outright loss of the plastome in some nonphotosynthetic eukaryotes. Several (nonsporozoan) myzozoans are known to have lost their plastomes but retained their plastid organelles for metabolic functions (e.g. fatty acid and isoprenoid biosynthesis) sustained by plastid-targeted nuclear genes (see Fig. 1). The colpodellids *Alphamonas*, *Colpodella*, and *Voromonas* (Apicomonada; see Fig. 1) seem to have lost their plastomes (Gile & Slamovits, 2014; Janouškovec et al., 2015). Among dinozoans, the perkisozoan *Perkinsus* is also reported to have lost its plastome, and no trace of a plastome has been found in the early-diverging nonphotosynthetic dinoflagellates *Oxyrrhis*, *Noctiluca*, and *Crypthecodinium* (Janouškovec et al., 2017; Sanchez-Puerta, Lippmeier, Apt, & Delwiche, 2007; Slamovits & Keeling, 2008). The more derived dinophyte *Dinophysis* has also retained the ancestral myzozoan plastid, but without its plastome (Janouškovec et al., 2017). The other cases of reported plastome losses among eukaryotes are the green alga *Polytomella* (Smith & Lee, 2014), and the parasitic land plant *Rafflesia lagascae* (Molina et al., 2014), both in the green plastid lineage.

4.8 Plastid Loss Among Myzozoans

The strong metabolic dependency that myzozoan cells have on their plastids makes plastid loss a rare evolutionary event. Only one case on plastid loss has been fully confirmed among apicomplexans: that of the intestinal parasite

Cryptosporidium (more closely related to gregarines; Zhu, Marchewka, & Keithly, 2000). The only other case of outright plastid loss documented among all eukaryotes happened in the dinozoan *Hematodinium* (and by extension also in its sister *Amoebophrya* (Syndina); Gornik et al., 2015, and also see Janouškovec et al., 2017). This dinoflagellate seems to have lost its plastid organelle (and biosynthetic pathways therein) before losing any cytosolic pathways for haeme, isoprenoid, or fatty acid biosynthesis. *Hematodinium* retained an ancestral metabolic redundancy by preserving the cytosolic pathways for fatty acid and haeme biosynthesis that allowed this crustacean parasite to dispense with its plastid organelle. And this was complemented by the evolution of isoprenoid scavenging from its animal host. The apicomplexan *Cryptosporidium*, in contrast, appears to have only conserved the cytosolic pathway for fatty acid biosynthesis, but evolved means to steal haeme and isoprenoids from its animal host cells. Because of this, *Cryptosporidium* was able to lose its plastid organelle. Knowledge is scarce about the very diverse gregarines, but similarly to *Cryptosporidium*, *Gregarina niphandrodes* might have lost its plastid organelles and genomes (Toso & Omoto, 2007). All studied colpodellids and classical intracellular sporozoans have retained plastid organelles, with or without plastomes (Fig. 1).

5. THE PLASTOMES OF MYZOZOANS

5.1 The Plastid Genomes of Apicomonad Algae

Myzozoans exhibit a great diversity of plastomes. Among apicomplexans, only apicomonad plastomes (those of chromerid algae) support photosynthesis. The plastomes of the chromerids *C. velia* (Moore et al., 2008) and *V. brassicaformis* (CCMP3315; Oborník et al., 2012) have the largest sizes and gene repertoires among all myzozoans (with the exception of tertiarily acquired plastids by some kareniacean dinophytes; Figs 2 and 3). Their gene repertoires encompass the reduced set of 12 photosynthetic genes found in most dinophyte plastomes, but also the translation and transcription genes found in sporozoan plastomes. But the gene content of apicomonad algae is still smaller than those of other red meta-algae such as haptophytes, cryptophytes, and ochrophytes (Fig. 3; see Janouškovec et al., 2010). This suggests that some degree of genome reduction through gene loss had already happened before the diversification of modern myzozoans. Because of their relatively big sizes and gene repertoires, chromerid plastomes are the most similar to the ancestral myzozoan plastome.

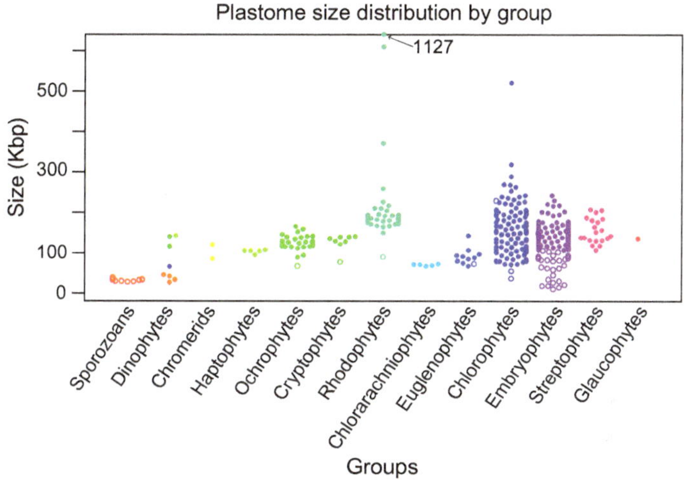

Fig. 2 The distribution of plastome sizes among all eukaryotic algal groups. The rhodophytes have recently been found to comprise the most extraordinarily large plastomes known, but this is a derived condition. In stark contrast, myzozoans, i.e., sporozoans and dinophytes, possess some of the smallest plastomes across eukaryotic algae. But some embryophytes (land plants) hold the record for the most reduced plastomes. The ancestral peridinin plastid of dinophytes can reach sizes smaller than those of the Sporozoa (e.g. in *Symbiodinium*). Derived plastids among dinophytes have larger plastomes and are *coloured* according to their provenance; the plastomes of peridinin dinophytes are coloured in orange. The plastomes of nonphotosynthetic plastids, or leucoplasts, are represented by *empty circles*. The database of plastome sizes used to make this figure can be found at: https://doi.org/10.17632/frxt79djmr.1.

The plastomes of the apicomonad algae *Chromera* and *Vitrella* are also considerably divergent relative to each other. Whereas *Vitrella* has a compact plastome with a size of only 85.5 Kbp, *Chromera*'s plastome is 121.2 Kbp in size. Despite this difference in size, *Vitrella*'s plastome encodes more genes than *Chromera*'s (81 vs 74 genes; Janouškovec et al., 2010; Oborník & Lukeš, 2015). The plastome of *Chromera* is also unusually divergent in (i) being considerably rearranged in comparison to those of sporozoans and *Vitrella*, (ii) possessing genes with long extensions, (iii) having split genes encoding for separately translated protein fragments, and (iv) being a noncircular-mapping linear ptDNA with terminal repeats (Janouškovec et al., 2013). *Vitrella*'s plastome, in contrast, lacks all these divergent oddities seen in *Chromera*'s plastome. The plastome of *Vitrella* has a canonical quadripartite organization shared with most apicomplexans (see below), has retained a 5S rRNA gene (unlike *Chromera*'s, dinoflagellate, and sporozoan plastomes), and also has a one of the highest GC

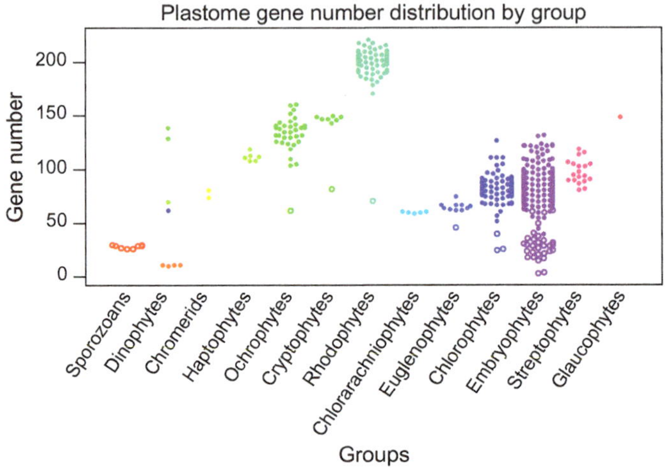

Fig. 3 The distribution of plastome gene repertoires among all eukaryotic groups. Myzozoans ancestrally have plastomes with smaller gene repertoires than other red meta-algae, as seen in *Chromera* and *Vitrella*. The origin of the Dinozoa and Apicomplexa further led to independent episodes of gene loss in the plastomes of each lineage. Derived plastids among dinophytes have larger gene repertoires than peridinin plastids and are *coloured* according to their provenance (see main text for discussion); the plastomes of peridinin dinophytes are coloured in orange. The plastomes of non-photosynthetic plastids, or leucoplasts, are represented by *empty circles*. Gene repertoires reported here are only based on the number protein-coding genes as reported primarily by NCBI (and some other minor sources). RNA-specifying genes and pseudogenes were ignored. Because the gene number reported depends on annotations and these are not unified, there is some variability in the numbers, but all obvious outliers were manually removed. The database of plastome gene numbers used to make this figure can be found at: https://doi.org/10.17632/frxt79djmr.1.

contents (47.7%) among eukaryotes (Fig. 4; Janouškovec et al., 2010, 2013). These features point to the less divergent nature of *Vitrella*'s plastome and suggest that the myzozoan cenancestor had a plastome more similar to *Vitrella*'s than to any other modern myzozoan currently known.

5.2 The Plastid Genomes of Sporozoans and Dinophytes

Dinophyte and sporozoan plastomes are considerably divergent from those of chromerids. Both groups have quite reduced plastomes, but in very different ways. Sporozoan plastomes have dispensed with all photosynthetic genes, whereas dinophyte plastomes have essentially only retained photosynthetic genes. The sporozoan plastome is contained within a single DNA molecule that is circular, but the dinophyte peridinin plastid genome has been fragmented into several 'minicircles', most of which encode one single gene.

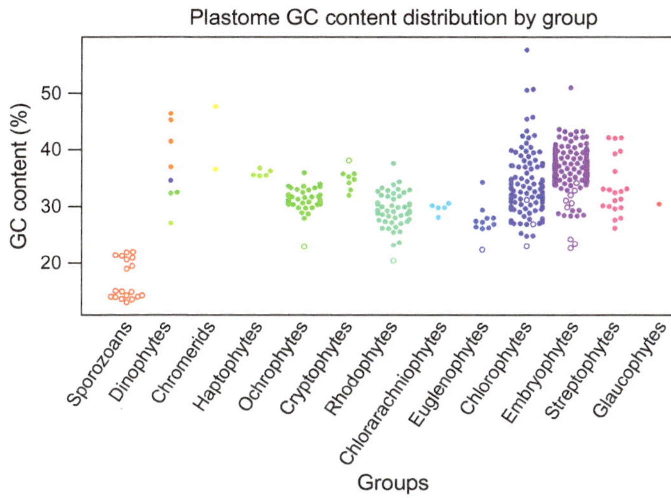

Fig. 4 The distribution of plastome GC content among all eukaryotic algal groups. Sporozoans have plastomes heavily compositionally biased towards low GC contents, and constitute the most extreme example among all eukaryotic algae. Among sporozoans, haemosporidians (e.g. *Plasmodium* and *Leucocytozoon*) have the lowest GC contents, whereas the plastomes of piroplasmids and coccidians have slightly higher GC contents. Among dinophytes, peridinin plastid plastomes have higher GC contents than those more recently acquired through serial secondary or tertiary endosymbiosis. The plastomes of nonphotosynthetic plastids, or leucoplasts, are represented by *empty circles*. Derived plastids among dinophytes are *coloured* according to their provenance; the plastomes of peridinin dinophytes are coloured in orange. The database of plastome GC contents used to make this figure can be found at: https://doi.org/10.17632/frxt79djmr.1.

5.2.1 Plastome Size and Gene Repertoires

Myzozoans are also extraordinary in having some of the smallest plastomes. Sporozoans have apicoplast plastomes that range from 28.6 to 39.5 Kbp in size (see Fig. 2). But even though apicoplast plastomes are incredibly small, some nonphotosynthetic land plants (or embryophytes) have reduced their plastomes even further (Fig. 2). For example, the plastomes of *Pilostyles* and *Epigogium* have sizes of just 11.4 Kbp (Bellot & Renner, 2015) and 19 Kbp (Schelkunov et al., 2015), respectively. The parasitic green alga *Helicosporidium* is another example of plastome reduction within the green plastid lineage (37.4 Kbp in size) and represents another interesting case of convergent evolution with the plastomes of sporozoans (de Koning & Keeling, 2006). These examples are the most extreme, but they are found within the green plastid lineage. Sporozoans, on the other hand, have the smallest plastomes for the red plastid lineage (compare to green plastids in Fig. 2).

Plastome expansion, the opposite to plastome reduction in non-photosynthetic parasites like sporozoans, is seen among primary plastids. Some green and red algae have massively expanded their plastomes by the accumulation of different kinds of noncoding DNA (introns, insertion sequences, or repetitions) and have then reached sizes of up to 1.13 Mbp in the case of the red algal unicell *Corynoplastis japonica* (see distant outlier for rhodophytes in Fig. 2; Muñoz-Gómez et al., 2017). One recent example also shows that leucoplast plastomes (like those of apicoplast's) are not immune to expansion. Even though it has lost all photosynthetic genes, the plastome of the heterotrophic green alga *Polytoma uvella* has inflated to a size of 230 Kbp, 75% of which is noncoding DNA (Fig. 2; Figueroa-Martinez, Nedelcu, Smith, & Reyes-Prieto, 2017). It has been suggested that the reason for this lies in that *Polytoma* is a free-living unicell, and so it does not necessarily experience the evolutionary forces that drive genome compaction in parasites like sporozoans (Figueroa-Martinez, Nedelcu, Reyes-Prieto, & Smith, 2017).

The current sizes and gene repertoires of dinoflagellate and apicomplexan plastomes seem to have been achieved through at least four episodic events of plastome reduction, the first two of which are associated to symbiogeneses (Oborník, Janouškovec, Chrudimský, & Lukeš, 2009). The largest and most ancestral plastome gene repertoires are found among the red algae (Fig. 3), but they only represent a fraction of the total gene number of their cyanobacterial genome progenitors. The progenitor of all plastomes was probably a cyanobacterial genome of only about 3.05 Mbp in size (2929 protein genes; Ponce-Toledo et al., 2017), and the ancestral plastome was about 200 Kbp in size ($\approx$200 protein genes; Figs 2 and 3), most similar to those of modern red algae like bangiophyceans and florideophyceans. The symbiogenetic origin of primary plastids was then the first episode of drastic plastome reduction. The gene repertoires of most red meta-algal groups reflect their red algal ancestry: they have, on average, larger gene repertoires than most green plastids (both primary and secondary), but still smaller than those of red algae (Fig. 3). Thus, the secondary symbiogenesis that led to the origin of red meta-algae was the second episode of plastome reduction. Chromerid plastomes most resemble the ancestral myzozoan plastome (74–81 protein genes), but they are notoriously reduced in comparison to those of other red meta-algae, i.e., cryptophytes ($\approx$147 protein genes), haptophytes ($\approx$111 protein genes), and ochrophytes ($\approx$134 protein genes); the third episode of plastome reduction (see Fig. 3). Dinophyte plastomes have the smallest gene set for any algal group. Sporozoan plastomes are also considerably reduced in terms of gene repertoires ($\approx$29 protein genes), but are still larger than their dinophyte sisters (≈11 protein genes). Both dinophyte and sporozoan plastomes greatly reduced after their divergence from a common ancestor, but they followed quite different evolutionary trajectories; the fourth episode of genome reduction.

5.2.2 Plastome GC Content

Another extreme feature of apicoplast plastomes is their strong nucleotide compositional bias. Indeed, their GC content is the lowest among all plastomes (see Fig. 4). Some apicoplast genes, like *rpl11*, are 95% AT. There is a general correlation between plastome size and GC content: the smallest plastome sizes have the lowest GC contents (compare leucoplast plastomes in Figs 2 and 3). It is a little surprising then that apicoplast plastomes have such strong compositional bias towards AT given that other plastomes, such as those of some heterotrophic land plants, are more reduced in size (see embryophytes in Fig. 2). Indeed, the correlation between plastome size and GC content is not perfect. Some exceptions are the leucoplasts of the cryptomonad *Cryptomonas paramecium* (38.14% GC) and the chlorophyte *Prototheca wickerhamii* (31.2% GC), whose plastomes are smaller but have higher GC contents than some of their photosynthetic relatives. Some peridinin dinophytes, which have the smallest plastomes known, have minicircles whose GC composition is relatively high (Fig. 4). In the case of myzozoans, evolutionary forces driving plastome reduction and high AT bias have been linked for apicoplast genomes, but it appears that the two trends have been unlinked during the evolution of peridinin plastomes. The comparatively high GC contents of the plastomes of apicomonad algae (*Vitrella* in particular) and many 'basal' dinophytes (as suggested by the third codon position in protein-coding genes; Dorrell et al., 2017) suggest that the ancestral myzozoan plastome was GC-rich.

5.2.3 Plastome Organization in Sporozoans

The most common and therefore ancestral plastome architecture among sporozoans corresponds to a circular-mapping DNA molecule with a quadripartite organization (Fig. 5). Inverted repeats (IRs) containing the small subunit (SSU) and large subunit (LSU) rRNA genes, as well as many tRNAs, divide the plastomes into a large single copy (LSC) region which virtually encompasses all other genes, and an extremely shrunk small single copy (SSC) region that contains no genes at all (Arisue & Hashimoto, 2015). This genome architecture is shared among haemosporidians, coccidians, and *Nephromyces*. A quadripartite organization of the plastome seems to be an

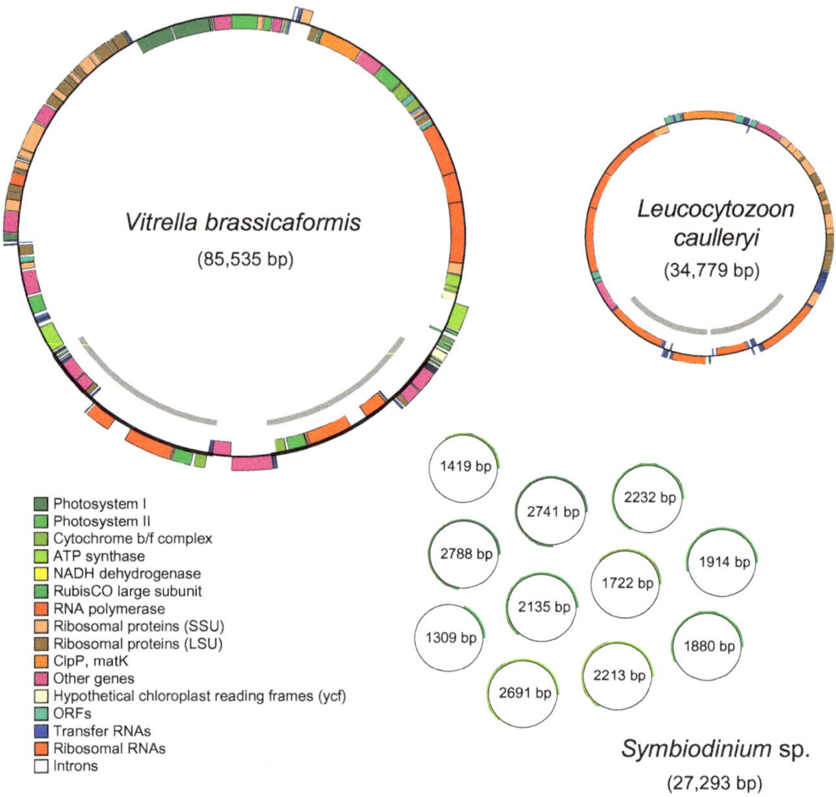

Fig. 5 Plastome organizations and structures among myzozoans. The plastomes of the apicomonad alga *Vitrella brassicaformis*, the haemosporidian sporozoan *Leucocytozoon caulleryi*, and the dinophycean *Symbiodinium* sp., are used as representatives for their groups. Apicomonad algae have ancestral-like plastomes that support photosynthesis and have the largest gene repertoires among myzozoans. Dinophytes have highly divergent plastomes that still support photosynthesis but are fragmented into plasmid-like minicircles, each encoding one to few genes. Sporozoans have small plastomes that do not support photosynthetic plastids but metabolic apicoplasts that make fatty acids, isoprenoids, and haeme.

ancestral feature to all plastids, although it also seems to be a feature prone to be lost or evolved convergently. There is also a strong strand polarity in the apicoplast plastomes of haemosporidians, coccidians, and *Nephromyces*, with half of the plastome having genes on one strand, whereas the other half having genes on the opposite strand (Fig. 5).

Sporozoan plastomes ancestrally support nonphotosynthetic plastids and therefore have reduced by losing all genes for photosynthetic proteins.

The protein-coding gene content of the sporozoan plastome is reduced to a set of translation (*rps*, *rpl*, and *tufA*) and transcription (*rpo*) genes, the chaperone *clpC*, the iron–sulfur cluster biogenesis protein *sufB*, and the unknown but conserved *ycf93* gene. ClpC is a plastid chaperone required to properly deliver unfolded proteins to the ClpP proteases. SufB is a protein required for the biogenesis of iron–sulfur-containing proteins (like fatty acid and isoprenoid biosynthetic enzymes). Ycf93 seems not to be a ribosomal protein, but a membrane protein whose exact function remains unknown (Goodman & McFadden, 2014). There are also some ORFs encoded by apicoplast plastome whose functions remained unknown, but many of them might be divergent ribosomal proteins—our own searches reveal that most of the unknown ORFs have remote similarities to ribosomal protein genes, namely *rps13*, *rps16*, *rps17*, *rps18*, *rpl11*, *rpl19*, and *rpl20*. The gene repertoires of apicoplast plastomes are fairly stable and comprise about 30 protein-coding genes with only sporadic gene losses in some species (Fig. 3). The RNA-specifying gene content of apicoplast plastomes includes 24 tRNAs and 2 rRNA genes (the '16S' SSU and '23S' LSU rRNA genes); there is no trace of a '5S' rRNA gene. Apicoplast plastomes are also extremely compact with insignificant intergenic regions (i.e. gene dense), and many instances of overlapping genes. Another intriguing property of the apicoplast plastomes of coccidians and *Nephromyces* is that they use the stop codon UGA for tryptophan instead (Oborník & Lukeš, 2015). This alternative genetic code is also observed in *Chromera*, but not in *Vitrella*, and is therefore assumed to be ancestral to all apicomplexans but to have been lost in the plastomes of *Vitrella*, haemosporidians and piroplasmids (which together form a clade, see Fig. 1).

Piroplasmids, like *Babesia* and *Theileria*, possess the most divergent apicoplast genomes. Even though their apicoplast genomes are circular mapping, they have no IRs and the rRNA genes are thus found as single copy. Piroplasmid plastomes are considerably more rearranged than those of other sporozoans and have repetitive unknown ORFs with varying degrees of similarity to each other. All protein-coding genes are encoded on the same strand, i.e., there is absolute strand polarity in their plastomes (Sato, 2011). They also have duplicated *clpC* genes and have lost *sufB*—the latter suggests plastome loss would be easier in piroplasms. All of these divergent features were gained early in the evolution of piroplasmids and therefore are also derived among sporozoans.

5.2.4 *Nephromyces* Is the Deepest-Branching Sporozoans With a Plastome

We have recently performed a genomic survey of *Nephromyces*, a probably mutualistic apicomplexan. *Nephromyces* is an endosymbiont of molgulid tunicates, where it is found infecting the tunicate's renal sac (Saffo, McCoy, Rieken, & Slamovits, 2010). This unusual apicomplexan has a complex life cycle composed by a succession of disparate stages, several of which are extracellular (Saffo & Nelson, 1983). Our survey revealed that *Nephromyces* contains a typical sporozoan plastome, quite similar in structure to those of coccidians. In preliminary trees, *Nephromyces* appears as the deepest-branching sporozoan known with an apicoplast genome.

5.2.5 Plastome Organization in Dinophytes

The plastomes of dinophyte peridinin plastids are unlike any other plastome known. On one side, these plastomes encode fewer proteins than any other photosynthetic or nonphotosynthetic plastome; they are also the smallest photosynthetic plastomes known, being only between 27.2 and 45.8 Kbp in size (Barbrook, Voolstra, & Howe, 2014; Howe, Nisbet, & Barbrook, 2008). The genes encoded in 'peridinin' plastomes are considerably divergent in comparison to their homologues in other eukaryotic algae, having accumulated many nonsynonymous substitutions, indels, unusual codon-usage preferences and alternative translation initiation codons (Dorrell et al., 2017). On the other side, the genes are not arranged collinearly in a circular-mapping molecule as usual, but they are split into very small circular DNA molecules termed minicircles (Zhang, Green, & Cavalier-Smith, 1999). Most minicircles contain one gene (protein-coding, tRNA-, or rRNA-specifying), and a few have been found to contain two genes. The largest number of genes in a single minicircle was recorded for *Amphidinium carterae* (Sulcodinea), where the largest minicircle carries three identified (*psbD*, *psbE*, and *psbI*) and one unknown ORF (Barbrook, Santucci, Plenderleith, Hiller, & Howe, 2006). This, however, appears to be an exceptional situation, likely resulting from fusion of otherwise single-gene minicircles (Howe et al., 2008). In addition to the coding region, minicircles include a noncoding element termed 'core' (Howe et al., 2008) or 'conserved noncoding sequence' (CNS) (Mungpakdee et al., 2014). This element is found in all minicircles and it is likely to have a regulatory function by driving transcription of the gene (Mungpakdee et al., 2014). While highly similar among the minicircles in one given species, CNSs are species specific, although some similarity between strains of the C phylotype of *Symbiodinium*

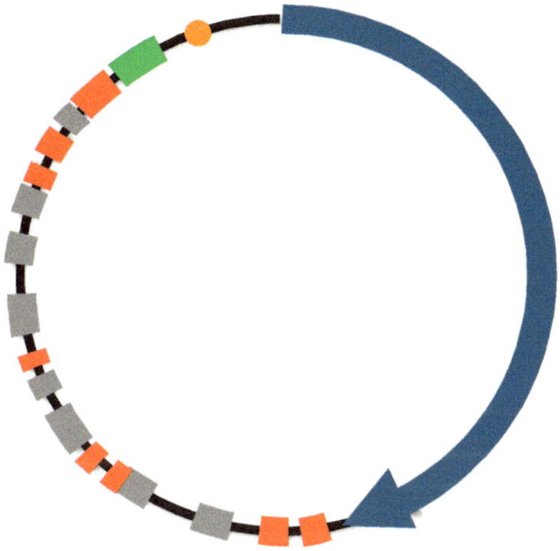

Fig. 6 Structure of a minicircle from the dinophyte *Symbiodinium* sp. The schematic representation of a single-gene minicircle shows the organization of the various elements found in most minicircles as determined in the most detailed analysis of a peridinin plastid genome conducted to date (Mungpakdee et al., 2014). The *blue arrow* represents either an ORF (if a protein-coding minicircle) or an rRNA gene. Upstream of the gene is the regulatory region consisting of a promoter (*green*) and a putative site for a pentatricopeptide RNA-binding protein (*orange circle*). *Minicircles* also contain a high density of conserved noncoding elements (*red*) and short repeats (*grey*).

sp. has been observed (Barbrook et al., 2014; Howe et al., 2003; Mungpakdee et al., 2014; Zhang et al., 1999). Aside from the coding region and the CNS core, small blocks of inverted and direct repeats are found throughout (Fig. 6; Barbrook et al., 2014; Mungpakdee et al., 2014). No function has been assigned or suggested for these small elements. It is possible that they constitute 'hot spots' of recombination. Several studies noticed certain level of heterogeneity in the composition of minicircles encoding a particular gene, and often the differences between different variants are due to small deletions spanning a few dozen base pairs (Santos, Gutierrez-Rodriguez, & Coffroth, 2003; Zhang et al., 1999). Though not yet experimentally studied, it is easy to envision that the abundance of small repeats throughout the minicircles can promote intermolecular recombination, resulting in a variety of rearranged forms. Other types of minicircle variants consistent with the occurrence of recombinational exchanges have been observed, including empty minicircles (Barbrook, Symington, Nisbet, Larkum, & Howe, 2001; Hiller, 2001), jumbled minicircles (Zhang, Cavalier-Smith, & Green, 2001), and microcircles (Nisbet, Koumandou, Barbrook, & Howe, 2004). Under closer scrutiny, some 'empty' circles were found to encode tRNA genes, although very few have been identified. In *A. carterae* and *Amphidinium operculatum* only one tRNA (formyl-methionine) appears to be encoded in the plastome. The tRNAs for proline and tryptophan (but not formyl-methionine) were found in *Heterocapsa triquetra* and *Heterocapsa pygmaea*. In contrast, no tRNAs were found in *Symbiodinium* spp., in spite of thorough examination in several species or isolates (Barbrook & Howe, 2000; Mungpakdee et al., 2014; Nelson et al., 2007; Nisbet et al., 2004; Zhang et al., 2001, 1999), therefore, the plastid must rely on tRNA molecules imported from the cytosol for protein synthesis.

Other unusual features reported for peridinin dinophytes include the possible nuclear localization (rather than plastidic) of the plastome minicircles in *Ceratium horridum* (Laatsch, Zauner, Stoebe-Maier, Kowallik, & Maier, 2004; plastome minicircles have nevertheless been experimentally shown to localize to the peridinin plastid stroma in *Amphidinium massartii*; Owari, Hayashi, & Ishida, 2014), and the possible lateral transfer of genes from nonphotosynthetic eubacteria to the plastomes of *Ceratium horridum* and *Pyrocystis lunula* (Mackiewicz, Bodył, & Moszczyński, 2013; Moszczyński, Mackiewicz, & Bodył, 2012; these reported laterally transferred genes are likely contaminants because they are not found in the close relatives of *Ceratium* and *Pyrocystis* (Dorrell et al., 2017).

5.2.6 The Plastomes of Derived Plastids in Dinophytes

There are currently four plastomes sequenced for dinophyte plastids that have replaced the ancestral peridinin plastid. The plastome of the tertiarily acquired plastid of the fucoxanthin dinophyte *Karlodinium veneficum* (Kareniacea) is considerably divergent relative to that of their haptophyte progenitor. *Karlodinium*'s plastome is larger than that of all haptophytes because of the expansion of its intergenic spacers (172.9 vs and average of 103.6 Kbp in size for haptophytes; see Fig. 2). It also encodes for considerably fewer protein-coding genes, only 70 rather than ≈111 like most haptophytes (Fig. 3). Furthermore, the 'fucoxanthin' plastome of *Karlodinium*'s is considerably rearranged and its gene sequences are fast evolving (as seen in phylogenies; Gabrielsen et al., 2011). Most interestingly, this plastome seems to also encode genes in extra-chromosomal elements that possibly resemble the minicircles of 'peridinin' plastomes—this points to convergence in plastome organization/structure due to constraints imposed by the genetic environment of the host (Espelund et al., 2012; Richardson, Dorrell, & Howe, 2014). The secondarily acquired plastid of the green dinophyte *L. chlorophorum* is also divergent relative to its chlorophyte ancestor, but to a lesser degree than in the Kareniaceae. The 'green' plastome of *Lepidodinium*'s is smaller (66.2 vs 98.3 Kbp; Fig. 2), more compact (shorter intergenic spacers; 13% vs 25% of the plastome) and has fewer genes (62 vs 82 protein genes; Fig. 3) than that of its closest chlorophyte relative, *Pedinomonas minor* (Kamikawa et al., 2015). Unlike green and fucoxanthin dinophytes, the dinotoms *Kriptoperidinium* and *Durisnkia* have plastomes which largely fall within the range of variation seen among their ochrophyte progenitors, i.e., they are not considerably divergent (see Figs 2–4; Imanian, Pombert, & Keeling, 2010). The state of dinotome plastomes probably reflects their most recent acquisition. In all the replacing plastids, however, plastome genes appear to now evolve faster relative to their progenitors (or donor groups). In the case of green and fucoxanthin dinophytes, their plastomes have undergone yet another episode of reduction that is associated with new symbiogeneses (on top of the preceding four; Figs 2 and 3).

5.2.7 Expression of Peridinin Plastome Genes in Dinophytes

The fragmented nature of the peridinin plastome is not the only unusual feature of these organelles. Unlike any other plastidic system, the transcripts of protein-coding genes in minicircles are polyuridylylated at their 3′ ends, resulting in a poly(U) tail spanning between 24 and 40 U residues in the mature mRNAs (Nelson et al., 2007; Wang & Morse, 2006). But such postranscriptional modification is also seen in the fucoxanthin plastids of *Karlenia mikimotoi* (Dorrell & Howe, 2012) and *Karlodinium veneficum* (Richardson et al., 2014), and in *Chromera velia* (Janouškovec et al., 2010). The poly(U) tail is not encoded in the minicircle DNA, and therefore it is inferred to be added posttranscriptionally by a yet to be identified enzyme. Transcription of minicircle-encoded genes involves synthesis of a primary RNA via a 'rolling circle' mechanism, which results in a long RNA spanning the minicircle up to several times. This long RNA is then cleaved into smaller pieces (pre-mRNAs) which are then subject to processing at both ends to produce a translatable monocistronic mRNA. Processing of the 3′ terminus involves trimming to a short 3′-UTR and polyuridylylation. Like the poly(A) tail of nuclear transcripts, the poly(U) tail is thought to contribute to stability and protection of the transcripts. The 5′ terminus of the pre-mRNA is also trimmed to a ~40 residue untranslated region (Fig. 7; Barbrook et al., 2012; Dang & Green, 2010). In some species, transcripts of plastid-encoded genes are subject to substitutional editing (i.e. a kind of RNA editing). This is, again, also seen in the

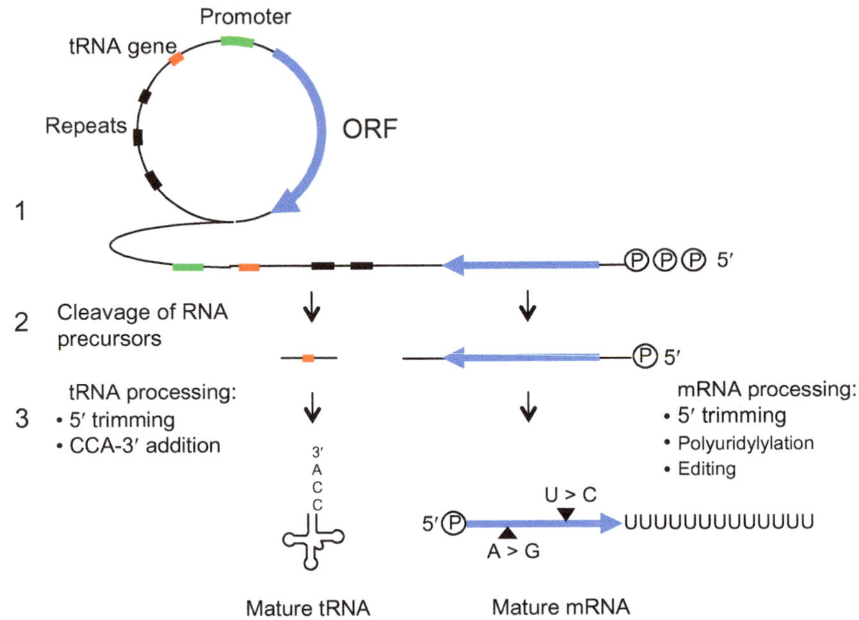

Fig. 7 A summary of the current understanding of gene expression in these organelles. Schematic representation of the steps involved in transcription of plastid-encoded genes in the peridinin plastids: (1) transcription by a yet not identified DNA-dependent RNA polymerase initiates from the 'core' or CNS (promoter) and presumably, RNA synthesis proceeds continuously spanning the entire minicircle more than once (*rolling circle*); (2) it was proposed that RNase Z-type RNases cleave the nascent transcript into 'pre-RNAs', each containing a gene; and (3) each gene is further processed into the mature forms (i.e. tRNA or mRNA).

fucoxanthin plastids of the Kareniaceae, further exemplifying convergent evolution with peridinin plastids (Jackson, Gornik, & Waller, 2013; Richardson et al., 2014). The extent of editing varies from species to species but also between genes. In *Ceratium horridum*, the genes encoding for PsbA, PsbB, and PsbE suffer editing in about 7% of their nucleotide positions, whereas in the SSU rRNA gene the proportion of edited sites was 3.3%. The most frequent substitutions were transitions, being A-to-G and U-to-C the most numerous substitutions, but all possible editing interconversions are known in dinophyte plastome transcripts (Dorrell & Howe, 2015). A similar pattern and proportion of substitutions were observed in *Lingulodinium polyedrum* (Wang & Morse, 2006), *H. triquetra* (Dang & Green, 2010), and in *Symbiodinium minutum* (Mungpakdee et al., 2014). No evidence of editing was found in *A. carterae* (Barbrook et al., 2001). Although editing affects a small proportion of nucleotides, the process seems to be critical for maintaining the proper performance of the encoded proteins. In *S. minutum*, a large majority of the substitutions resulting from editing (88%) caused amino acid changes (Mungpakdee et al., 2014). Like in other systems were editing happens, such as the mitochondria of plants, kinetoplastids, and other organisms (including dinoflagellates), translation of the DNA-encoded sequence results in loss of conserved and otherwise functionally important amino acid positions, or even in premature stop codons. Editing, therefore, is essential to overcome deleterious substitutions.

5.3 Availability of Myzozoan Plastomes

To date, as of September 2017, numerous apicoplast plastomes have been sequenced (see Table 1). However, many of them have been incompletely sequenced or assembled and are therefore found as partial in public databases. The presence of nearly identical IRs in many apicomplexan plastomes difficult their final assembly and circularization. Only two chromerid plastomes have been sequenced so far (for the only two culturable species), and there are currently eight plastomes sequenced for dinophytes, four for peridinin dinophytes, one for a green dinophyte, one for a fucoxanthin dinophyte, and two for dinotoms. In addition to the four fully (or almost fully) sequenced plastomes for peridinin dinophytes (*Amphidinium carterae* CCAP1102/6, *Amphidinium carterae* CS21, *Heterocapsa triquetra*, and *Symbiodinium* sp. clade C3), there are some few plastome minicircles/genes reported for *Adenoides eludens* (*psbA, psbD*), *Ceratium horridum* (*psaA, psaB, psbB, psbC, psbD, petB, ycf24, ycf16, psbE, psaB, psbC, psbD*), *Heterocapsa niei* (*psbA*, 23S rRNA), *Heterocapsa pygmaea* (*psbA*, 23S rRNA), *Heterocapsa rotundata* (23S rRNA, *psbA, trnW, trnP*), *Protoceratium reticulatum* (23S rRNA), *Pyrocystis lunula* (*rpl28* (?), *rpl33* (?), *psbC, psbC*), *Symbiodinium* sp. clade A (*psbA*), and *Symbiodinium* sp. clade B (*psbA*) (see Howe et al., 2008; Moszczyński, Mackiewicz, & Bodył, 2012).

6. GENE TRANSFER IN MYZOZOANS

Gene transfer from organelle to nuclear genomes, or EGT, is a well-known phenomenon (Martin, 2003). The main evolutionary function of EGT in the evolution of plastids has been to transfer genes from the ancestral plastome to the nuclear genome of its host. This has served to integrate plastids within their host cells, and to relieve plastome genes from mutational meltdown due to Muller's ratchet (the accumulation of deleterious mutations in asexual lineages). EGT has been well studied in plants (all eukaryotes

Table 1 Availability of Sequenced Plastomes for Species in the Myzozoa

Species	Accession Number	Source	Completeness	Reference
Sporozoa				
Babesia bovis T2Bo	AAXT01000007	NCBI GenBank	Complete	Brayton et al. (2007)
Babesia microti RI	LK028575	NCBI GenBank	Complete	Garg et al. (2014)
Babesia orientalis Wuhan	KT428643	NCBI GenBank	Complete	Huang et al. (2015)
Babesia sp. Lintan	KX881915.1	NCBI GenBank	Complete	Wang et al. (2017)
Babesia sp. Xinjiang	KX881914.1	NCBI GenBank	Complete	Wang et al. (2017)
Cyclospora cayetanensis CHN HEN01	KP866208	NCBI GenBank	Complete	Tang et al. (2015)
Eimeria tenella Penn State	AY217738	NCBI GenBank	Complete	Cai et al. (2003)
Leucocytozoon caulleryi Niigata	AP013071	NCBI GenBank	Complete	Imura et al. (2014)
Plasmodium berghei ANKA	LK023130	NCBI GenBank	Partial	Aslett et al. (unpublished)
Plasmodium berghei ANKA	AB649421	NCBI GenBank	Partial	Arisue et al. (2012)
Plasmodium berghei NK65	NC_030892.1	NCBI GenBank	Partial	GenBank (unpublished)
Plasmodium brasilianum Bolivian I	CM007351	NCBI GenBank	Partial	Talundzic (unpublished)
Plasmodium chabaudi AS	AB649423	NCBI GenBank	Partial	Arisue et al. (2012)
Plasmodium chabaudi chabaudi	HF563595	NCBI GenBank	Partial	Sato, Sesay, and Holder (2013)
Plasmodium coatyeni CDC	AB649420	NCBI GenBank	Partial	Arisue et al. (2012)

Plasmodium falciparum	X95275–X95276	NCBI GenBank	Partial	Arisue et al. (2012)
Plasmodium falciparum HB3	DQ642846	NCBI GenBank	Partial	Birren et al. (unpublished)
Plasmodium gaboni SY75	CM003884	NCBI GenBank	Partial	Sundararaman et al. (2016)
Plasmodium gallinaceum A8	AB649424	NCBI GenBank	Partial	Arisue et al. (2012)
Plasmodium malariae Kisii67	AB649418	NCBI GenBank	Partial	Arisue et al. (2012)
Plasmodium ovale NIGERIA II	AB649417	NCBI GenBank	Partial	Arisue et al. (2012)
Plasmodium reichenowi SY75	CM003883	NCBI GenBank	Partial	Sundararaman et al. (2016)
Plasmodium vivax Brazil I	JQ437257	NCBI GenBank	Partial	Neafsey et al. (unpublished)
Plasmodium vivax Mauritania I	JQ437258	NCBI GenBank	Partial	Neafsey et al. (unpublished)
Plasmodium vivax North Korean	JQ437259	NCBI GenBank	Partial	Neafsey et al. (unpublished)
Plasmodium vivax Salvador I	AB649419	NCBI GenBank	Partial	Arisue et al. (2012)
Plasmodium yoelii 17NXL	AB649422	NCBI GenBank	Partial	Arisue et al. (2012)
Plasmodium yoelii 17X	LM993669	NCBI GenBank	Partial	Aslett et al. (unpublished)
Plasmodium yoelii YM	LK934643	NCBI GenBank	Partial	Aslett et al. (unpublished)
Theileria parva Muguga	AAGK01000009	NCBI GenBank	Complete	Gardner et al. (2005)
Toxoplasma gondii ME49 RH	U87145	NCBI GenBank	Complete	Kissinger et al. (unpublished)
Apicomanda				
Chromera velia CCMP2878	NC_014340.2	NCBI GenBank	Complete	Janouškovec et al. (2010)
Vitrella brassicaformis CCMP3155	HM222968.1	NCBI GenBank	Complete	Janouškovec et al. (2010)

Continued

Table 1 Availability of Sequenced Plastomes for Species in the Myzozoa—cont'd

Species	Accession Number	Source	Completeness	Reference
Dinokaryota				
Amphidinium carterae CCAP1102/6	Many nonconsecutive GenBank entries. See reference for accession numbers.	NCBI GenBank	'Complete'	Barbrook and Howe (2000), Barbrook et al. (2001), Nisbet et al. (2004), and Barbrook, Santucci, et al. (2006)
Amphidinium carterae CS21	Many nonconsecutive GenBank entries. See reference for accession numbers.	NCBI GenBank	'Complete'	Hiller (2001) and Barbrook, Santucci, et al. (2006)
Durinskia baltica CS-38	NC_014287.1	NCBI GenBank	Complete	Imanian et al. (2010)
Heterocapsa triquetra	Many nonconsecutive GenBank entries. See reference for accession numbers.	NCBI GenBank	'Complete'	Zhang et al. (2001, 1999) and Nelson et al. (2007)
Karlodinium veneficum	JN039300.1	NCBI GenBank	Partial	Gabrielsen et al. (2011)
Kryptoperidinium foliaceum CCMP1326	NC_014267.1	NCBI GenBank	Complete	Imanian et al. (2010)
Lepidodinium chlorophorum	NC_027093.1	NCBI GenBank	Complete	Kamikawa et al. (2015)
Symbiodinium sp. clade C3	HG515015–HG515025, HG515027, and HG515028	NCBI GenBank	'Complete'	Barbrook et al. (2014)

that belong to the clade Archaeplastida) where it has been inferred that nuclear genomes have 200–600 genes of cyanobacterial origin (Moustafa & Bhattacharya, 2008; Price et al., 2012). Most of these genes likely have plastid functions (Reyes-Prieto, Hackett, Soares, Bonaldo, & Bhattacharya, 2006). In the case of myzozoans, which acquired their plastid from a red alga, most genes were transferred from the nucleus of the endosymbiotic red alga to the host nucleus (of either an ancestral chromalveolate or myzozoan; see discussion above). Because myzozoan plastomes have smaller gene repertoires than those of other red meta-algae (Fig. 3), some direct gene transfer from the plastome to the host nucleus also happened. The proteome of a photosynthetic plastid in red meta-algae is estimated to be composed of about 800–1000 proteins (Dorrell et al., 2017; Gruber et al., 2007), which means that about 700–900 genes might have been transferred and now reside in the myzozoan nucleus (ancestral myzozoan plastomes encoded only 80 genes). During dinozoan evolution even more genes, about 69 (all ribosomal and many photosynthetic proteins), were transferred from the plastome to the 'dinokaryon'. Some reports have attempted to estimate the impact of EGT in some dinophyte nuclear genomes (e.g. see Hackett et al., 2004; Hehenberger et al., 2016; Nosenko et al., 2006; Patron, Waller & Keeling, 2006; Minge et al., 2010; Burki et al., 2014). Apicomplexans, on the other hand, greatly reduced their plastid proteome when they lost photosynthesis. It is estimated that the apicoplast proteomes has 500 proteins (Ralph et al., 2004), and thus about 470 ancestrally plastome genes now reside in the apicomplexan nucleus. Of course, these are just rough estimates because some ancestral plastid genes could simply have been lost (instead of transferred, i.e. the plastid proteome simplified), and the host could also have retargeted its own new proteins to the plastid.

EGT has also contributed to the accumulation of nonfunctional and noncoding DNA in nuclear genomes. These are called 'NUPT' for nuclear plastid DNA. Analyses of genomes have concluded that sporozoans have relatively low amounts of NUPTs (Smith, Crosby, & Lee, 2011). This is expected according to the 'limited window transfer' hypothesis which proposes low rates of DNA transfer from the apicoplast to the nucleus (because all sporozoans have one single apicoplast). The relatively small nuclear genomes of parasitic sporozoans seem to primarily evolve in a reductive fashion and therefore also make them less prone to accumulate noncoding DNA-like NUPTs (Smith et al., 2011). In some sporozoans, like the piroplasmids *Babesia* and *Theileria*, no NUPTs were found at all; they also have the smallest nuclear genomes among sporozoans. The coccidians

Eimeria and *Toxoplasma*, which have larger nuclear genomes, have 31 and 77 NUPTs reported, respectively. But these numbers are insignificant in comparison to land plants which harbour many plastids per cell, have incredibly bloated nuclear genomes, and can have as many as 2036 NUPTs in the case of *Oryza sativa*. Unfortunately, rates of gene transfer from plastids to the nuclei of dinozoans have not been studied yet. The reason is that those dinozoans for which we have nuclear genomes have lost their plastomes (*Perkinsus* and *Hematodonium*), whereas dinozoans for which we have their plastomes (*Heterocapsa* and *Amphidinium*) do not have their nuclear genomes sequenced (because of their massive proportions). The only exception would be the coral endosymbiont *Symbiodinium* for which there are now both nuclear and plastid genomes available (Aranda et al., 2016; Barbrook et al., 2014; Lin et al., 2015; Shoguchi et al., 2013); however, no search for NUPTs has been done yet. It is expected for dinophytes to have large number of NUPTs because they usually possess numerous peridinin-containing plastids and have easily expandable genomes. Such analyses are also wanting for the apicomonads *Chromera* and *Vitrella*, for which both nuclear and plastid genomes are now available (Janouškovec et al., 2010; Woo et al., 2015).

7. CONCLUSIONS AND FUTURE DIRECTIONS

This chapter has provided a general description of the main features observed among the diversity of myzozoan plastomes. It has also attempted to generally describe the evolutionary trajectories that plastome-bearing myzozoans have followed. We aimed to do both things within a general framework where some of the eccentricities observed among myzozoan plastomes can be compared to all other plastid-bearing eukaryotes.

To summarize, myzozoan plastomes most likely have a most recent common ancestor. But it is more uncertain whether this ancestral myzozoan inherited its plastid vertically from a distant ancestor or laterally from an unrelated alga. Myzozoan diversification produced a great diversity of plastids. Some preserved the ancestral property of performing photosynthesis (like in some apicomonads and dinophytes). But leucoplasts evolved repeatedly among myzozoans. One lineage turned plastids into leucoplasts that retained a plastome (the Sporozoa), whereas many others repeatedly lost the plastome altogether (some apicomonads and dinozoans). The only two examples of outright plastid loss known to date are myzozoans, one dinoflagellate and one sporozoan. The ancestral myzozoan plastome most closely resembled that of *Vitrella*'s among the sampled diversity of modern myzozoans. The other two plastome-bearing lineages (sporozoans and dinophytes) have followed divergent evolutionary lines and their plastome now virtually have non-overlapping gene repertoires. Peridinin dinophytes have the smallest plastomes among eukaryotes, and yet they support photosynthesis. They are also fragmented into plasmid-like minicircles that generally contain one single gene. Sporozoans have a more typical leucoplast plastome that generally conserves a classical quadripartite organization. Myzozoans plastomes hold records as the smallest plastomes (for dinophytes) and the most GC-rich (for sporozoans). The small sizes of myzozoan plastomes seems to have been achieved through four episodes of genome reduction.

Future sampling will undoubtedly expand the known diversity of myzozoan plastomes. We will most likely find new chromerid plastomes (like ARLs), as well as 'deeply diverging' apicoplast plastomes (like environmental lineages VI–X), and perhaps dinozoan plastomes that are less reduced and fragmented. These will allow us to better reconstruct the changes that gave rise to the reduced plastomes of sporozoans and dinophytes. For instance, the fine-grain sampling among parasitic land plants has unravelled the gradual mode of plastome reduction in different embryophytic lineages. The field of metagenomics promises to make these discoveries soon and to greatly improve our knowledge of plastome diversity in the Myzozoa.

ACKNOWLEDGEMENTS

We are indebted to Jeff Palmer for providing valuable feedback on a late version of this manuscript. S.A.M.-G. is supported by a Killam Predoctoral Scholarship and a Nova Scotia Graduate Scholarship. C.H.S. is supported by NSERC (Discovery Grant RGPIN/05754-2015).

REFERENCES

Adl, S. M., et al. (2012). The revised classification of eukaryotes. *The Journal of Eukaryotic Microbiology*, *59*, 429–493.

Aranda, M., et al. (2016). Genomes of coral dinoflagellate symbionts highlight evolutionary adaptations conducive to a symbiotic lifestyle. *Scientific Reports*, *6*, 39734.

Arisue, N., & Hashimoto, T. (2015). Phylogeny and evolution of apicoplasts and apicomplexan parasites. *Parasitology International*, *64*, 254–259.

Arisue, N., Hashimoto, T., Mitsui, H., Palacpac, N. M. Q., Kaneko, A., Kawai, S., et al. (2012). The Plasmodium apicoplast genome: Conserved structure and close relationship of P. ovale to rodent malaria parasites. *Molecular Biology and Evolution*, *29*, 2095–2099.

Barbrook, A. C., Dorrell, R. G., Burrows, J., Plenderleith, L. J., Nisbet, R. E. R., & Howe, C. J. (2012). Polyuridylylation and processing of transcripts from multiple gene minicircles in chloroplasts of the dinoflagellate Amphidinium carterae. *Plant Molecular Biology*, *79*, 347–357.

Barbrook, A. C., & Howe, C. J. (2000). Minicircular plastid DNA in the dinoflagellate Amphidinium operculatum. *Molecular & General Genetics*, *263*, 152–158.

Barbrook, A. C., Howe, C. J., & Purton, S. (2006). Why are plastid genomes retained in non-photosynthetic organisms? *Trends in Plant Science*, *11*, 101–108.
Barbrook, A. C., Santucci, N., Plenderleith, L. J., Hiller, R. G., & Howe, C. J. (2006). Comparative analysis of dinoflagellate chloroplast genomes reveals rRNA and tRNA genes. *BMC Genomics*, 7, 297.
Barbrook, A. C., Symington, H., Nisbet, R. E., Larkum, A., & Howe, C. J. (2001). Organisation and expression of the plastid genome of the dinoflagellate Amphidinium operculatum. *Molecular Genetics and Genomics*, *266*, 632–638.
Barbrook, A. C., Voolstra, C. R., & Howe, C. J. (2014). The chloroplast genome of a Symbiodinium sp. clade C3 isolate. *Protist*, *165*, 1–13.
Bellot, S., & Renner, S. S. (2015). The plastomes of two species in the Endoparasite genus Pilostyles (Apodanthaceae) each retain just five or six possibly functional genes. *Genome Biology and Evolution*, *8*, 189–201.
Blanchard, J. L., & Hicks, J. S. (1999). The non-photosynthetic plastid in malarial parasites and other apicomplexans is derived from outside the green plastid lineage. *The Journal of Eukaryotic Microbiology*, *46*, 367–375.
Bodył, A. (2017). Did some red alga-derived plastids evolve via kleptoplastidy? A hypothesis. *Biological Reviews of the Cambridge Philosophical Society*. https://doi.org/10.1111/brv.12340. [Epub ahead of print].
Bodyl, A., Stiller, J. W., & Mackiewicz, P. (2009). Chromalveolate plastids: Direct descent or multiple endosymbioses. *Trends in Ecology & Evolution*, *24*, 119–121.
Brayton, K. A., et al. (2007). Genome sequence of Babesia bovis and comparative analysis of apicomplexan hemoprotozoa. *PLoS Pathogens*, *3*, 1401–1413.
Burki, F., Imanian, B., Hehenberger, E., Hirakawa, Y., Maruyama, S., & Keeling, P. J. (2014). Endosymbiotic gene transfer in tertiary plastid-containing dinoflagellates. *Eukaryotic Cell*, *13*, 246–255.
Cai, X., Fuller, A. L., McDougald, L. R., & Zhu, G. (2003). Apicoplast genome of the coccidian Eimeria tenella. *Gene*, *321*, 39–46.
Cavalier-Smith, T. (1999). Principles of protein and lipid targeting in secondary symbiogenesis: Euglenoid, dinoflagellate, and sporozoan plastid origins and the eukaryote family tree. *The Journal of Eukaryotic Microbiology*, *46*, 347–366.
Cavalier-Smith, T. (2017). Kingdom Chromista and its eight phyla: A new synthesis emphasising periplastid protein targeting, cytoskeletal and periplastid evolution, and ancient divergences. *Protoplasma*. https://doi.org/10.1007/s00709-017-1147-3. [Epub ahead of print].
Cavalier-Smith, T., & Chao, E. E. (2004). Protalveolate phylogeny and systematics and the origins of Sporozoa and dinoflagellates (phylum Myzozoa nom. nov.). *European Journal of Protistology*, *40*, 185–212.
Dang, Y., & Green, B. R. (2010). Long transcripts from dinoflagellate chloroplast minicircles suggest "rolling circle" transcription. *Journal of Biological Chemistry*, *285*, 5196–5203.
de Koning, A. P., & Keeling, P. J. (2006). The complete plastid genome sequence of the parasitic green alga Helicosporidium sp. is highly reduced and structured. *BMC Biology*, *4*, 12.
Dorrell, R. G., & Howe, C. J. (2012). Functional remodeling of RNA processing in replacement chloroplasts by pathways retained from their predecessors. *Proceedings of the National Academy of Sciences of the United States of America*, *109*, 18879–18884.
Dorrell, R. G., & Howe, C. J. (2015). Integration of plastids with their hosts: Lessons learned from dinoflagellates. *Proceedings of the National Academy of Sciences of the United States of America*, *112*, 10247–10254.
Dorrell, R. G., Klinger, C. M., Newby, R. J., Butterfield, E. R., Richardson, E., Dacks, J. B., et al. (2017). Progressive and biased divergent evolution underpins the origin and diversification of peridinin dinoflagellate plastids. *Molecular Biology and Evolution*, *34*, 361–379.

Dorrell, R. G., & Smith, A. G. (2011). Do red and green make brown?: Perspectives on plastid acquisitions within chromalveolates. *Eukaryotic Cell*, *10*, 856–868.

Dorrell, R. G., et al. (2017). Chimeric origins of ochrophytes and haptophytes revealed through an ancient plastid proteome. *eLife*, *6*, e23717.

Espelund, M., Minge, M. A., Gabrielsen, T. M., Nederbragt, A. J., Shalchian-Tabrizi, K., Otis, C., et al. (2012). Genome fragmentation is not confined to the peridinin plastid in dinoflagellates. *PloS One*, *7*, e38809.

Fast, N. M., Kissinger, J. C., Roos, D. S., & Keeling, P. J. (2001). Nuclear-encoded, plastid-targeted genes suggest a single common origin for apicomplexan and dinoflagellate plastids. *Molecular Biology and Evolution*, *18*, 418–426.

Fast, N. M., Xue, L., Bingham, S., & Keeling, P. J. (2002). Re-examining alveolate evolution using multiple protein molecular phylogenies. *Journal of Eukaryotic Microbiology*, *49*, 30–37.

Fawcett, R. C., & Parrow, M. W. (2014). Mixotrophy and loss of phototrophy among geographic isolates of freshwater Esoptrodinium/Bernardinium sp. (Dinophyceae). *Journal of Phycology*, *50*, 55–70.

Figueroa-Martinez, F., Nedelcu, A. M., Reyes-Prieto, A., & Smith, D. R. (2017). The plastid genomes of nonphotosynthetic algae are not so small after all. *Communicative & Integrative Biology*, *10*, e1283080.

Figueroa-Martinez, F., Nedelcu, A. M., Smith, D. R., & Reyes-Prieto, A. (2017). The plastid genome of Polytoma uvella is the largest known among colorless algae and plants and reflects contrasting evolutionary paths to nonphotosynthetic lifestyles. *Plant Physiology*, *173*, 932–943.

Funes, S., Davidson, E., Reyes-Prieto, A., Magallón, S., Herion, P., King, M. P., et al. (2002). A green algal apicoplast ancestor. *Science*, *298*, 2155.

Funes, S., Reyes-Prieto, A., Pérez-Martínez, X., & González-Halphen, D. (2004). On the evolutionary origins of apicoplasts: Revisiting the rhodophyte vs. chlorophyte controversy. *Microbes and Infection*, *6*, 305–311.

Füssy, Z., & Oborník, M. (2017). Chromerids and their plastids. *Advances in Botanical Research*, *84*, 187–218, Academic Press.

Gabrielsen, T. M., et al. (2011). Genome evolution of a tertiary dinoflagellate plastid. *PLoS One*, *6*, e19132.

Gardner, M. J., Feagin, J. E., Moore, D. J., Spencer, D. F., Gray, M. W., Williamson, D. H., et al. (1991). Organisation and expression of small subunit ribosomal RNA genes encoded by a 35-kilobase circular DNA in Plasmodium falciparum. *Molecular and Biochemical Parasitology*, *48*, 77–88.

Gardner, M. J., Williamson, D. H., & Wilson, R. J. M. (1991). A circular DNA in malaria parasites encodes an RNA polymerase like that of prokaryotes and chloroplasts. *Molecular and Biochemical Parasitology*, *44*, 115–123.

Gardner, M. J., et al. (2005). Genome sequence of Theileria parva, a bovine pathogen that transforms lymphocytes. *Science (New York, NY)*, *309*, 134–137.

Garg, A., Stein, A., Zhao, W., Dwivedi, A., Frutos, R., Cornillot, E., et al. (2014). Sequence and annotation of the apicoplast genome of the human pathogen Babesia microti. *PLoS One*, *9*, e107939.

Gavelis, G. S., Hayakawa, S., Iii, R. A. W., Gojobori, T., Suttle, C. A., Keeling, P. J., & Leander, B. S. (2015). Eye-like ocelloids are built from different endosymbiotically acquired components. *Nature*, *523*, 204.

Gile, G. H., & Slamovits, C. H. (2014). Transcriptomic analysis reveals evidence for a cryptic plastid in the colpodellid Voromonas pontica, a close relative of chromerids and apicomplexan parasites. *PLoS One*, *9*, e96258.

Goodman, C. D., & McFadden, G. I. (2014). Ycf93 (Orf105), a small apicoplast-encoded membrane protein in the relict plastid of the malaria parasite Plasmodium falciparum that is conserved in Apicomplexa. *PLoS One*, *9*, e91178.

Gornik, S. G., Ford, K. L., Mulhern, T. D., Bacic, A., McFadden, G. I., & Waller, R. F. (2012). Loss of nucleosomal DNA condensation coincides with appearance of a novel nuclear protein in dinoflagellates. *Current Biology*, *22*, 2303–2312.

Gornik, S. G., et al. (2015). Endosymbiosis undone by stepwise elimination of the plastid in a parasitic dinoflagellate. *Proceedings of the National Academy of Sciences of the United States of America*, *112*, 5767–5772.

Gruber, A., Vugrinec, S., Hempel, F., Gould, S. B., Maier, U.-G., & Kroth, P. G. (2007). Protein targeting into complex diatom plastids: Functional characterisation of a specific targeting motif. *Plant Molecular Biology*, *64*, 519–530.

Hackett, J. D., Yoon, H. S., Soares, M. B., Bonaldo, M. F., Casavant, T. L., Scheetz, T. E., et al. (2004). Migration of the plastid genome to the nucleus in a peridinin dinoflagellate. *Current Biology*, *14*, 213–218.

Harper, J. T., & Keeling, P. J. (2003). Nucleus-encoded, plastid-targeted glyceraldehyde-3-phosphate dehydrogenase (GAPDH) indicates a single origin for chromalveolate plastids. *Molecular Biology and Evolution*, *20*, 1730–1735.

Harper, J. T., Waanders, E., & Keeling, P. J. (2005). On the monophyly of chromalveolates using a six-protein phylogeny of eukaryotes. *International Journal of Systematic and Evolutionary Microbiology*, *55*, 487–496.

Hehenberger, E., Burki, F., Kolisko, M., & Keeling, P. J. (2016). Functional relationship between a dinoflagellate host and its diatom endosymbiont. *Molecular Biology and Evolution*, *33*, 2376–2390.

Hiller, R. G. (2001). "Empty" minicircles and petB/atpA and psbD/psbE (cytb559 α) genes in tandem in Amphidinium carterae plastid DNA11The sequences reported in this paper have been deposited in the EMBL data base under numbers AJ311628–AJ311633, AJ307009–AJ3079016 and AJ318067. *FEBS Letters*, *505*, 449–452.

Hoppenrath, M. (2017). Dinoflagellate taxonomy—A review and proposal of a revised classification. *Marine Biodiversity*, *47*, 381–403.

Howe, C. J., Barbrook, A. C., Koumandou, V. L., Nisbet, R. E. R., Symington, H. A., & Wightman, T. F. (2003). Evolution of the chloroplast genome. *Philosophical Transactions of the Royal Society of London Series B: Biological Sciences*, *358*, 99–107.

Howe, C. J., Nisbet, R. E. R., & Barbrook, A. C. (2008). The remarkable chloroplast genome of dinoflagellates. *Journal of Experimental Botany*, *59*, 1035–1045.

Huang, Y., et al. (2015). Characterization and annotation of Babesia orientalis apicoplast genome. *Parasites & Vectors*, *8*, 543.

Imanian, B., Pombert, J.-F., & Keeling, P. J. (2010). The complete plastid genomes of the two "Dinotoms" Durinskia baltica and Kryptoperidinium foliaceum. *PLoS One*, *5*, e10711.

Imura, T., Sato, S., Sato, Y., Sakamoto, D., Isobe, T., Murata, K., et al. (2014). The apicoplast genome of Leucocytozoon caulleryi, a pathogenic apicomplexan parasite of the chicken. *Parasitology Research*, *113*, 823–828.

Jackson, C. J., Gornik, S. G., & Waller, R. F. (2013). A tertiary plastid gains RNA editing in its new host. *Molecular Biology and Evolution*, *30*, 788–792.

Janouškovec, J., Horák, A., Barott, K. L., Rohwer, F. L., & Keeling, P. J. (2013). Environmental distribution of coral-associated relatives of apicomplexan parasites. *The ISME Journal*, 7, 444–447.

Janouškovec, J., Horák, A., Oborník, M., Lukes, J., & Keeling, P. J. (2010). A common red algal origin of the apicomplexan, dinoflagellate, and heterokont plastids. *Proceedings of the National Academy of Sciences of the United States of America*, *107*, 10949–10954.

Janouškovec, J., Tikhonenkov, D. V., Burki, F., Howe, A. T., Kolísko, M., Mylnikov, A. P., et al. (2015). Factors mediating plastid dependency and the origins of parasitism in apicomplexans and their close relatives. *Proceedings of the National Academy of Sciences of the United States of America*, *112*, 10200–10207.

Janouškovec, J., et al. (2013). Split photosystem protein, linear-mapping topology, and growth of structural complexity in the plastid genome of Chromera velia. *Molecular Biology and Evolution*, *30*, 2447–2462.

Janouškovec, J., et al. (2017). Major transitions in dinoflagellate evolution unveiled by phylotranscriptomics. *Proceedings of the National Academy of Sciences of the United States of America*, *114*, E171–E180.

Kamikawa, R., Tanifuji, G., Kawachi, M., Miyashita, H., Hashimoto, T., & Inagaki, Y. (2015). Plastid genome-based phylogeny pinpointed the origin of the green-colored plastid in the dinoflagellate Lepidodinium chlorophorum. *Genome Biology and Evolution*, *7*, 1133–1140.

Kilejian, A. (1975). Circular mitochondrial DNA from the avian malarial parasite Plasmodium lophurae. *Biochimica et Biophysica Acta*, *390*, 276–284.

Köhler, S., Delwiche, C. F., Denny, P. W., Tilney, L. G., Webster, P., Wilson, R. J. M., et al. (1997). A plastid of probable green algal origin in Apicomplexan parasites. *Science*, *275*, 1485–1489.

Laatsch, T., Zauner, S., Stoebe-Maier, B., Kowallik, K. V., & Maier, U.-G. (2004). Plastid-derived single gene minicircles of the dinoflagellate Ceratium horridum are localized in the nucleus. *Molecular Biology and Evolution*, *21*, 1318–1322.

Lau, A. O. T., McElwain, T. F., Brayton, K. A., Knowles, D. P., & Roalson, E. H. (2009). Babesia bovis: A comprehensive phylogenetic analysis of plastid-encoded genes supports green algal origin of apicoplasts. *Experimental Parasitology*, *123*, 236–243.

Lin, S., et al. (2015). The Symbiodinium kawagutii genome illuminates dinoflagellate gene expression and coral symbiosis. *Science*, *350*, 691–694.

Mackiewicz, P., Bodył, A., & Moszczyński, K. (2013). The case of horizontal gene transfer from bacteria to the peculiar dinoflagellate plastid genome. *Mobile Genetic Elements*, *3*, e25845.

Mahé, F., et al. (2017). Parasites dominate hyperdiverse soil protist communities in Neotropical rainforests. *Nature Ecology & Evolution*, *1*, 91.

Martin, W. (2003). Gene transfer from organelles to the nucleus: Frequent and in big chunks. *Proceedings of the National Academy of Sciences*, *100*, 8612–8614.

Matsumoto, T., Kawachi, M., Miyashita, H., & Inagaki, Y. (2012). Prasinoxanthin is absent in the green-colored dinoflagellate Lepidodinium chlorophorum strain NIES-1868: Pigment composition and 18S rRNA phylogeny. *Journal of Plant Research*, *125*, 705–711.

Matsumoto, T., et al. (2011). Green-colored plastids in the dinoflagellate genus Lepidodinium are of core chlorophyte origin. *Protist*, *162*, 268–276.

McFadden, G. I., Reith, M. E., Munholland, J., & Lang-Unnasch, N. (1996). Plastid in human parasites. *Nature*, *381*, 482.

Minge, M. A., Shalchian-Tabrizi, K., Tørresen, O. K., Takishita, K., Probert, I., Inagaki, Y., et al. (2010). A phylogenetic mosaic plastid proteome and unusual plastid-targeting signals in the green-colored dinoflagellate Lepidodinium chlorophorum. *BMC Evolutionary Biology*, *10*, 191.

Molina, J., et al. (2014). Possible loss of the chloroplast genome in the parasitic flowering plant Rafflesia lagascae (Rafflesiaceae). *Molecular Biology and Evolution*, *31*, 793–803.

Moore, R. B., et al. (2008). A photosynthetic alveolate closely related to apicomplexan parasites. *Nature*, *451*, 959–963.

Moszczyński, K., Mackiewicz, P., & Bodył, A. (2012). Evidence for horizontal gene transfer from bacteroidetes bacteria to dinoflagellate minicircles. *Molecular Biology and Evolution*, *29*, 887–892.

Moustafa, A., & Bhattacharya, D. (2008). PhyloSort: A user-friendly phylogenetic sorting tool and its application to estimating the cyanobacterial contribution to the nuclear genome of Chlamydomonas. *BMC Evolutionary Biology*, *8*, 6.

Mungpakdee, S., et al. (2014). Massive gene transfer and extensive RNA editing of a symbiotic dinoflagellate plastid genome. *Genome Biology and Evolution, 6*, 1408–1422.
Muñoz-Gómez, S. A., Mejía-Franco, F. G., Durnin, K., Colp, M., Grisdale, C. J., Archibald, J. M., et al. (2017). The new red algal subphylum Proteorhodophytina comprises the largest and most divergent plastid genomes known. *Current Biology, 27*, 1677–1684.e4.
Nelson, M. J., Dang, Y., Filek, E., Zhang, Z., Yu, V. W. C., Ishida, K., et al. (2007). Identification and transcription of transfer RNA genes in dinoflagellate plastid minicircles. *Gene, 392*, 291–298.
Nisbet, R. E. R., Koumandou, L. V., Barbrook, A. C., & Howe, C. J. (2004). Novel plastid gene minicircles in the dinoflagellate Amphidinium operculatum. *Gene, 331*, 141–147.
Nosenko, T., Lidie, K. L., Van Dolah, F. M., Lindquist, E., Cheng, J.-F., & Bhattacharya, D. (2006). Chimeric plastid proteome in the Florida "red tide" dinoflagellate Karenia brevis. *Molecular Biology and Evolution, 23*, 2026–2038.
Oborník, M., Janouškovec, J., Chrudimský, T., & Lukeš, J. (2009). Evolution of the apicoplast and its hosts: From heterotrophy to autotrophy and back again. *International Journal for Parasitology, 39*, 1–12.
Oborník, M., & Lukeš, J. (2015). The organellar genomes of Chromera and Vitrella, the phototrophic relatives of apicomplexan parasites. *Annual Review of Microbiology, 69*, 129–144.
Oborník, M., et al. (2012). Morphology, ultrastructure and life cycle of Vitrella brassicaformis n. sp., n. gen., a novel chromerid from the great barrier reef. *Protist, 163*, 306–323.
Okamoto, N., & Keeling, P. J. (2014). The 3D structure of the apical complex and association with the flagellar apparatus revealed by serial TEM tomography in Psammosa pacifica, a distant relative of the Apicomplexa. *PloS One, 9*, e84653.
Owari, S., Hayashi, A., & Ishida, K. (2014). Subcellular localization of minicircle DNA in the dinoflagellate Amphidinium massartii. *Phycological Research, 62*, 1–8.
Patron, N. J., Waller, R. F., & Keeling, P. J. (2006). A tertiary plastid uses genes from two endosymbionts. *Journal of Molecular Biology, 357*, 1373–1382.
Petersen, J., Ludewig, A.-K., Michael, V., Bunk, B., Jarek, M., Baurain, D., et al. (2014). Chromera velia, endosymbioses and the rhodoplex hypothesis—Plastid evolution in cryptophytes, alveolates, stramenopiles, and haptophytes (CASH lineages). *Genome Biology and Evolution, 6*, 666–684.
Ponce-Toledo, R. I., Deschamps, P., López-García, P., Zivanovic, Y., Benzerara, K., & Moreira, D. (2017). An early-branching freshwater Cyanobacterium at the origin of plastids. *Current Biology, 27*, 386–391.
Price, D. C., et al. (2012). Cyanophora paradoxa genome elucidates origin of photosynthesis in algae and plants. *Science, 335*, 843–847.
Ralph, S. A., van Dooren, G. G., Waller, R. F., Crawford, M. J., Fraunholz, M. J., Foth, B. J., et al. (2004). Tropical infectious diseases: Metabolic maps and functions of the *Plasmodium falciparum* apicoplast. *Nature Reviews Microbiology, 2*, 203.
Reñé, A., Alacid, E., Ferrera, I., & Garcés, E. (2017). Evolutionary trends of Perkinsozoa (Alveolata) characters based on observations of two new genera of parasitoids of dinoflagellates, Dinovorax gen. nov. and Snorkelia gen. nov. *Frontiers in Microbiology, 8*.
Reyes-Prieto, A., Hackett, J. D., Soares, M. B., Bonaldo, M. F., & Bhattacharya, D. (2006). Cyanobacterial contribution to algal nuclear genomes is primarily limited to plastid functions. *Current Biology, 16*, 2320–2325.
Richardson, E., Dorrell, R. G., & Howe, C. J. (2014). Genome-wide transcript profiling reveals the coevolution of plastid gene sequences and transcript processing pathways in the fucoxanthin dinoflagellate Karlodinium veneficum. *Molecular Biology and Evolution, 31*, 2376–2386.

Saffo, M. B., McCoy, A. M., Rieken, C., & Slamovits, C. H. (2010). Nephromyces, a beneficial apicomplexan symbiont in marine animals. *Proceedings of the National Academy of Sciences*, *107*, 16190–16195.

Saffo, M. B., & Nelson, R. (1983). The cells of nephromyces: Developmental stages of a single life cycle. *Canadian Journal of Botany*, *61*, 3230–3239.

Saldarriaga, J. F., & Taylor, F. J. R. (2017). Dinoflagellata. In J. M. Archibald, A. G. B. Simpson, C. H. Slamovits, L. Margulis, M. Melkonian, D. J. Chapman, & J. O. Corliss (Eds.), *Handbook of the protists* (pp. 1–54). Springer International Publishing.

Saldarriaga, J. F., Taylor, F. J., Keeling, P. J., & Cavalier-Smith, T. (2001). Dinoflagellate nuclear SSU rRNA phylogeny suggests multiple plastid losses and replacements. *Journal of Molecular Evolution*, *53*, 204–213.

Sanchez-Puerta, M. V., & Delwiche, C. F. (2008). A hypothesis for plastid evolution in chromalveolates. *Journal of Phycology*, *44*, 1097–1107.

Sanchez-Puerta, M. V., Lippmeier, J. C., Apt, K. E., & Delwiche, C. F. (2007). Plastid genes in a non-photosynthetic dinoflagellate. *Protist*, *158*, 105–117.

Santos, S. R., Gutierrez-Rodriguez, C., & Coffroth, M. A. (2003). Phylogenetic identification of symbiotic dinoflagellates via length heteroplasmy in domain V of chloroplast large subunit (cp23S)-ribosomal DNA sequences. *Marine Biotechnology (New York, NY)*, *5*, 130–140.

Sato, S. (2011). The apicomplexan plastid and its evolution. *Cellular and Molecular Life Sciences*, *68*, 1285–1296.

Sato, S., Sesay, A. K., & Holder, A. A. (2013). The unique structure of the apicoplast genome of the rodent malaria parasite Plasmodium chabaudi chabaudi. *PLoS One*, *8*, e61778.

Schelkunov, M. I., Shtratnikova, V. Y., Nuraliev, M. S., Selosse, M.-A., Penin, A. A., & Logacheva, M. D. (2015). Exploring the limits for reduction of plastid genomes: A case study of the mycoheterotrophic orchids Epipogium aphyllum and Epipogium roseum. *Genome Biology and Evolution*, 7, 1179–1191.

Schnepf, E., & Deichgräber, G. (1984). "Myzocytosis", a kind of endocytosis with implications to compartmentation in endosymbiosis. *Naturwissenschaften*, *71*, 218–219.

Ševčíková, T., et al. (2015). Updating algal evolutionary relationships through plastid genome sequencing: Did alveolate plastids emerge through endosymbiosis of an ochrophyte? *Scientific Reports*, *5*, 10134.

Shoguchi, E., et al. (2013). Draft assembly of the Symbiodinium minutum nuclear genome reveals dinoflagellate gene structure. *Current Biology*, *23*, 1399–1408.

Slamovits, C. H., & Keeling, P. J. (2008). Plastid-derived genes in the nonphotosynthetic alveolate Oxyrrhis marina. *Molecular Biology and Evolution*, *25*, 1297–1306.

Smith, D. R., Crosby, K., & Lee, R. W. (2011). Correlation between nuclear plastid DNA abundance and plastid number supports the limited transfer window hypothesis. *Genome Biology and Evolution*, *3*, 365–371.

Smith, D. R., & Lee, R. W. (2014). A plastid without a genome: Evidence from the nonphotosynthetic green algal genus Polytomella. *Plant Physiology*, *164*, 1812–1819.

Sundararaman, S. A., et al. (2016). Genomes of cryptic chimpanzee *Plasmodium* species reveal key evolutionary events leading to human malaria. *Nature Communications*, 7, 11078.

Suplick, K., Akella, R., Saul, A., & Vaidya, A. B. (1988). Molecular cloning and partial sequence of a 5.8 kilobase pair repetitive DNA from Plasmodium falciparum. *Molecular and Biochemical Parasitology*, *30*, 289–290.

Tang, K., et al. (2015). Genetic similarities between Cyclospora cayetanensis and cecum-infecting avian Eimeria spp. in apicoplast and mitochondrial genomes. *Parasites & Vectors*, *8*, 358.

Tengs, T., Dahlberg, O. J., Shalchian-Tabrizi, K., Klaveness, D., Rudi, K., Delwiche, C. F., et al. (2000). Phylogenetic analyses indicate that the 19′Hexanoyloxy-fucoxanthin-containing dinoflagellates have tertiary plastids of haptophyte origin. *Molecular Biology and Evolution*, *17*, 718–729.
Toso, M. A., & Omoto, C. K. (2007). Gregarina niphandrodes may lack both a plastid genome and organelle. *The Journal of Eukaryotic Microbiology*, *54*, 66–72.
Vaidya, A. B., Akella, R., & Suplick, K. (1989). Sequences similar to genes for two mitochondrial proteins and portions of ribosomal RNA in tandemly arrayed 6-kilobase-pair DNA of a malarial parasite. *Molecular and Biochemical Parasitology*, *35*, 97–107.
van Dooren, G. G., & Hapuarachchi, S. V. (2017). The dark side of the chloroplast: Biogenesis, metabolism and membrane biology of the apicoplast. *Advances in Botanical Research*, *84*, 145–185, Academic Press.
Votýpka, J., Modrý, D., Oborník, M., Šlapeta, J., & Lukeš, J. (2016). Apicomplexa. In J. M. Archibald, A. G. B. Simpson, C. H. Slamovits, L. Margulis, M. Melkonian, D. J. Chapman, & J. O. Corliss (Eds.), *Handbook of the protists* (pp. 1–58). Springer International Publishing.
Waller, R. F., Gornik, S. G., Koreny, L., & Pain, A. (2016). Metabolic pathway redundancy within the apicomplexan-dinoflagellate radiation argues against an ancient chromalveolate plastid. *Communicative & Integrative Biology*, *9*, e1116653.
Waller, R. F., & Keeling, P. J. (2006). Alveolate and chlorophycean mitochondrial cox2 genes split twice independently. *Gene*, *383*, 33–37.
Waller, R. F., Keeling, P. J., van Dooren, G. G., & McFadden, G. I. (2003). Comment on "a green algal apicoplast ancestor" *Science*, *301*, 49.
Waller, R. F., & Kořený, L. (2017). Plastid complexity in dinoflagellates: A picture of gains, losses, replacements and revisions. *Advances in Botanical Research*, *84*, 105–143, Academic Press.
Wang, T., Guan, G., Korhonen, P. K., Koehler, A. V., Hall, R. S., Young, N. D., et al. (2017). The apicoplast genomes of two taxonomic units of Babesia from sheep. *Veterinary Parasitology*, *233*, 123–128.
Wang, Y., & Morse, D. (2006). Rampant polyuridylylation of plastid gene transcripts in the dinoflagellate Lingulodinium. *Nucleic Acids Research*, *34*, 613–619.
Woo, Y. H., et al. (2015). Chromerid genomes reveal the evolutionary path from photosynthetic algae to obligate intracellular parasites. *eLife*, *4*, e06974.
Yamada, N., Sym, S. D., & Horiguchi, T. (2017). Identification of highly divergent diatom-derived chloroplasts in dinoflagellates, including a description of Durinskia kwazulunatalensis sp. nov. (Peridiniales, Dinophyceae). *Molecular Biology and Evolution*, *34*, 1335–1351.
Yoon, H. S., Hackett, J. D., Dolah Van, M. F., Nosenko, T., Lidie, K. L., & Bhattacharya, D. (2005). Tertiary endosymbiosis driven genome evolution in dinoflagellate algae. *Molecular Biology and Evolution*, *22*, 1299–1308.
Zhang, Z., Cavalier-Smith, T., & Green, B. R. (2001). A family of selfish Minicircular chromosomes with jumbled chloroplast gene fragments from a dinoflagellate. *Molecular Biology and Evolution*, *18*, 1558–1565.
Zhang, Z., Green, B. R., & Cavalier-Smith, T. (1999). Single gene circles in dinoflagellate chloroplast genomes. *Nature*, *400*, 155–159.
Zhang, Z., Green, B. R., & Cavalier-Smith, T. (2000). Phylogeny of ultra-rapidly evolving dinoflagellate chloroplast genes: A possible common origin for sporozoan and dinoflagellate plastids. *Journal of Molecular Evolution*, *51*, 26–40.
Zhu, G., Marchewka, M. J., & Keithly, J. S. (2000). Cryptosporidium parvum appears to lack a plastid genome. *Microbiology (Reading, England)*, *146*(Pt. 2), 315–321.

CHAPTER FOUR

Comparative Plastid Genomics of Glaucophytes

Adrian Reyes-Prieto*,[1], Sarah Russell*, Francisco Figueroa-Martinez[†], Christopher Jackson[‡]

*University of New Brunswick, Fredericton, NB, Canada
[†]CONACyT-Universidad Autónoma Metropolitana, México City, México
[‡]School of Biosciences, University of Melbourne, Melbourne, VIC, Australia
[1]Corresponding author: e-mail address: areyes@unb.ca

Contents

Abstract

Diverse studies of plastid data suggest that the photosynthetic organelles of red algae, viridiplants, and glaucophytes, the three lineages comprising the Archaeplastida supergroup, share a common ancestor. Glaucophyte plastids are unique among archaeplastidians due to the presence of a vestigial peptidoglycan wall and the

Advances in Botanical Research, Volume 85
ISSN 0065-2296
https://doi.org/10.1016/bs.abr.2017.11.012

accumulation of RuBisCO in the stroma that resembles cyanobacterial carboxysomes. These ancestral traits, typically observed in cyanobacteria, have led to suggestions that glaucophytes are the earliest branching Archaeplastida lineage. Plastid phylogenomic surveys recover Glaucophyta as the earliest-diverging branch, but tree topology tests have not rejected the placement of red algae or viridiplants as the first splitting group. Resolving the branching history of the primary plastids might rely on both the implementation of phylogenetic methods that cope better with systematic errors and further expansion of the taxonomic sampling. The paucity of the Glaucophyta genome data has been a limitation when contrasting different hypotheses about the diversification of the Archaeplastida. The plastome of *Cyanophora paradoxa* was the only available from Glaucophyta for almost 20 years, until recently when plastomes of *Glaucocystis*, *Cyanoptyche*, and *Gloeochaete* and other *Cyanophora* species were sequenced. Comparative analyses show that the plastid gene repertoire of glaucophytes is highly conserved, and that the size and gene content of their plastomes do not differ drastically from those of other archaeplastidians. In fact, in terms of gene content, red algal plastomes are likely more similar to the repertoire of the ancestral primary plastid. Studies of plastomes have expanded our perspective about the diversity within Glaucophyta, but such studies are still based on limited taxonomic samples. The further inclusion of data from novel glaucophyte taxa will be critical to obtain more solid answers about the evolution and diversity of these rare algae.

1. INTRODUCTION

The Glaucophyta (Skuja, 1954) is one of the three major photosynthetic lineages comprising the supergroup Archaeplastida (Adl et al., 2012). The blue-green plastids of glaucophytes, historically known as cyanelles or muroplasts, are the most conspicuous trait of this algal group. The peculiar colouration of the glaucophyte photosynthetic organelles, similar to some cyanobacteria, is the result of combining the accessory blue photopigments allophycocyanin and C-phycocyanin with chlorophyll *a*. As in the case of the other two members of the Archaeplastida, red algae (Rhodoplantae or Rhodophyta) and viridiplants (Chloroplastida or Viridiplantae), the particular plastid pigmentation inspired the composite name of the lineage: *glaukos* (γλαυκός), blue-green or bluish grey, and *phyton* (φυτόν), plant. In addition to their distinctive combination of photopigments, glaucophyte plastids are exceptional among eukaryotes due to the presence of a vestigial peptidoglycan wall between the organelle membranes and the stromal accumulations of RuBisCO resembling the appearance of cyanobacterial carboxysomes (Hall & Claus, 1963; Kies & Kremer, 1986; Löffelhardt & Bohnert, 2001). These latter traits are shared with some modern cyanobacteria and likely reflect ancestral characteristics present in the photosynthetic organelles of the first archaeplastidians.

Diverse plastid-derived data suggest that the photosynthetic organelles of the Archaeplastida evolved from a single endosymbiotic event between a eukaryote and a cyanobacterium (i.e. primary endosymbiosis) (Cavalier-Smith, 1982; Palmer, 2003). Evidence supporting a unique origin of the primary plastids comes from gene clusters conserved in plastid genomes (plastomes) across the three Archaeplastida groups (Stoebe & Kowallik, 1999), shared enzyme replacements in plastid-localized biochemical pathways (Reyes-Prieto & Bhattacharya, 2007; Reyes-Prieto & Moustafa, 2012), the common origin of key components of the plastid protein import machinery (McFadden & van Dooren, 2004; Steiner, Yusa, Pompe, & Löffelhardt, 2005), and plastid phylogenomics (Criscuolo & Gribaldo, 2011; Deschamps & Moreira, 2009; Ponce-Toledo et al., 2017).

In contrast to plastid-derived inferences, recent phylogenomic analyses based on nuclear sequences have consistently failed to recover the Archaeplastida groups as a monophyletic assemblage (Burki et al., 2016; Burki, Okamoto, Pombert, & Keeling, 2012; Derelle et al., 2015; Yabuki et al., 2014). It is important to emphasize here that the hypothetical common origin of the Archaeplastida host (the nucleo-cytoplasm core) and their plastids are not necessarily coincident events. If the nucleo-cytoplasm components and the primary plastids were present in the hypothetical last common ancestor of the Archaeplastida, then we would expect that phylogenetic analyses of nuclear and plastome data should produce similar results. However, this latter scenario has rarely been the outcome of most phylogenetic studies (Rodríguez-Ezpeleta et al., 2005), and incongruent phylogenetic results from nuclear vs plastid data are a recurrent theme (Mackiewicz & Gagat, 2014). One of the major constraints when investigating and contrasting alternative hypotheses regarding Archaeplastida evolution (e.g. separate establishment of plastids in different eukaryote hosts via serial endosymbiosis; Kim & Maruyama, 2014; Stiller, 2014) has been the limited amount of glaucophyte genomic data analysed. With few exceptions (Burki et al., 2016; Zhao et al., 2012), most nuclear and plastid phylogenomic studies have included data from only *Cyanophora paradoxa*, by far the most-studied glaucophyte species, with some analyses also including data from *Glaucocystis nostochinearum*. Further investigations of glaucophyte diversity at a genomic scale will play an important role in our capabilities to untangle the early evolution of photosynthetic eukaryotes.

In recent years, the use of plastome data to investigate the evolution and diversity of major algal groups has increased noticeably (Lee, Cho, et al., 2016; Leliaert et al., 2016; Lemieux, Otis, & Turmel, 2014, 2016; Lemieux, Vincent, Labarre, Otis, & Turmel, 2015; Muñoz-Gómez et al., 2017; Sun et al., 2016), with complete plastomes from rare and understudied algal groups becoming available at a relatively high rate. Given the recognized importance of glaucophytes when investigating the early evolution of photosynthetic eukaryotes, it is surprising that the sequencing frenzy has not positively impacted the study of the group. For instance, until 2015 just a single glaucophyte plastome, sequenced more than 20 years ago, was available in public repositories (Stirewalt, Michalowski, Löffelhardt, Bohnert, & Bryant, 1995). The recent sequencing of mitochondrial genomes and plastomes from diverse glaucophyte species has for the first time allowed investigations of species diversity and evolution within this algal lineage using information from complete organelle genomes (F. Figueroa-Martinez et al., unpublished; Jackson & Reyes-Prieto, 2014; Price et al., 2012; S. Russell et al., unpublished; Smith, Jackson, & Reyes-Prieto, 2014).

2. THE GLAUCOPHYTA PLASTIDS, ORGANELLES WITH ANCESTRAL MORPHOLOGICAL TRAITS

Besides the blue-green colouration (Fig. 1A), the peptidoglycan wall and the carboxysome-like bodies (CLBs) (Fig. 1B) are the main causes of the "cyanobacterial" appearance of the glaucophyte plastids. These conspicuous characteristics led some authors to originally describe and classify the glaucophyte plastids as discrete cyanobacterial species. For example, the plastids of *C. paradoxa* were initially recognized as endosymbiotic "blue-green algae", named *Cyanocyta korschikoffiana*, living inside a "cryptomonad" host (Hall & Claus, 1963). Later, these ancestral traits led other authors to propose that the glaucophyte "blue-green insertions" were an intermediate stage between free-living cyanobacteria and actual organelles (Herdman, 1977; Kies, 1979), and to suggest that the glaucophytes represent the earliest-diverging lineage within the Archaeplastida (Fathinejad et al., 2008; Steiner & Löffelhardt, 2011). The sequence of the *C. paradoxa* plastome demonstrated that this glaucophyte species has a genome similar in length and gene content to those from viridiplants, red alga, and lineages with secondary plastids (Douglas & Turner, 1991; Stirewalt et al., 1995). In fact, the gene repertoire of the glaucophyte plastomes does not include several

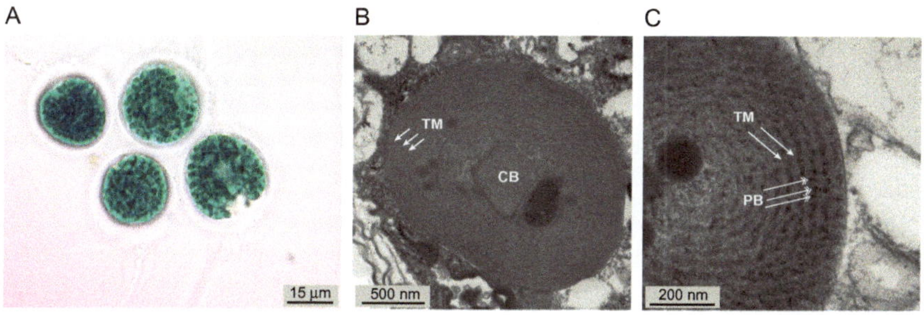

Fig. 1 Microscopy images of diverse glaucophytes. (A) The light microscopy photography of a cluster of four *Gloeochaete wittrockiana* (SAG 46.84) cells shows the typical blue-green colour of the glaucophyte plastids. It is possible to distinguish numerous individual plastids inside each cell. The scanning electron micrography (SEM) of (B) *Cyanoptyche gloeocystis* (SAG 4.97) reveals the carboxysome-like body (CB) localized in the center of the plastid. The nonstacked concentric disposition of the thylakoidal membranes (TM) is indicated with *arrowheads*. In the SEM (C) of *Glaucocystis incrassata* (SAG 229-2) the concentric organization of the TM is more evident. The electron-dense structures arranged along the TMs indicated with *double-headed arrows* are the phycobilisome-like inclusions (PB). *Scale bars* are indicated in each panel.

proteins responsible for the ancestral traits of these photosynthetic organelles. For instance, enzymes involved in peptidoglycan biosynthesis and the nonpigmented components of the phycobilisomes (Fig. 1C) are nuclear encoded in *C. paradoxa* (Bhattacharya et al., 2014; Price et al., 2012; Sato, Nishikawa, Kajitani, & Kawano, 2007). The nuclear location of these coding sequences revealed that the cyanobacteria-like appearance of the *C. paradoxa* plastids does not rely on an unusually rich set of plastid-encoded proteins.

2.1 The Glaucophyte Phycobilisomes: Source of the Blue Colour

Phycobilisomes are membrane-anchored multimeric (up 16×10^6 Da) light-harvesting complexes present in cyanobacteria, red algae, and glaucophytes (Grossman, Schaefer, Chiang, & Collier, 1993; Watanabe, Sato, Kondo, Narikawa, & Ikeuchi, 2012; Zhang et al., 2017). The structure of phycobilisomes resembles a fan-like arrangement composed of rods connected to a central core (Watanabe et al., 2012). The central core and the rods are made up from a series of multimeric disks of phycobiliproteins (PBPs; proteins with covalently linked tetrapyrrole chromophores), organized in cylindrical structures and stabilized by linker proteins (David, Marx, & Adir, 2011). Most red algal phycobilisomes contain phycoerythrin, a red accessory photopigment, as the main PBP and minor amounts of the blue PBPs allophycocyanin (APC) and C-phycocyanin (PC) (Chapman, 1966). In contrast, glaucophyte phycobilisomes include only APC and PC as accessory photopigments. The different pigment-binding protein subunits that constitute the allophycocyanin (ApcA-B-D-E-F) and C-phycocyanin (and CpcA-B) multimeric discs (Price, Steiner, Yoon, Bhattacharya, & Löffelhardt, 2017; Watanabe et al., 2012) are encoded in all known glaucophyte plastomes (Fig. 2; F. Figueroa-Martinez et al., unpublished; S. Russell et al., unpublished), but the nonpigmented proteins with core linker (ApcC1, ApcC2), rod linker (CpcK1 and CpcK2), and rod–core linker (CpcG1 and CpcG2) roles are located in the nuclear genome of *C. paradoxa* (Price et al., 2012, 2017; Watanabe et al., 2012) and have not been detected in plastomes of other glaucophyte species. The presumed unique origin of primary plastids suggests that phycobilisomes were part of the light-harvesting assembly in the plastids of the last Archaeplastida common ancestor and were subsequently lost in viridiplants (Tomitani et al., 1999).

2.2 The Vestigial Peptidoglycan Plastid Wall

Cyanobacteria, and many other bacterial lineages, usually possess a 20–40-nm mesh of peptidoglycan (a heteropolymer network of monosaccharides cross-linked by short peptide chains) surrounding the plasma membrane (Vollmer, Blanot, & de Pedro, 2008). The peptidoglycan mesh, called the cell wall, preserves cell integrity and actively participates in the cell division process (Vollmer et al., 2008). The glaucophyte plastids have retained vestiges of the cyanobacterial cell wall between the outer and inner membranes of the organelle. It is unclear if the plastid peptidoglycan wall has an osmotic function in glaucophytes, but it certainly plays an important role during the early stages of organelle division by forming a dividing septum that leads the invagination of the plastid membranes (Miyagishima, Kabeya, Sugita, Sugita, & Fujiwara, 2014). Detailed analyses of plastid division in diverse Archaeplastida indicate that all lineages share a common mechanism involving the formation of an annular structure composed of diverse proteins (e.g. FtsA, ARC6) on the stromal side (stromal plastid-dividing ring) of the organelle (Miyagishima, Suzuki, Okazaki, & Kabeya, 2012). However, the peptidoglycan-dividing septum of glaucophytes is a major difference between the plastid division mechanisms of these algae and those of red algae and viridiplants. During plastid division in

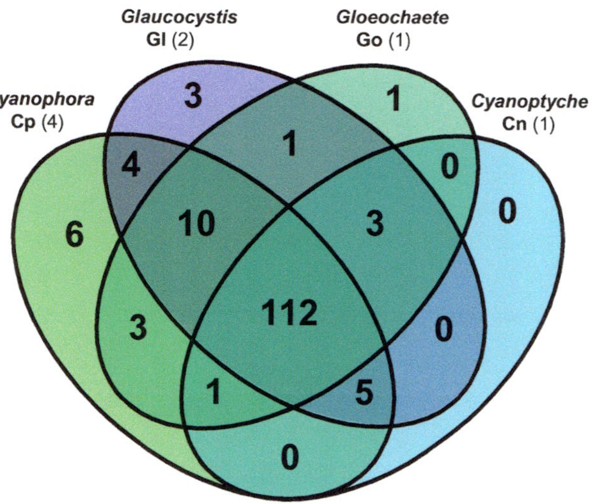

Genera	Shared protein-coding genes
Cn, Cp, Gl, Go	apcA, apcB, apcD, apcE, apcF, atpA, atpB, atpD, atpE, atpF, atpG, atpH, ccsA, chlB, chlI, chlL, chlN, clpP1, **clpP2**, cpcA, cpcB, dnaK, ftsW, groEL, **hemA**, **nadA**, petA, petB, petD, petF, petG, petL, petM, petN, preA, psaA, psaB, psaC, psaE, psaF, psaI, psaJ, psb30, psbA, psbB, psbC, psbD, psbE, psbF, psbH, psbI, psbJ, psbK, psbL, psbM, psbN, psbT, psbV, psbW, psbX, rbcL, rbcR, rbcS, rpl1, rpl11, rpl14, rpl16, rpl18, rpl19, rpl2, rpl20, rpl21, rpl22, rpl27, rpl28, rpl3, rpl35, rpl36, rpl5, rpl6, rpl7/12, rpoA, rpoB, rpoC1, rpoC2, rps10, rps11, rps12, rps13, rps14, rps16, rps17, rps19, rps2, rps20, rps3, rps4, rps5, rps6, rps7, rps8, secY, tufA, ycf16, ycf24, ycf3, ycf36, ycf38, ycf39, ycf4, **ycf48**, **ycf51**
Cp, Gl, Go	bioY, **ftsQ**, psaM, psbY, **rbrA**, rpl33, trpG, ycf29, **ycf49**, ycf9 (psbZ)
Cn, Cp, Gl	**groES**, **recO**, ycf17, ycf23, ycf34
Cn, Cp, Go	rps18
Cn, Gl, Go	psaK, rpl9, rpl23
Cp, Gl	ycf21, ycf27, ycf33, ycf35
Cp, Go	**hisH**, rpl34, **sepF**
Gl, Go	rps15
	Exclusive protein-coding genes
Cp	acpP, **crtE**, **mntA**, **mntB**, rps9, ycf37
Gl	**secE**, ycf26, ycf52
Go	tpx (bas1)

Fig. 2 Shared protein-coding genes in glaucophyte plastomes. The Venn diagram illustrates the number genes shared between the plastomes of *Cyanophora* (Cp), *Glaucocystis* (Gl), *Gloeochaete* (Go), and *Cyanoptyche* (Cn). The numbers in parenthesis indicate the number of species investigated. The table to right of the Venn diagram lists the names of the shared genes. Gene names in *red* are apparently exclusive of Glaucophyta.

nonglaucophyte archaeplastidians, an annular multiprotein complex is also formed on the cytosolic side of the plastid outer membrane. This cytosolic plastid-dividing ring, only present in red algae and viridiplants, presumably evolved as a mechanical replacement for the peptidoglycan septum after the loss of the heteropolymer wall (Iino & Hashimoto, 2003; Sato et al., 2007).

All glaucophyte plastomes sequenced to date (Fig. 2) encode only one enzyme thought to participate in the synthesis of peptidoglycan, the putative lipid flipase FtsW (*fts*W). Several other *C. paradoxa* proteins (e.g. MurA-G, MraY, and various penicillin-binding proteins) involved in peptidoglycan biosynthesis are encoded in the nuclear genome. The origin of the *Cyanophora* plastid-targeted enzymes participating in peptidoglycan biosynthesis is not entirely clear because phylogenetic estimations are not conclusive in most cases. Some genes have a putative cyanobacterial origin, but others are more similar to noncyanobacterial counterparts, suggesting that multiple enzyme replacements have occurred during the evolution of the glaucophyte plastid proteome (Bhattacharya et al., 2014; Löffelhardt, Bohnert, & Bryant, 1997; Plaimauer, Pfanzagl, Berenguer, de Pedro, & Löffelhardt, 1991; Price et al., 2012).

2.3 The Plastid RuBisCO Inclusions: Carboxysomes or Pyrenoids?

Carboxysomes are microcompartments found in diverse bacterial groups that accumulate both RuBisCO and carbonic anhydrase inside a semipermeable polyhedral protein shell. The protein shell is a barrier to gas diffusion (O_2 and CO_2), but allows the exchange of ribulose-1,5-bisphosphate, phosphoglycerate, and bicarbonate (HCO_3^-). Carboxysomes increase the CO_2 concentration inside the protein shell by promoting the activity of the carbonic anhydrase ($HCO_3^- + H^+ \rightarrow H_2CO_3 \rightarrow H_2O + CO_2$). Then, the released CO_2 favours the carboxylation reaction of the RuBisCO that fixes CO_2 into organic molecules (3-phosphoglycerate) (Burey et al., 2005; Mangeney & Gibbs, 1987; Rae, Long, Badger, & Price, 2013). The plastids of glaucophytes have stromal electron-dense bodies that resemble the carboxysomes of free-living cyanobacteria (Fig. 1B; Hall & Claus, 1963, 1967; Kies, 1989; Mangeney & Gibbs, 1987). However, genes encoding typical components of carboxysomes, such as proteins of the semipermeable shell and plastid-targeted homologues of cyanobacterial carbonic anhydrases, have not been identified in glaucophytes (Fathinejad et al., 2008; Price et al., 2012). Only mitochondrial and cytosolic carbonic anhydrases are encoded in the *Cyanophora* nuclear genome (Bhattacharya et al., 2014).

In fact, the only components of the putative carboxysomes detected in glaucophytes are the two subunits of RuBisCO (*rbc*S and *rbc*L; both plastid-encoded) and the RuBisCO activase (nuclear-encoded). Regardless of the apparent differences in protein composition between the CLBs of glaucophytes and bona fide cyanobacterial carboxysomes, experimental evidence indicates that the CLBs of *Cyanophora* are part of a plastid CO_2-concentrating mechanism (CCM) (Burey et al., 2005).

The presence of CCMs is not limited to cyanobacteria and glaucophyte plastids, and several eukaryote photosynthetic groups have analogous plastid CO_2-concentrating microcompartments called pyrenoids. The distinction between carboxysomes and pyrenoids is based on ultrastructural characteristics: pyrenoids lack a protein shell and are usually traversed by thylakoids, whereas carboxysomes are not penetrated by any membranous structure and always present a proteinaceous cover (Fathinejad et al., 2008). The CLBs of glaucophytes are not traversed by thylakoidal membranes, but the absence of key carboxysome components raises questions about the actual nature of the Glaucophyta CCM compartments (Bhattacharya et al., 2014; Price et al., 2017). Are they carboxysomes inherited directly from the plastid ancestor or are they just pyrenoids similar to those observed in other algae and plants? A recent survey of the *Cyanophora* genome revealed some nuclear genes (e.g. LCIA, LCIB, LCIC) encoding plastid-targeted proteins homologous to pyrenoidal components in *Chlamydomonas reinhardtii* (Bhattacharya et al., 2014). Overall, the glaucophyte CLBs seem to be in an intermediate ultrastructural state between cyanobacterial carboxysomes and algal pyrenoids, but the list of proteins apparently participating in the CCM resembles more a pyrenoid-like system (Price et al., 2017). Plastid microcompartments involved in CCMs have evolved and been independently lost numerous times during algal and plant evolution (Badger et al., 1998; Silberfeld et al., 2011; Villarreal & Renner, 2012). It should be investigated if the glaucophyte CLBs are evolutionary remnants of carboxysomes present in the plastid ancestor, or are pyrenoids that evolved independently.

3. THE GENETIC REPERTOIRE OF THE BLUE-GREEN PLASTIDS

3.1 Genome Size, Noncoding Regions, and RNA-Coding Genes

The plastome of *C. paradoxa* (strain UTEX LB 555, equivalent of SAG 29.80 and CCMP 329) was the only glaucophyte plastome available in public repositories for many years (Stirewalt et al., 1995). Recently, complete plastome sequences from three *Cyanophora* species, two different *Glaucocystis* isolates, *Cyanoptyche gloeocystis* and *Gloeochaete wittrockiana* became available (Table 1; F. Figueroa-Martinez et al., unpublished; Price et al., 2012; S. Russell et al., unpublished; Smith, Jackson et al., 2014). All glaucophyte plastomes assembled to date are circular mapping with similar G + C contents (30%–33%) and lengths (~130–150 kb) (Fig. 3 and Table 1; F. Figueroa-Martinez et al., unpublished). Glaucophyte plastomes are relatively compact, with no evidence of introns in *Glaucocystis* and *Gloeochaete* and just a single group IB intron identified in the plastomes of *Cyanoptyche gloeocystis* and *Cyanophora* species (Table 1). Interestingly, this intron is located in the same plastid gene (*trn*L; UAA anticodon) of the two latter taxa.

The proportion of noncoding regions in glaucophyte plastomes varies from 15% in *Glaucocystis* species to 30% in *Cyanophora sudae*. How do glaucophyte plastomes compare with other Archaeplastida in terms of sizes and the proportion of noncoding (NC) regions? In viridiplants we find the tiny plastome of the prasinophyte *Ostreococcus tauri* (71.6 kb, 1 intron, 15% NC) and the extreme case of *Volvox carteri* (525 kb, 9 introns, 80% NC). Within red algae, the extremophile *Cyanidioschyzon merolae* has a plastome that is much more compact (149.9 kb, no introns, 6% NC) than the gigantic plastomes of *Flintiella sanguinaria* (370.6 kb, 179 introns, 61% NC), *Bulboplastis apyrenoidosa* (610 kb, 220 introns, 79% NC), and *Corynoplastis japonica* (1127 kb, 310 introns, 88% NC). Hence, in comparison to both size variability and proportion of noncoding DNA found in other archaeplastidians, the glaucophyte plastomes stand somewhere in the middle, with no major differences between the diverse glaucophyte taxa compared.

All glaucophyte plastomes present a quadripartite structure with three ribosomal RNAs genes (*rrf*A [5S rRNA], *rrsA* [16S rRNA], and *rrlA* [23S rRNA]) located in the inverted repeat (IR) region. The length of the IR and the number of genes contained within it varies among lineages. The *G. wittrockiana* IR is 24.6 kb in length and contains 21 protein-coding genes, 4 open reading frames (ORFs), and 9 tRNAs, tripling at least the coding capacity of IRs from other glaucophyte plastomes (Fig. 3). The majority of tRNAs encoded in all glaucophyte plastomes are shared (33/43) between all species, and in the case of those completely sequenced the tRNA collection (≥31 in each case) is sufficient to decode all amino acids used in plastid-encoded proteins (Table 1; F. Figueroa-Martinez et al., unpublished).

Table 1 General Characteristics of Glaucophyte Plastomes

	Glaucocystis sp. Strain BBH	*Glaucocystis incrassata* (SAG 229-2)	*Gloeochaete wittrockiana* (SAG 46.84)	*Cyanoptyche gloeocystis* (SAG 4.97)	*Cyanophora paradoxa* (UTEX LB 555)	*Cyanophora kugrensii* (NIES-763)	*Cyanophora sudae* (NIES-764)	*Cyanophora biloba* (UTEX LB 2766)
GenBank accession	MF167424	MF167425	MF167426	MF167427	U30821	KM198929	MG601102	MG601103
Length (bp)	130,276	137,017	143,342	130,047[a]	135,599	142,028	150,128	130,509
Inverted repeat length (bp) (genes)[d]	10,582 (7)	10,538 (7)	24,788 (21)	9348 (5)	11,285 (4)	12,876 (4)	13,247 (1)	7637 (1)
Noncoding DNA (bp [%])	20,257 [15.5]	21,243 [15.5]	26,439 [18.44]	24,431 [18.7]	26,951 [19.9]	36,072 [25.4]	45,712 [30.4]	25,600 [19.6]
GC content (%)	33.4	33.6	29.6	30.6	30.5	30.3	29.8	30.2
Mean intergenic size (bp)[b]	148	152	199	179	199	263	339	189
Protein-coding genes[c]	137	137	129	121	136	138	136	136
Uknown ORFs	11	20	18	20	8	7	18	10
RNA-coding genes								
rRNA genes	3	3	3	3	3	3	3	3
tRNA genes	32	31	31	29	31	35	35	36
tmRNA genes	ND	ND	ND	1	1	1	1	1
rnpB gene	1	1	1	1	1	1	1	1
Introns	ND	ND	ND	1	1	1	1	1

[a]Partial sequence.
[b]tRNA genes were not considered in the estimation.
[c]Excluding unknown ORFs.
[d]Protein coding.
ND, not detected.

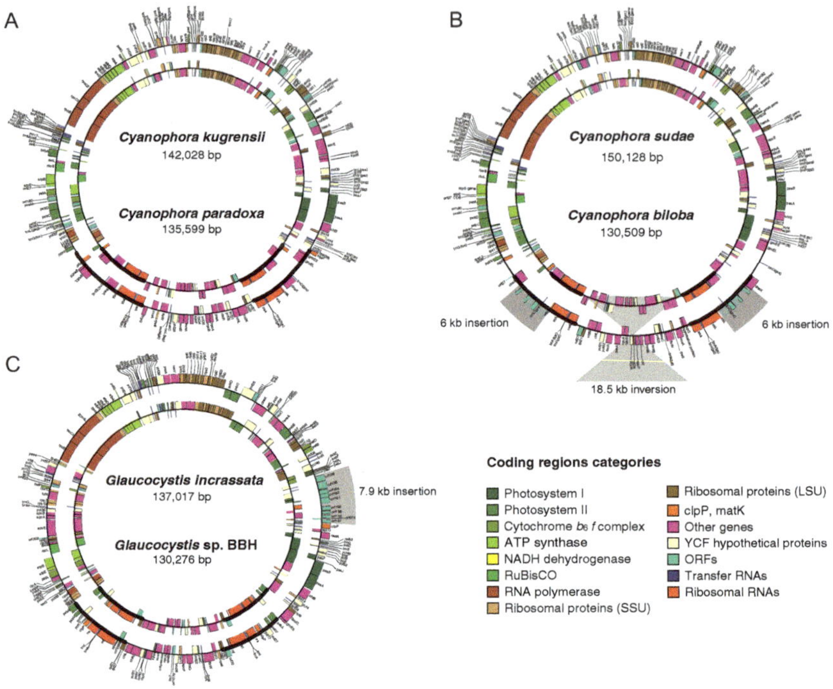

Fig. 3 Circular maps of diverse glaucophyte plastomes. The gene maps of (A) *Cyanophora paradoxa* (UTEX LB 555; *inner map*), *Cyanophora kugrensii* (NIES-763; *outer map*); (B) *Cyanophora biloba* (UTEX LB 2766; *inner map*), *Cyanophora sudae* (NIES-764; *outer map*); and (C) *Glaucocystis* sp. stain BBH (*inner map*) and *Glaucocystis incrassata* (SAG 229-2; *outer map*). Each pair of circular maps represents species that appear closely related in molecular phylogenetic studies (see Fig. 5). Inverted repeats are indicated with *thick lines* and detected insertions with *grey boxes*. The colour of the loci across the chromosome correspond to the functional categories of OGDRAW default values (Lohse, Drechsel, Kahlau, & Bock, 2013).

The gene *ssrA*, which encodes a transfer-messenger RNA, is present in the plastomes of *Cyanophora* species and *C. gloeocystis*, but no homologues were detected in *Glaucocystis* or *Gloeochaete* sequences (F. Figueroa-Martinez et al., unpublished; S. Russell et al., unpublished). Transfer-messenger RNAs (tmRNA), also present in the plastomes of some other algae (Gueneau de Novoa & Williams, 2004), are key mediators of the *trans*-translation process that rescues ribosomes arrested (e.g. stalled at end of truncated mRNAs with no stop codon) during protein translation (Janssen & Hayes, 2012). An RNA-coding sequence present in all known glaucophyte plastomes is the gene *rnp*B that encodes the RNA component of the Ribonuclease P. This enzyme is a ribonucleoprotein responsible for the maturation of the 5′ end of tRNA molecules. The gene *smp*B, encoding the proteinaceous section of the Ribonuclease P, has not been identified in glaucophyte plastomes, but the plastid *rnp*B transcript of *C. paradoxa* presents endonuclease catalytic activity and is able to process the 5′ ends of tRNA molecules in the absence of the protein component (Li, Willkomm, Schön, & Hartmann, 2007).

3.2 The Protein-Coding Gene Complement

If we count duplicated loci once, the number of protein-coding genes in completely sequenced glaucophyte plastomes varies from 129 in *G. wittrockiana* to 137 in *Glaucocystis* species (F. Figueroa-Martinez et al., unpublished). These plastid repertoires are larger than most viridiplant counterparts (between 50 and 100 genes; only three known cases with >130 genes), but smaller than the majority of red algal plastomes (between 160 and 210 genes). The glaucophyte plastomes share a set of 112 protein-coding genes, including sequences encoding subunits of photosystem I (7), photosystem II (17), the phycobilisome (7), ATPase (7), the cytochrome b_6f complex (8), the RNA polymerase (4), RuBisCO (2), proteins involved in chlorophyll biosynthesis (4), 34 ribosomal proteins, and diverse hypothetical and miscellaneous proteins. The content of the glaucophyte plastomes is largely conserved, with only 10 genes exclusively present in a single genus, including 6 of them only present in *Cyanophora* (Fig. 2; F. Figueroa-Martinez et al., unpublished; S. Russell et al., unpublished).

If we collate all the protein-coding sequences of the glaucophyte plastomes, we can identify an *all-glaucophyte* plastid collection of 149 genes (see the inset table of Fig. 2). The majority (133 genes) of the *all-glaucophyte* collection have homologues in plastomes of red algae or viridiplants, but only 68 of them are universally shared by the three Archaeplastida lineages. Glaucophytes share more plastid genes exclusively with red algae (57) than with viridiplants (8), while red algae and viridiplants share 23 genes not detected in glaucophytes (F. Figueroa-Martinez et al., unpublished). This three-way comparison reveals a set of 16 protein-coding sequences exclusive to glaucophytes plastomes, but only 5 of them are shared between all the species analysed (Fig. 2). The set of glaucophyte-exclusive genes includes sequences encoding the subunit A (*nad*A) of the quinolinate synthetase (an enzyme that is part of the de novo synthesis of pyridine nucleotides), a glutamyl-tRNA reductase (*hem*A, involved in protoporphyrin biosynthesis), a sequence (*clp*P2) similar to the proteolytic subunit of the CLP protease system encoded by the gene *clp*P1 (the latter is also encoded in glaucophyte plastomes), and the hypothetical proteins Ycf48 (putative assembly factor of photosystem II) and Ycf51 (DUF2518 family of unknown function). Other glaucophyte-exclusive plastid genes that are not present in all genera include the cochaperone GroES (*gro*ES), symerythrin (*rbr*A; a putative rubrerythrin-like FNR-dependent peroxidase; Cooley, Arp, & Karplus, 2011), the hypothetical membrane protein Ycf49 (a putative distant homologue is also present in *Nannochloropsis oceanica*) (Wei et al., 2013), the glutamine amidotransferase (*his*H) subunit of the imidazole glycerol phosphate synthase (involved in histidine biosynthesis), the "cell division" protein FtsQ (*fts*Q), the subunit SecE of the Sec-translocase (*sec*E), the DNA repair protein RecO (*rec*O), the geranylgeranyl diphosphate synthase (*crt*E; carotenoid production), and the putative protein SepF (*sep*F) involved in the formation of the *Z*-ring during cell division. Additionally, the plastomes of *Cyanophora* species have two genes (*mnt*A and *mnt*B) that encode subunits of a putative manganese/zinc ABC-transporter.

3.3 The Genome of the Last Common Ancestor of the Primary Plastids

Based on the ancestral features (i.e. peptidoglycan wall and carboxysomes-like structures) conserved by the glaucophyte plastids, it could have been hypothesized that their genomes might retain a transitional "primitive" state. That is, they might resemble the genome of a free-living cyanobacterium more closely than the plastomes of red and green algae, exhibiting features such as a larger genome size with a higher number of genes. Such a "transitional" state is evident in the photosynthetic organelles (i.e. chromatophores) of some species of the euglyphid genus *Paulinella* (Cercozoa, Rhizaria). The *Paulinella* chromatophores evolved via endosymbiosis with cyanobacteria more recently (90–140 million years ago; Delaye, Valadez-Cano, & Pérez-Zamorano, 2016) than the Archaeplastida plastids (1.2–1.9 billion years old) (Sánchez-Baracaldo, Raven, Pisani, & Knoll, 2017; Yoon, Hackett, Ciniglia, Pinto, & Bhattacharya, 2004) and have been studied thoroughly to investigate cellular and genomic mechanisms associated with endosymbiosis and early organelle evolution (Marin, Nowack, & Melkonian, 2005; Nowack, Melkonian, & Glöckner, 2008; Singer et al., 2017; Yoon et al., 2009; Yoon, Reyes-Prieto, Melkonian, & Bhattacharya, 2006). The three *Paulinella* chromatophore genomes sequenced to date have similar sizes (~1 Mb) and encode circa one-third (~870 genes) of the genes presumed to have been present in their free-living ancestor (~3300 genes in a 3-Mb genome), which is closely related to some species of the genus *Synechococcus* (Lhee et al., 2017; Nowack et al., 2008; Reyes-Prieto et al., 2010).

If we consider that the cyanobacterial ancestor of the Archaeplastida primary plastids likely had a similar repertoire of ~3300 genes, then the 149 *all-glaucophyte* plastid collection accounts for only 4% of the original cyanobacterial gene collection vs the ~30% encoded in the *Paulinella* photosynthetic organelle. In fact, the ~200 gene complement of red algal plastomes seems slightly more similar (6% of the hypothetical original cyanobacterial set) to the genetic repertoire of the last common ancestor of primary plastids. This comparison clearly indicates that the glaucophyte plastomes, together with those from other archaeplastidians, are not in a transitional state comparable to the genome of the *Pualinella* chromatophore. Moreover, the new data corroborate that the ancestral ultrastructural characteristics retained by the glaucophyte photosynthetic organelle are not associated with a copious ancestral gene collection in the plastome.

3.4 Few Gene Clusters Are Widely Conserved

Alignments of complete glaucophyte plastomes revealed few conserved gene groups. Those detected include two clusters of ribosomal proteins 5′-*rps*12-*rps*7-*tuf*A-*rps*10-3′ and 5′-*rpl*3-*rpl*23-*rpl*2-*rps*19-*rpl*22-*rps*3-*rpl*16-*rps*17-*rpl*14-*rpl*5-*rps*8-*rpl*6-*rpl*18-*rps*5-3′ (Michalowski, Pfanzagl, Löffelhardt, & Bohnert, 1990) that are almost identical in all glaucophytes, apart from a few missing genes in *Cyanophora* (F. Figueroa-Martinez et al., unpublished). The cluster 5′-*rpoB*-*rpoC1*-*rpoC2*-*rps2*-*atpH*-*atpG*-*atpF*-*atpD*-*atpA*-3′, conserved in most red algae and viridiplants, is also present in all known glaucophyte plastomes. The high conservation of this latter cluster in the vast majority of archaeplastidian plastomes, and the fact that the same syntenic arrangement has not been found in extant cyanobacterial genomes, has been suggested as evidence of the common origin of primary plastids (Löffelhardt, 2014; Stoebe & Kowallik, 1999).

Multiple alignments of complete glaucophyte plastomes indicate that several genomic rearrangements have occurred during diversification of glaucophyte genera, but there is no evidence of significant expansions, compactions, or major architectural changes (F. Figueroa-Martinez et al., unpublished). The relatively simple architecture of the sequenced glaucophyte plastomes contrasts with the complexity observed throughout the evolution of plastomes in green (Lemieux et al., 2016; Smith & Keeling,

2015; Turmel, Otis, & Lemieux, 2015) and early-diverging red algae (Muñoz-Gómez et al., 2017).

4. INTERSPECIFIC COMPARATIVE GENOMICS

Recent investigations using individual molecular markers from nuclear, mitochondrial, and plastid genomes revealed cryptic diversity in *Cyanophora* and *Glaucocystis*, leading to revaluations of the species boundaries in these two glaucophyte genera (Chong, Jackson, Kim, Yoon, & Reyes-Prieto, 2014; Takahashi et al., 2016, 2014). The new taxonomic schemes derived from the analysis of organelle genomic data have modified our perspective on diversity within Glaucophyta and validated the utility of organelle sequences for future studies of the group (Chong et al., 2014; Smith, Jackson, et al., 2014).

4.1 Plastomes in the Genus *Cyanophora*

Phylogenetic and distance analyses using markers from diverse genomic compartments resolved discrete genetic groups within *Cyanophora* (Chong et al., 2014; Takahashi et al., 2014), and later analyses based on complete mitochondrial (Jackson & Reyes-Prieto, 2014) and plastid genomes (S. Russell et al., unpublished) confirmed that *C. paradoxa* and *C. kugrensii* are sister taxa, separated from a clade formed by *C. sudae* and *C. biloba* (Fig. 5). The close affiliation between both duos of *Cyanophora* species is consistent with overall similarities shared between their plastomes. Gene synteny is almost identical in *C. paradoxa* (135.6 kb) and *C. kugrensii* (142 kb) with no differences in coding capacity (Fig. 3A). The gene order between *C. biloba* (130.5 kb) and *C. sudae* (150.1 kb) is very similar, but plastome alignments revealed an 18.5 kb inversion in *C. sudae*. These latter two *Cyanophora* species contain *dna*K as the only protein-coding gene within the IR region because *gro*EL, *gro*ES, and *clp*P1, encoded in the IRs of *C. paradoxa* and *C. kugrensii*, appear as singletons. The IR of *C. sudae* (13.3 kb) is double the size of the *C. biloba* IR (7.6 kb) due to a ~6-kb insertion that contains eight ORFs (at least 100 bp long) absent in other *Cyanophora* species (Fig. 3B).

Comparisons of nucleotide substitutions between *C. paradoxa* and *C. kugrensii* revealed that the rates of both synonymous (d_S) and nonsynonymous (d_N) substitutions in plastid protein-coding sequences ($d_S = 1.01 \pm 1.2$ and $d_N = 0.03 \pm 0.04$) are circa five times slower than rates in mitochondrial genes ($d_S = 5.3 \pm 3.2$ and $d_N = 0.13 \pm 0.11$; Smith,

Jackson, et al., 2014). Preliminary comparisons between the organelle genomes of *Glaucocystis* sp. BBH and *Glaucocystis incrassata* (SAG 229-2) have produced similar results, with plastid genes accumulating nucleotide substitutions ($d_S = 1.03 \pm 0.8$; $d_N = 0.03 \pm 0.03$) at a lower pace than mitochondrial sequences ($d_S = 6.01 \pm 2.5$; $d_N = 0.12 \pm 0.13$), suggesting that this pattern is common among glaucophytes. These results in glaucophytes are consistent with analyses in other photosynthetic groups, which have reported lower substitution rates in plastid sequences than mitochondrial counterparts (Smith, Arrigo, Alderkamp, & Allen, 2014; Smith & Keeling, 2012). With few exceptions, such as land plants and certain dinoflagellates, it seems that there is a widespread tendency of plastids to accumulate nucleotide substitutions at slower rates than mitochondrial and nuclear sequences (Smith, 2015). If the relatively slow substitution rate has been broadly conserved throughout plastid evolution, which is apparently the case, then it is reasonable to assume that plastid sequences are more suitable to elucidate ancient evolutionary relationships among photosynthetic eukaryotes than faster-evolving sequences (e.g. nuclear or mitochondrial genes). In other words, the high multiple substitution rates of fast-evolving sequences tend to increase the levels of homoplasy and saturation, with concurrent dilution of the phylogenetic signal (Klopfstein, Kropf, & Quicke, 2010).

4.2 A Case of HGT in One *Glaucocystis* Plastome

The plastomes of *G. incrassata* (137 kb) and *Glaucocystis* sp. BBH (130.2 kb) are the only available for this genus in public databases (F. Figueroa-Martinez et al., unpublished). Genetic distances estimated between *G. incrassata* and *Glaucocystis* sp. BBH using both plastid and mitochondrial genes strongly suggest that these two isolates are representatives of different species (Chong et al., 2014; Takahashi et al., 2016). Regardless of the genetic distance, gene synteny is highly conserved between both taxa with no evidence of major genomic rearrangements (Fig. 3C). However, the alignment of both complete plastomes revealed that *G. incrassata* possesses a 7.9-kb stretch, between the genes *clp*P1 and *psa*I, that is not present in the *Glaucocystis* sp. BBH plastome (Fig. 3C; F. Figueroa-Martínez et al., unpublished). The insertion in the *G. incrassata* plastome contains 10 ORFs, of which only 4 show similarity to known protein sequences. These four ORFs encode proteins similar to phage-type DNA primase/helicases (ORF 166), peptidoglycan aminohydrolases (ORF 163), and serine recombinases (ORFs 151 and 161). Maximum likelihood phylogenetic analyses of the four putative proteins suggest that the coding genes are of noncyanobacterial origin, but most nodes in the single-protein trees have weak bootstrap support and the results are not entirely conclusive (Fig. 4; F. Figueroa-Martinez et al., unpublished).

Most plastid genes in algae and plants have a cyanobacterial origin, presumably because such sequences were present in the genome of the plastid ancestor and have been inherited vertically as part of the reduced genome of the organelle. Considerable evidence indicates that plastomes are less prone than mitochondrial and nuclear genomes to recruit sequences via horizontal gene transfer (HGT) (Keeling & Palmer, 2008), but there are known cases of plastid sequences of noncyanobacterial origin that were likely acquired via HGT. Some examples are the genes of proteobacterial origin encoding the subunits of RuBisCO in red algae (Delwiche & Palmer, 1996), the gene *rpl36* in cryptophytes and haptophytes (Rice & Palmer, 2006), genes involved in the biosynthesis of vitamin K in cyanidiales red algae (Gross, Meurer, & Bhattacharya, 2008), ORFs of possible mitochondrial origin in the green alga *Oedogonium cardiacum* (Brouard, Otis, Lemieux, & Turmel, 2008), diverse genes in diatom plastomes acquired from plasmids resident in both the nucleus and plastids of the same diatoms (Ruck, Nakov, Jansen, Theriot, & Alverson, 2014), several bacterial-derived genes encoding enzymes involved in DNA replication and mobilization (e.g. DNA polymerases, transposases, integrases, and primases) in the green algae *Bryopsis plumosa* and *Tydemania expeditiones* (Leliaert & Lopez-Bautista, 2015), the DNA polymerase of the cryptophytes *Rhodomonas salina* and *Teleaulax amphioxeia* (Khan et al., 2007; Kim et al., 2015), genes involved in isoprenoid synthesis in the eustigmatophyte *Monodopsis* (Yurchenko, Ševčíková, Strnad, Butenko, & Eliáš, 2016), and intron sequences in the cryptophyte *R. salina* (Khan et al., 2007) and the diatom *Seminavis robusta* (Brembu et al., 2014). The ~8-kb insertion in the plastome of *G. incrassata* seems to be a new example of HGT occurring during plastome evolution.

4.3 Is the 7.9-kb Fragment Inserted Into the Plastome of *G. incrassata* Derived From a DNA Mobile Element?

DNA recombinases are defined as enzymes able to mediate site-specific excision and reintegration of DNA fragments. Based on characteristics of their active sites, two unrelated families of DNA recombinases are recognized: tyrosine recombinases (TR) and serine recombinases (SR). DNA recombinases are known to be involved in the insertion of phage genomes, transposons, and plasmids into foreign DNA regions (e.g. bacterial chromosomes) (Smith & Thorpe, 2002; Stark, 2014). Thus, the presence of two

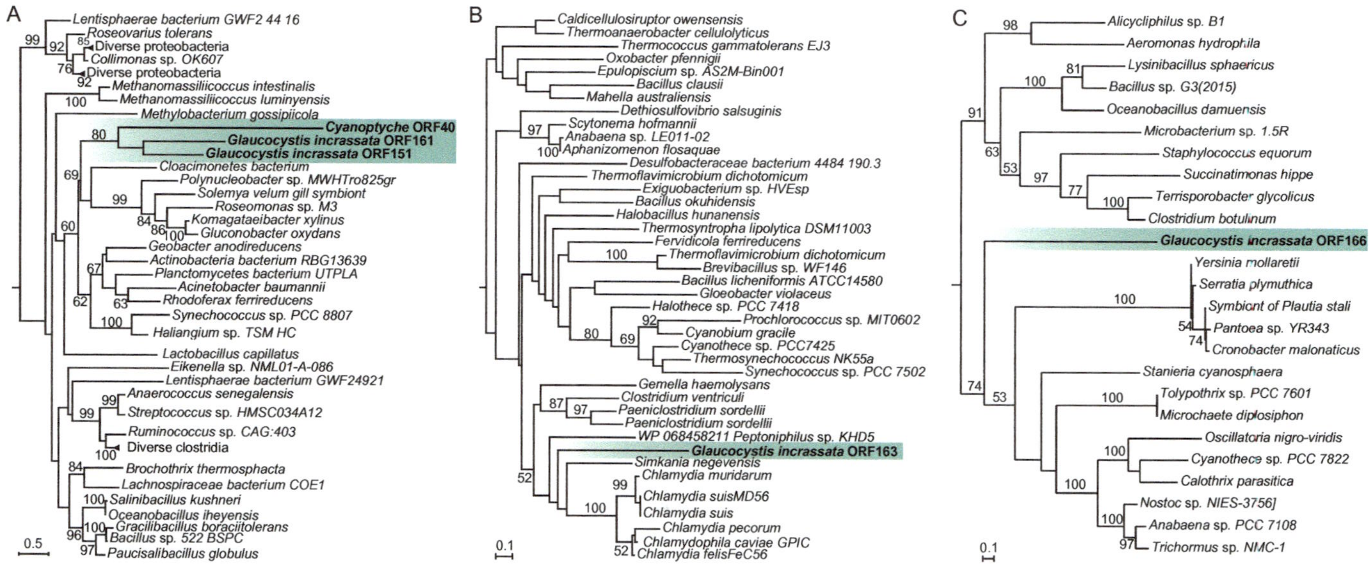

Fig. 4 Phylogenetic trees of four ORFs of the *Glaucocystis incrassata* plastome insertion. The maximum likelihood phylogenetic trees of (A) putative serine recombiases (ORFS 151, 161), (B) peptidoglycan aminohydrolase (ORF 163), and (C) DNA primase/helicase (ORF 166) of *Glaucocystis incrassata* were estimates with RAxML v8.2.6 (Stamatakis, 2014) considering the LG+G substitution model. Numbers near nodes indicate bootstrap values from 500 nonparametric bootstrap replicates. Branch lengths are proportional to the number of substitutions per site indicated by the *scale bars*.

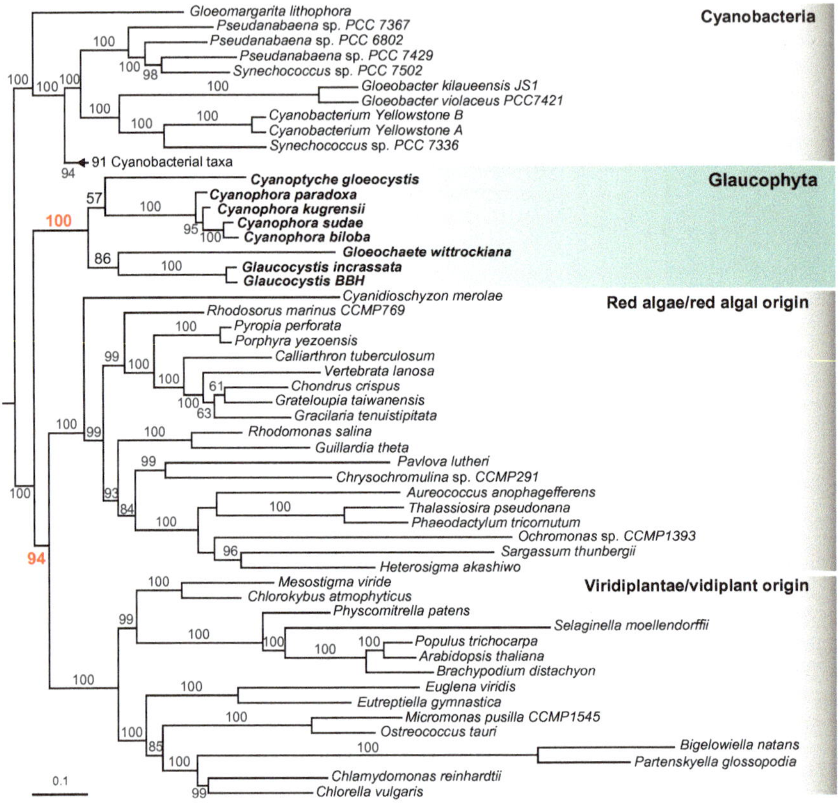

Fig. 5 Phylogenetic tree estimated from plastid sequences. The maximum likelihood phylogenetic tree was estimated from a set of 42 protein sequences using RAxML v8.2.6 (Stamatakis, 2014) and considering the LG + G substitution model. Numbers near nodes represent bootstrap proportion support values from 500 nonparametric bootstrap replicates. Branch lengths are proportional to the number of amino acid substitutions per site indicated by the *scale bar*. The amino acid sequences used were conceptual translations of the protein-coding genes *atp*A, *atp*E, *atp*H, *ccs*A, *chl*I, *pet*A, *pet*B, *pet*D, *psa*A, *psa*B, *psa*C, *psb*B, *psb*C, *psb*E, *psb*H, *psb*I, *psb*J, *psb*K, *psb*N, *rpl*14, *rpl*16, *rpl*19, *rpl*2, *rpl*20, *rpl*21, *rpl*22, *rpl*5, *rpl*6, *rpo*B, *rps*11, *rps*12, *rps*14, *rps*16, *rps*19, *rps*2, *rps*3, *rps*4, *rps*7, *rps*8, *tuf*A, *ycf*3, and *ycf*4.

ORFs encoding SRs in the plastome of *G. incrassata* raises questions about their potential participation in the integration of the ~8-kb fragment detected in the plastid chromosome. The plastome of *C. gloeocystis* also encodes a putative SR (ORF 40) of noncyanobacterial origin likely acquired via HGT. It is unknown if the SRs encoded in the *G. incrassata* and *C. gloeocystis* plastomes are transcribed and translated, but most active sites identified in bacterial homologues are present in the corresponding conceptual translations (F. Figueroa-Martinez et al., unpublished). It is yet to be investigated if these putative SRs are part of an active DNA recombination system in glaucophyte plastids.

The plastomes of some pennate diatoms and the "dinotom" (a dinoflagellate with a tertiary plastid of diatom origin) *Kryptoperidinium foliaceum* encode SRs recruited via HGT from plasmids localized in the same diatoms (Brembu et al., 2014; Hildebrand et al., 1992; Imanian, Pombert, Keeling, Schleiermacher, & Stoye, 2010; Ruck et al., 2014). Thus, SRs encoded in plastomes are not restricted to glaucophytes. Moreover, genes encoding TRs have been identified in plastomes of some green (Brouard et al., 2008; Civáň, Foster, Embley, Séneca, & Cox, 2014) and stramenopile algae (Brembu et al., 2014; Cattolico et al., 2008; Imanian et al., 2010). The presence of TRs and SRs in plastomes of some photosynthetic eukaryotes strongly suggests that DNA site-specific recombination has the potential to mediate, if only rarely, the integration of foreign sequences into plastid chromosomes.

In addition to DNA recombinases, the DNA primase/helicase and the peptidoglycan aminohydrolase, also encoded in the *G. incrassata* plastid insertion, represent other enzyme types frequently present in bacteriophage genomes and plasmids, where they play key roles in the replication and mobilization, respectively, of those DNA mobile elements (e.g. DeWitt & Grossman, 2014; Ilyina, Gorbalenya, & Koonin, 1992; Laverde Gomez, Bhatty, & Christie, 2014; Regamey & Karamata, 1998; Rutherford & Van Duyne, 2014). The *G. incrassata* insertion is relatively small and does not contain enough phylogenetic and architectural information to identify the origin of the entire ~8 kb segment, but the fact that four ORFs encode putative enzymes with recognized roles in DNA mobile element activity opens the possibility that the *G. incrassata* insertion originated from a phage or a plasmid sequence. If this is the case, it seems to be unique to *G. incrassata* among known glaucophytes. Cases of plastid sequences derived from DNA mobile elements have also been identified in diatoms (Ruck et al., 2014), green (Brouard, Turmel, Otis, & Lemieux, 2016; Leliaert & Lopez-Bautista, 2015), and red algae (Lee, Kim, et al., 2016; Muñoz-Gómez et al., 2017).

5. PHYLOGENOMICS, THE ORIGIN OF THE PRIMARY PLASTIDS AND THE ARCHAEPLASTIDA HYPOTHESIS

Most phylogenetic analyses based on plastid sequences strongly suggest a unique origin of the Archaeplastida photosynthetic organelles

(Criscuolo & Gribaldo, 2011; Ponce-Toledo et al., 2017; Qiu, Yang, Bhattacharya, & Yoon, 2012). However, these results do not directly support the common ancestry of the Archaeplastida nucleo-cytoplasm (the "host" component), which in contrast has rarely been supported in recent analyses of nuclear sequences. Regardless of these phylogenetic uncertainties between data from diverse genomic compartments, we can still separately investigate the evolutionary history of the host and the plastid components.

What do recent phylogenetic studies of nuclear data tell us about Archaeplastida evolution? Phylogenomic surveys based on sequences of nuclear-encoded proteins and considering broad eukaryote sampling have failed to recover the monophyly of the Archaeplastida nucleo-cytoplasm component (Burki et al., 2016, 2012; Derelle et al., 2015; Yabuki et al., 2014). In particular, the eukaryote group called Cryptista (cryptomonads, katablepharids, and palpitomonads) appears to have a phylogenetic connection with archaeplastidians that interrupts the Archaeplastida clade (Burki et al., 2016). These results directly challenge the hypothesis that red algae, viridiplants, and glaucophytes constitute a monophyletic group and leave open other alternatives to explain the evolution of the archaeplastidian host lineages and their plastids. For instance, some authors have suggested that the photosynthetic organelles of the Archaeplastida likely share a unique cyanobacterial ancestor, but that the plastid distribution we observe in modern archaeplastidians is the product of posterior independent plastid recruitments in different eukaryote hosts via cryptic secondary (i.e. eukaryote–eukaryote) endosymbiosis, rather than a vertical (phyletic) inheritance of the organelles (see Kim & Maruyama, 2014; Stiller, 2014; Stiller & Hall, 1997).

What do we know about the evolution of the primary plastid lineages? A recent phylogenetic reconstruction using a set of 42 plastid-encoded proteins, cyanobacterial homologues and the largest glaucophyte taxonomic sample to date (4 genera, 7 species), recovered a monophyletic Archaeplastida with Glaucophyta as the earliest-diverging branch (Fig. 5; S. Russell et al., unpublished). Previous investigations using plastid and cyanobacterial data have also recovered the Glaucophyta as the earliest-diverging archaeplastidian group (Deschamps & Moreira, 2009; Ponce-Toledo et al., 2017; Qiu et al., 2012), but other studies have alternatively resolved red algae (Criscuolo & Gribaldo, 2011; Janouskovec, Horák, Oborník, Lukes, & Keeling, 2010; Price et al., 2012) or viridiplants (Deschamps & Moreira, 2009; Janouskovec et al., 2010) as the earliest Archaeplastida branch. Moreover, tree topology tests based on multiscale bootstrap approximations (i.e. approximately unbiased tests;

Shimodaira & Goldman, 2002) do not reject topologies arbitrarily placing red algae or viridiplants as the first diverging archaeplastidian group (F. Figueroa-Martinez et al., unpublished). Overall, diverse independent phylogenomic approaches that differ in the amount of plastid loci analysed, the nature of the sequence data considered (nucleotide or protein) and taxon sampling, have not converged on a common answer identifying the earliest-diverging plastid lineage during algal evolution (see Mackiewicz & Gagat, 2014).

The phylogenetic history of the Archaeplastida and their plastids still require unambiguous answers, but inferences derived solely from plastid sequences have to be viewed with caution because the absence of data from plastid-lacking eukaryotes mostly limits conclusions to the origin and evolution of the organelles themselves. Additionally, the inclusion of novel plastid-lacking lineages (e.g. putative yet-unknown groups that have never possessed plastids but are related to Archaeplastida, or additional Cryptista representatives) in further analyses of nuclear data has the potential to completely dismantle the Archaeplastida monophyly hypothesis.

6. THE KNOWN GLAUCOPHYTE GROUPS REPRESENT LINEAGES OF PUTATIVE ANCIENT DIVERGENCE

Plastid phylogenomics recovers all Glaucophyta lineages in a single clade (F. Figueroa-Martinez et al., unpublished; S. Russell et al., unpublished). The same analyses recovered sister relationships, with weak to moderate support, between the genera *Cyanophora–Cyanoptyche* and *Gloeochaete–Glaucocystis*, respectively (Fig. 5). These intergenera relationships are consistent with previous phylogenetic estimations using mitochondrial data (Jackson & Reyes-Prieto, 2014). Further inclusion of additional species will be important to resolve phylogenetic relationships between glaucophyte genera, but the current data have provided some insights into genetic divergence within this algal group. For instance, the modest collection of glaucophyte plastomes has allowed exploration of the level of genetic distance between the known genera.

Pairwise genetic distances estimated from 17 different plastid loci revealed that divergence between certain glaucophyte genera is of the same magnitude as genetic distances estimated between some species belonging to different classes within the red and green algal lineages (Fig. 6). Unsurprisingly, each of plastid loci evaluated reflect different levels of sequence divergence, with some genes showing higher mean genetic distances than others (e.g. Kimura 2P distances ~0.45 for *chl*I vs ~0.2 for *psb*A). However,

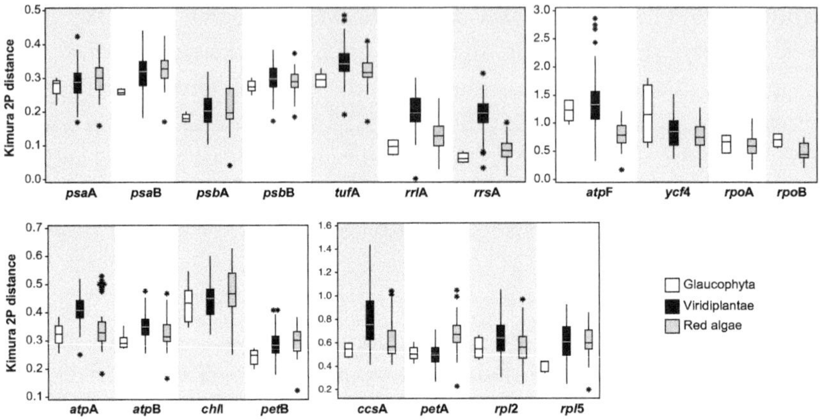

Fig. 6 Box and whisker plot of genetic distances estimated with diverse plastid protein-coding and ribosomal RNA genes. Nucleotide sequences of 17 protein-coding genes (*atp*A, *atp*B, *atp*F, *ccs*A, *chl*I, *pet*A, *pet*B, *psa*A, *psa*B, *psb*A, *psb*B, *rpl*2, *rpl*5, *rpo*A, *rpo*B, *tuf*A, and *ycf*4) and 2 ribosomal RNAs (*rrlA* [23S ribosomal RNA] and *rrsA* [16S ribosomal RNA]) were collected from representative species of the 4 glaucophyte genera (*Glaucocystis*, *Cyanophora*, *Gloeochaete*, and *Cyanoptyche*), and diverse species of red algae (17 species from 15 orders of 9 different classes/subclasses) and viridiplants (22 species from 21 orders of 18 different classes). *Boxes* represent 25th and 75th percentile with the median value, whiskers illustrate the 10th and 90th percentiles, and *asterisks* denote outlier comparisons. Nucleotide sequences of each Archaeplastida group were aligned independently considering codon positions. Pairwise Kimura 2-parameter distances were estimated independently for each of the three Archaeplastida lineages with MEGA v7 (Kumar, Stecher, & Tamura, 2016). The complete list of taxa investigated is included in the original analysis by F. Figueroa-Martinez et al. (unpublished).

in most cases the glaucophyte intergenus sequence divergence was higher than several pairwise comparisons within viridiplants or red algal groups of ancient divergence. For example, The Kimura 2P distances calculated with 13 (*atp*A, *atp*B, *atp*F, *chl*I, *pet*B, *psa*A, *psb*A, *psb*B, *rpl*2, *rpo*A, *rpo*B, *tuf*A, and *ycf*4) of the 17 plastid loci between representative species of the red algal classes Bangiophyceae and Florideophyceae were frequently smaller than distances calculated between different glaucophyte genera (Fig. 6; F. Figueroa-Martinez et al., unpublished). Comparisons of absolute distance between different Archaeplastida groups should be viewed with caution given the intrinsic subjectivity of higher taxonomic delimitations, but the relative comparison of genetic divergence values has revealed some patterns that deserve further investigation. For instance, if we consider that the Bangiophyceae and Florideophyceae node likely split 0.8–1.0 billion years ago (Yang et al., 2016), then the higher distances estimated between certain glaucophyte genera suggest that these subgroups represent lineages of ancient divergence within the Archaeplastida context. Another possibility is that glaucophyte plastomes have accumulated nucleotide changes more quickly than red algae. Unfortunately, the paucity of glaucophyte taxon sampling and the associated data analysed makes it difficult to discern between the two scenarios.

7. CONCLUSIONS

7.1 The Branching History of the Plastid Lineages

For many years the limited amount of glaucophyte genomic data has been a major constraint when investigating the origin of primary plastids and the presumed Archaeplastida monophyly. Fortunately, the scenario is changing and organelle genomes from poorly studied glaucophytes, such as *Glaucocystis*, *Cyanoptyche*, and *Gloeochaete*, have been recently sequenced. This expanded glaucophyte plastome sampling has allowed more robust investigations of the evolution of the primary plastid lineages. Phylogenies based only on plastome data cannot solve the Archaeplastida monophyly puzzle, but can certainly provide information about the tempo (evolutionary rates) and mode (branching patterns) of diversification in plastid lineages. Recent analyses of plastome data suggest that glaucophytes represent the earliest branching lineage within Archaeplastida. However, this latter scenario is not entirely conclusive yet, because alternative branching hypotheses are not rejected by tree topology tests. Solving the branching history of primary plastid groups might rely on further analyses with expanded taxon sampling, including plastomes of additional glaucophyte taxa and early branching red algae, but also on the development of phylogenetic methods that better cope with systematic errors, such as the use of inappropriate substitution models, and unequal nucleotide or amino acid frequencies between lineages.

An additional research avenue to solve the primary plastid branching history is the use of nonphylogenetic approaches, such as analyses based on comparative genomics. For example, besides the set of protein-coding genes (~68) universally present in all Archaeplastida plastomes, glaucophytes and red algae share an exclusive repertoire (57 genes) larger than the sets that each of them shares only with viridiplants (8 genes and 32 genes, respectively). If the number of shared genes is an indication of close phylogenetic relationships, then glaucophyte and red algal plastids seem to be sister groups given that they have more genes in common. If this latter scenario is true, then viridiplants must be the earliest-diverging lineage from the primary plastid stem. However, this conclusion should be tempered with some caveats: viridiplants possess the smallest plastid genetic complement of all archaeplastidians, indicating that the dissimilarity in gene content is, at least in part, a consequence of the highly derived state of the green algal and land plant plastomes.

The comparison of the plastid gene collections from the diverse Archaeplastida groups provides also some insights into the gene complement of the last common ancestor of the primary plastids. If we assume that the common ancestor of the Archaeplastida photosynthetic organelles had a larger gene complement than current extant plastids, then the numbers suggest that the red algal plastome set (~200 genes) may most closely resemble the gene collection of the last plastid common ancestor. Having the largest coding capacity does not mean red algae are the earliest-diverging plastid group, because there is no reason to assume that early-diverging plastid lineages have lost fewer genes than recently derived groups.

7.2 Glaucophyte Diversity

While plastome phylogenomics strongly supports the Glaucophyta monophyly, the intergenera phylogenetic relationships are not completely resolved. The nodes connecting the different genera in recent analyses are, at best, only moderately supported. In the case of studies within glaucophyte genera, analyses of single plastid markers (e.g. *psbA*) and complete plastome sequences have converged on similar conclusions regarding the phylogenetic relationships between the different *Cyanophora* and *Glaucocystis* species (Chong et al., 2014; Takahashi et al., 2016, 2014), suggesting that plastid data provide good phylogenetic resolution and can be used to further explore glaucophyte diversity. Even though these results have expanded our perspective on diversity within the Glaucophyta, they are still based on small taxonomic samples (e.g. only single isolates of *Gloeochaete* and *Cyanoptyche* have been analysed), and further inclusion of data from hypothetical novel species will be critical to obtain more solid answers about the intergenera relationships. Recent reports of new glaucophyte isolates, such as *Chalarodora azurea* (Hindak & Hindakova, 2012; no sequence data available), motivate further exploration of the environment for new representatives. The development of more efficient sampling methods (e.g. fluorescence-activated cell sorting; R. Calvaruso et al., unpublished), as well as DNA sequencing approaches that do not require cell cultures (e.g. single-cell genomics) to produce high-quality genomic data are promising avenues to continue the investigation of the rare and fascinating glaucophytes.

REFERENCES

Adl, S. M., Simpson, A. G. B., Lane, C. E., Lukeš, J., Bass, D., Bowser, S. S., et al. (2012). The revised classification of eukaryotes. *The Journal of Eukaryotic Microbiology, 59*, 429–493.

Badger, M. R., Andrews, T. J., Whitney, S. M., Ludwig, M., Yellowlees, D. C., Leggat, W., et al. (1998). The diversity and coevolution of Rubisco, plastids, pyrenoids, and chloroplast-based CO_2-concentrating mechanisms in algae. *Canadian Journal of Botany, 76*, 1052–1071.

Bhattacharya, D., Price, D. C., Gross, J., Chan, C. X., Steiner, J. M., & Loffelhardt, A. W. (2014). Analysis of the genome of *Cyanophora paradoxa*: An algal model for understanding primary endosymbiosis. In A. W. Loffelhardt (Ed.), *Endosymbiosis* (pp. 135–148). Vienna: Springer.

Brembu, T., Winge, P., Tooming-Klunderud, A., Nederbragt, A. J., Jakobsen, K. S., & Bones, A. M. (2014). The chloroplast genome of the diatom *Seminavis robusta*: New features introduced through multiple mechanisms of horizontal gene transfer. *Marine Genomics, 16*, 17–27.

Brouard, J.-S., Otis, C., Lemieux, C., & Turmel, M. (2008). Chloroplast DNA sequence of the green alga *Oedogonium cardiacum* (Chlorophyceae): Unique genome architecture, derived characters shared with the Chaetophorales and novel genes acquired through horizontal transfer. *BMC Genomics, 9*, 290.

Brouard, J.-S., Turmel, M., Otis, C., & Lemieux, C. (2016). Proliferation of group II introns in the chloroplast genome of the green alga *Oedocladium carolinianum* (Chlorophyceae). *PeerJ, 4*. e2627.

Burey, S. C., Fathi-Nejad, S., Poroyko, V., Steiner, J. M., Löffelhardt, W., & Bohnert, H. J. (2005). The central body of the cyanelles of *Cyanophora paradoxa*: A eukaryotic carboxysome? *Canadian Journal of Botany, 83*, 758–764.

Burki, F., Kaplan, M., Tikhonenkov, D. V., Zlatogursky, V., Minh, B. Q., Radaykina, L. V., et al. (2016). Untangling the early diversification of eukaryotes: A phylogenomic study of the evolutionary origins of Centrohelida, Haptophyta and Cryptista. *Proceedings of the Royal Society of London B, 283*. 20152802.

Burki, F., Okamoto, N., Pombert, J.-F., & Keeling, P. J. (2012). The evolutionary history of haptophytes and cryptophytes: Phylogenomic evidence for separate origins. *Proceedings of the Royal Society B, 279*, 2246–2254.

Cattolico, R., Jacobs, M. A., Zhou, Y., Chang, J., Duplessis, M., Lybrand, T., et al. (2008). Chloroplast genome sequencing analysis of *Heterosigma akashiwo* CCMP452 (West Atlantic) and NIES293 (West Pacific) strains. *BMC Genomics, 9*, 211.

Cavalier-Smith, T. (1982). The origins of plastids. *Biological Journal of the Linnean Society, 17*, 289–306.

Chapman, D. J. (1966). The pigments of the symbiotic algae (Cyanomes) of *Cyanophora paradoxa* and *Glaucocystis nostochinearum* and two Rhodophyceae, *Porphyridium aerugineum* and *Asteroeytis ramosa*. *Archiv für Mikrobiologie, 55*, 17–25.

Chong, J., Jackson, C., Kim, J. I., Yoon, H. S., & Reyes-Prieto, A. (2014). Molecular markers from different genomic compartments reveal cryptic diversity within glaucophyte species. *Molecular Phylogenetics and Evolution, 76C*, 181–188.

Civáň, P., Foster, P. G., Embley, M. T., Séneca, A., & Cox, C. J. (2014). Analyses of charophyte chloroplast genomes help characterize the ancestral chloroplast genome of land plants. *Genome Biology and Evolution, 6*, 897–911.

Cooley, R. B., Arp, D. J., & Karplus, P. A. (2011). Symerythrin structures at atomic resolution and the origins of rubrerythrins and the ferritin-like superfamily. *Journal of Molecular Biology*, *413*, 177–194.

Criscuolo, A., & Gribaldo, S. (2011). Large-scale phylogenomic analyses indicate a deep origin of primary plastids within Cyanobacteria. *Molecular Biology and Evolution*, *28*, 3019–3032.

David, L., Marx, A., & Adir, N. (2011). High-resolution crystal structures of trimeric and rod phycocyanin. *Journal of Molecular Biology*, *405*, 201–213.

Delaye, L., Valadez-Cano, C., & Pérez-Zamorano, B. (2016). How really ancient is *Paulinella chromatophora*? *PLoS Currents 8*, https://doi.org/10.1371/currents.tol.e68a099364bb1a1e129a17b4e06b0c6b.

Delwiche, C. F., & Palmer, J. D. (1996). Rampant horizontal transfer and duplication of rubisco genes in eubacteria and plastids. *Molecular Biology and Evolution*, *13*, 873–882.

Derelle, R., Torruella, G., Klimeš, V., Brinkmann, H., Kim, E., Vlček, Č., et al. (2015). Bacterial proteins pinpoint a single eukaryotic root. *Proceedings of the National Academy of Sciences of the United States of America*, *112*, E693–E699.

Deschamps, P., & Moreira, D. (2009). Signal conflicts in the phylogeny of the primary photosynthetic eukaryotes. *Molecular Biology and Evolution*, *26*, 2745–2753.

DeWitt, T., & Grossman, A. D. (2014). The bifunctional cell wall hydrolase CwlT is needed for conjugation of the integrative and conjugative element ICEBs1 in *Bacillus subtilis* and *B. anthracis*. *Journal of Bacteriology*, *196*, 1588–1596.

Douglas, S. E., & Turner, S. (1991). Molecular evidence for the origin of plastids from a cyanobacterium-like ancestor. *Journal of Molecular Evolution*, *33*, 267–273.

Fathinejad, S., Steiner, J. M., Reipert, S., Marchetti, M., Allmaier, G., Burey, S. C., et al. (2008). A carboxysomal carbon-concentrating mechanism in the cyanelles of the "coelacanth" of the algal world, *Cyanophora paradoxa*? *Physiologia Plantarum*, *133*, 27–32.

Gross, J., Meurer, J., & Bhattacharya, D. (2008). Evidence of a chimeric genome in the cyanobacterial ancestor of plastids. *BMC Evolutionary Biology*, *8*, 117.

Grossman, A. R., Schaefer, M. R., Chiang, G. G., & Collier, J. L. (1993). The phycobilisome, a light-harvesting complex responsive to environmental conditions. *Microbiological Reviews*, *57*, 725–749.

Gueneau de Novoa, P., & Williams, K. P. (2004). The tmRNA website: Reductive evolution of tmRNA in plastids and other endosymbionts. *Nucleic Acids Research*, *32*, D104–D108.

Hall, W. T., & Claus, G. (1963). Ultrastructural studies on the blue-green algal symbiont in *Cyanophora paradoxa* korschikoff. *The Journal of Cell Biology*, *19*, 551–563.

Hall, W. T., & Claus, G. (1967). Ultrastructural studies on the cyanelles of *Glaucocystis nostochinearum* itzigsohn. *Journal of Phycology*, *3*, 37–51.

Herdman, M. (1977). The cyanelle: Chloroplast or endosymbiotic prokaryote? *FEMS Microbiology Letters*, *1*, 7–11.

Hildebrand, M., Hasegawa, P., Ord, R. W., Thorpe, V. S., Glass, C. A., & Volcani, B. E. (1992). Nucleotide sequence of diatom plasmids: Identification of open reading frames with similarity to site-specific recombinases. *Plant Molecular Biology*, *19*, 759–770.

Hindak, F., & Hindakova, A. (2012). *Chalarodora azurea* Pascher 1929—A rare glaucophyte found in the peat-bog Klin (Orava, Northern Slovakia). In K. Wołowski, I. Kaczmarska, J. M. Ehrman, & A. Z. Wojtal (Eds.), *Current advances in algal taxonomy and its applications: Phylogenetic, ecological and applied perspective* (pp. 53–60). Kraków: Polish Academy of Sciences, W Szafer Institute of Botany.

Iino, M., & Hashimoto, H. (2003). Intermediate features of cyanelle division of *Cyanophora paradoxa* (glaucocystophyta) between cyanobacterial and plastid division. *Journal of Phycology*, *39*, 561–569.

Ilyina, T. V., Gorbalenya, A. E., & Koonin, E. V. (1992). Organization and evolution of bacterial and bacteriophage primase-helicase systems. *Journal of Molecular Evolution, 34*, 351–357.

Imanian, B., Pombert, J.-F., Keeling, P. J., Schleiermacher, C., & Stoye, J. (2010). The complete plastid genomes of the two "Dinotoms" Durinskia baltica and Kryptoperidinium foliaceum. ed A. Brandstaetter*PLoS One, 5*, e10711.

Jackson, C. J., & Reyes-Prieto, A. (2014). The mitochondrial genomes of the glaucophytes *Gloeochaete wittrockiana* and *Cyanoptyche gloeocystis*: Multilocus phylogenetics suggests a monophyletic Archaeplastida. *Genome Biology and Evolution, 6*, 2774–2785.

Janouskovec, J., Horák, A., Oborník, M., Lukes, J., & Keeling, P. J. (2010). A common red algal origin of the apicomplexan, dinoflagellate, and heterokont plastids. *Proceedings of the National Academy of Sciences of the United States of America, 107*, 10949–10954.

Janssen, B. D., & Hayes, C. S. (2012). The tmRNA ribosome-rescue system. *Advances in Protein Chemistry and Structural Biology, 86*, 151–191.

Keeling, P. J., & Palmer, J. D. (2008). Horizontal gene transfer in eukaryotic evolution. *Nature Reviews. Genetics, 9*, 605–618.

Khan, H., Parks, N., Kozera, C., Curtis, B. A., Parsons, B. J., Bowman, S., et al. (2007). Plastid genome sequence of the cryptophyte alga *Rhodomonas salina* CCMP1319: Lateral transfer of putative DNA replication machinery and a test of chromist plastid phylogeny. *Molecular Biology and Evolution, 24*, 1832–1842.

Kies, L. (1979). Zur systematischen Einordnung von *Cyanophora paradoxa, Gloeochaete wittrockiana* und *Glaucocystis nostochinearum*1. *Berichte der Deutschen Botanischen Gesellschaft, 92*, 445–454.

Kies, L. (1989). Ultrastructure of *Cyanoptyche gloeocystis* f.dispersa (Glaucocystophyceae). *Plant Systematics and Evolution, 164*, 65–73.

Kies, L., & Kremer, B. P. (1986). Typification of the Glaucocystophyta. *Taxon, 35*, 128–133.

Kim, E., & Maruyama, S. (2014). A contemplation on the secondary origin of green algal and plant plastids. *Acta Societatis Botanicorum Poloniae, 83*, 331–336.

Kim, J. I., Yoon, H. S., Yi, G., Kim, H. S., Yih, W., & Shin, W. (2015). The plastid genome of the cryptomonad *Teleaulax amphioxeia. PLoS One, 10*, e0129284.

Klopfstein, S., Kropf, C., & Quicke, D. L. J. (2010). An evaluation of phylogenetic informativeness profiles and the molecular phylogeny of diplazontinae (Hymenoptera, Ichneumonidae). *Systematic Biology, 59*, 226–241.

Kumar, S., Stecher, G., & Tamura, K. (2016). MEGA7: Molecular evolutionary genetics analysis version 7.0 for bigger datasets. *Molecular Biology and Evolution, 33*, 1870–1874.

Laverde Gomez, J. A., Bhatty, M., & Christie, P. J. (2014). PrgK, a multidomain peptidoglycan hydrolase, is essential for conjugative transfer of the pheromone-responsive plasmid pCF10. *Journal of Bacteriology, 196*, 527–539.

Lee, J., Cho, C. H., Park, S. I., Choi, J. W., Song, H. S., West, J. A., et al. (2016). Parallel evolution of highly conserved plastid genome architecture in red seaweeds and seed plants. *BMC Biology, 14*, 75.

Lee, J., Kim, K. M., Yang, E. C., Miller, K. A., Boo, S. M., Bhattacharya, D., et al. (2016). Reconstructing the complex evolutionary history of mobile plasmids in red algal genomes. *Scientific Reports, 6*, 23744.

Leliaert, F., & Lopez-Bautista, J. M. (2015). The chloroplast genomes of *Bryopsis plumosa* and *Tydemania expeditiones* (Bryopsidales, Chlorophyta): Compact genomes and genes of bacterial origin. *BMC Genomics, 16*, 204.

Leliaert, F., Tronholm, A., Lemieux, C., Turmel, M., DePriest, M. S., Bhattacharya, D., et al. (2016). Chloroplast phylogenomic analyses reveal the deepest-branching lineage of the Chlorophyta, Palmophyllophyceae class. nov. *Scientific Reports, 6*, 25367.

Lemieux, C., Otis, C., & Turmel, M. (2014). Chloroplast phylogenomic analysis resolves deep-level relationships within the green algal class Trebouxiophyceae. *BMC Evolutionary Biology, 14*, 211.

Lemieux, C., Otis, C., & Turmel, M. (2016). Comparative chloroplast genome analyses of streptophyte green algae uncover major structural alterations in the Klebsormidiophyceae, Coleochaetophyceae and Zygnematophyceae. *Frontiers in Plant Science*, *7*, 697.

Lemieux, C., Vincent, A. T., Labarre, A., Otis, C., & Turmel, M. (2015). Chloroplast phylogenomic analysis of chlorophyte green algae identifies a novel lineage sister to the Sphaeropleales (Chlorophyceae). *BMC Evolutionary Biology*, *15*, 264.

Lhee, D., Yang, E. C., Kim, J. I., Nakayama, T., Zuccarello, G., Andersen, R. A., et al. (2017). Diversity of the photosynthetic *Paulinella* species, with the description of *Paulinella micropora* sp. nov. and the chromatophore genome sequence for strain KR01. *Protist*, *168*, 155–170.

Li, D., Willkomm, D. K., Schön, A., & Hartmann, R. K. (2007). RNase P of the *Cyanophora paradoxa* cyanelle: A plastid ribozyme. *Biochimie*, *89*, 1528–1538.

Löffelhardt, W. (2014). The single primary endosymbiotic event. In A. W. Loffelhardt (Ed.), *Endosymbiosis* (pp. 39–52). Vienna: Springer.

Löffelhardt, W., & Bohnert, H. J. (2001). The cyanelle (muroplast) of *Cyanophora paradoxa*: A paradigm for endosymbiotic organelle evolution. In J. Seckbach (Ed.), *Cellular origin, life in extreme habitats and astrobiology*: *Vol. 4. Symbiosis. Mechanisms and model systems* (pp. 111–130). Dordrecht: Springer.

Löffelhardt, W., Bohnert, H. J., & Bryant, D. A. (1997). The complete sequence of the *Cyanophora paradoxa* cyanelle genome (Glaucocystophyceae). In D. Bhattacharya (Ed.), *Plant systematics and evolution*: *Vol. 11. Origins of algae and their plastids* (pp. 149–162). Vienna: Springer.

Lohse, M., Drechsel, O., Kahlau, S., & Bock, R. (2013). OrganellarGenomeDRAW—A suite of tools for generating physical maps of plastid and mitochondrial genomes and visualizing expression data sets. *Nucleic Acids Research*, *41*, W575–W581.

Mackiewicz, P., & Gagat, P. (2014). Monophyly of Archaeplastida supergroup and relationships among its lineages in the light of phylogenetic and phylogenomic studies. Are we close to a consensus? *Acta Societatis Botanicorum Poloniae*, *83*, 263–280.

Mangeney, E., & Gibbs, S. P. (1987). Immunocytochemical localization of ribulose-1,5-bisphosphate carboxylase/oxygenase in the cyanelles of *Cyanophora paradoxa* and *Glaucocystis nostochinearum*. *European Journal of Cell Biology*, *43*, 65–70.

Marin, B., Nowack, E. C. M., & Melkonian, M. (2005). A plastid in the making: Evidence for a second primary endosymbiosis. *Protist*, *156*, 425–432.

McFadden, G. I., & van Dooren, G. G. (2004). Evolution: Red algal genome affirms a common origin of all plastids. *Current Biology*, *14*, R514–516.

Michalowski, C. B., Pfanzagl, B., Löffelhardt, W., & Bohnert, H. J. (1990). The cyanelle S10 spc ribosomal protein gene operon from *Cyanophora paradoxa*. *Molecular & General Genetics: MGG*, *224*, 222–231.

Miyagishima, S., Kabeya, Y., Sugita, C., Sugita, M., & Fujiwara, T. (2014). DipM is required for peptidoglycan hydrolysis during chloroplast division. *BMC Plant Biology*, *14*, 57.

Miyagishima, S.-Y., Suzuki, K., Okazaki, K., & Kabeya, Y. (2012). Expression of the nucleus-encoded chloroplast division genes and proteins regulated by the algal cell cycle. *Molecular Biology and Evolution*, *29*, 2957–2970.

Muñoz-Gómez, S. A., Mejía-Franco, F. G., Durnin, K., Colp, M., Grisdale, C. J., Archibald, J. M., et al. (2017). The new red algal subphylum proteorhodophytina comprises the largest and most divergent plastid genomes known. *Current Biology*, *27*, 1677–1684.

Nowack, E. C. M., Melkonian, M., & Glöckner, G. (2008). Chromatophore genome sequence of *Paulinella* sheds light on acquisition of photosynthesis by eukaryotes. *Current Biology*, *18*, 410–418.

Palmer, J. D. (2003). The symbiotic birth and spread of plastids: How many times and whodunit? *Journal of Phycology*, *39*, 4–12.

Plaimauer, B., Pfanzagl, B., Berenguer, J., de Pedro, M. A., & Löffelhardt, W. (1991). Subcellular distribution of enzymes involved in the biosynthesis of cyanelle murein in the protist *Cyanophora paradoxa*. *FEBS Letters*, *284*, 169–172.

Ponce-Toledo, R. I., Deschamps, P., López-García, P., Zivanovic, Y., Benzerara, K., Moreira, D., et al. (2017). An early-branching freshwater cyanobacterium at the origin of plastids. *Current Biology*, *27*, 386–391.

Price, D. C., Chan, C. X., Yoon, H. S., Yang, E. C., Qiu, H., Weber, A. P. M., et al. (2012). *Cyanophora paradoxa* genome elucidates the origin of photosynthesis in algae and plants. *Science*, *335*, 843–847.

Price, D. C., Steiner, J. M., Yoon, H. S., Bhattacharya, D., & Löffelhardt, W. (2017). Glaucophyta. In J. M. Archibald, A. G. B. Simpson, & C. H. Slamovits (Eds.), *Handbook of the protists* (pp. 23–87). Springer International Publishing.

Qiu, H., Yang, E. C., Bhattacharya, D., & Yoon, H. S. (2012). Ancient gene paralogy may mislead inference of plastid phylogeny. *Molecular Biology and Evolution*, *29*, 3333–3343.

Rae, B. D., Long, B. M., Badger, M. R., & Price, G. D. (2013). Functions, compositions, and evolution of the two types of carboxysomes: Polyhedral microcompartments that facilitate CO_2 fixation in cyanobacteria and some proteobacteria. *Microbiology and Molecular Biology Reviews*, *77*, 357–379.

Regamey, A., & Karamata, D. (1998). The N-acetylmuramoyl-L-alanine amidase encoded by the *Bacillus subtilis* 168 prophage SP? *Microbiology*, *144*, 885–893.

Reyes-Prieto, A., & Bhattacharya, D. (2007). Phylogeny of Calvin cycle enzymes supports Plantae monophyly. *Molecular Phylogenetics and Evolution*, *45*, 384–391.

Reyes-Prieto, A., & Moustafa, A. (2012). Plastid-localized amino acid biosynthetic pathways of Plantae are predominantly composed of non-cyanobacterial enzymes. *Scientific Reports*, *2*, 955.

Reyes-Prieto, A., Yoon, H. S., Moustafa, A., Yang, E. C., Andersen, R. A., Boo, S. M., et al. (2010). Differential gene retention in plastids of common recent origin. *Molecular Biology and Evolution*, *27*, 1530–1537.

Rice, D. W., & Palmer, J. D. (2006). An exceptional horizontal gene transfer in plastids: Gene replacement by a distant bacterial paralog and evidence that haptophyte and cryptophyte plastids are sisters. *BMC Biology*, *4*, 31.

Rodríguez-Ezpeleta, N., Brinkmann, H., Burey, S. C., Roure, B., Burger, G., Löffelhardt, W., et al. (2005). Monophyly of primary photosynthetic eukaryotes: Green plants, red algae, and glaucophytes. *Current Biology*, *15*, 1325–1330.

Ruck, E. C., Nakov, T., Jansen, R. K., Theriot, E. C., & Alverson, A. J. (2014). Serial gene losses and foreign DNA underlie size and sequence variation in the plastid genomes of diatoms. *Genome Biology and Evolution*, *6*, 644–654.

Rutherford, K., & Van Duyne, G. D. (2014). The ins and outs of serine integrase site-specific recombination. *Current Opinion in Structural Biology*, *24*, 125–131.

Sánchez-Baracaldo, P., Raven, J. A., Pisani, D., & Knoll, A. H. (2017). Early photosynthetic eukaryotes inhabited low-salinity habitats. *Proceedings of the National Academy of Sciences of the United States of America*, *114*, E7737–E7745.

Sato, M., Nishikawa, T., Kajitani, H., & Kawano, S. (2007). Conserved relationship between FtsZ and peptidoglycan in the cyanelles of *Cyanophora paradoxa* similar to that in bacterial cell division. *Planta*, *227*, 177–187.

Shimodaira, H., & Goldman, N. (2002). An approximately unbiased test of phylogenetic tree selection. *Systematic Biology*, *51*, 492–508.

Silberfeld, T., Racault, M.-F. L. P., Fletcher, R. L., Couloux, A., Rousseau, F., & De Reviers, B. (2011). Systematics and evolutionary history of pyrenoid-bearing taxa in brown algae (Phaeophyceae). *European Journal of Phycology*, *46*, 361–377.

Singer, A., Poschmann, G., Mühlich, C., Valadez-Cano, C., Hänsch, S., Hüren, V., et al. (2017). Massive protein import into the early-evolutionary-stage photosynthetic organelle of the amoeba *Paulinella chromatophora*. *Current Biology: CB, 27*. 2763–2773.e5.
Skuja, H. (1954). Glaucophyta. In H. Melchior & E. Werdermann (Eds.), *A. Engler's Syllabus der Pflanzenfamilien* (pp. 56–57). Berlin: Borntraeger.
Smith, D. R. (2015). Mutation rates in plastid genomes: They are lower than you might think. *Genome Biology and Evolution, 7*, 1227–1234.
Smith, D. R., Arrigo, K. R., Alderkamp, A.-C., & Allen, A. E. (2014). Massive difference in synonymous substitution rates among mitochondrial, plastid, and nuclear genes of Phaeocystis algae. *Molecular Phylogenetics and Evolution, 71*, 36–40.
Smith, D. R., Jackson, C. J., & Reyes-Prieto, A. (2014). Nucleotide substitution analyses of the glaucophyte *Cyanophora* suggest an ancestrally lower mutation rate in plastid vs mitochondrial DNA for the Archaeplastida. *Molecular Phylogenetics and Evolution, 79*, 380–384.
Smith, D. R., & Keeling, P. J. (2012). Twenty-fold difference in evolutionary rates between the mitochondrial and plastid genomes of species with secondary red plastids. *Journal of Eukaryotic Microbiology, 59*, 181–184.
Smith, D. R., & Keeling, P. J. (2015). Mitochondrial and plastid genome architecture: Reoccurring themes, but significant differences at the extremes. *Proceedings of the National Academy of Sciences of the United States of America, 112*, 10177–10184.
Smith, M. C. M., & Thorpe, H. M. (2002). Diversity in the serine recombinases. *Molecular Microbiology, 44*, 299–307.
Stamatakis, A. (2014). RAxML version 8: A tool for phylogenetic analysis and post-analysis of large phylogenies. *Bioinformatics, 30*, 1312–1313.
Stark, W. M. (2014). The serine recombinases. *Microbiology Spectrum, 2*, 1–16. https://doi.org/10.1128/microbiolspec.MDNA3-0046-2014.
Steiner, J. M., & Löffelhardt, W. (2011). The photosynthetic apparatus of the living fossil, *Cyanophora paradoxa*. In G. Peschek, C. Obinger, & G. Renger (Eds.), *Bioenergetic processes of cyanobacteria* (pp. 71–87). Dordrecht: Springer.
Steiner, J. M., Yusa, F., Pompe, J. A., & Löffelhardt, W. (2005). Homologous protein import machineries in chloroplasts and cyanelles. *The Plant Journal: For Cell and Molecular Biology, 44*, 646–652.
Stiller, J. W. (2014). Toward an empirical framework for interpreting plastid evolution. ed C. Lane. *Journal of Phycology, 50*, 462–471.
Stiller, J. W., & Hall, B. D. (1997). The origin of red algae: Implications for plastid evolution. *Proceedings of the National Academy of Sciences of the United States of America, 94*, 4520–4525.
Stirewalt, V. L., Michalowski, C. B., Löffelhardt, W., Bohnert, H. J., & Bryant, D. A. (1995). Nucleotide sequence of the cyanelle genome from *Cyanophora paradoxa*. *Plant Molecular Biology Reporter, 13*, 327–332.
Stoebe, B., & Kowallik, K. V. (1999). Gene-cluster analysis in chloroplast genomics. *Trends in Genetics, 15*, 344–347.
Sun, L., Fang, L., Zhang, Z., Chang, X., Penny, D., & Zhong, B. (2016). Chloroplast phylogenomic inference of green algae relationships. *Scientific Reports, 6*. 20528.
Takahashi, T., Nishida, T., Tuji, A., Saito, C., Matsuzaki, R., Sato, M., et al. (2016). Delineation of six species of the primitive algal genus *Glaucocystis* based on in situ ultrastructural characteristics. *Scientific Reports, 6*. 29209.
Takahashi, T., Sato, M., Toyooka, K., Matsuzaki, R., Kawafune, K., Kawamura, M., et al. (2014). Five *Cyanophora* (Cyanophorales, Glaucophyta) species delineated based on morphological and molecular data. *Journal of Phycology, 50*, 1058–1069.
Tomitani, A., Okada, K., Miyashita, H., Matthijs, H. C., Ohno, T., & Tanaka, A. (1999). Chlorophyll b and phycobilins in the common ancestor of cyanobacteria and chloroplasts. *Nature, 400*, 159–162.

Turmel, M., Otis, C., & Lemieux, C. (2015). Dynamic evolution of the chloroplast genome in the green algal classes Pedinophyceae and Trebouxiophyceae. *Genome Biology and Evolution*, 7, 2062–2082.
Villarreal, J. C., & Renner, S. S. (2012). Hornwort pyrenoids, carbon-concentrating structures, evolved and were lost at least five times during the last 100 million years. *Proceedings of the National Academy of Sciences of the United States of America*, *109*, 18873–18878.
Vollmer, W., Blanot, D., & de Pedro, M. A. (2008). Peptidoglycan structure and architecture. *FEMS Microbiology Reviews*, *32*, 149–167.
Watanabe, M., Sato, M., Kondo, K., Narikawa, R., & Ikeuchi, M. (2012). Phycobilisome model with novel skeleton-like structures in a glaucocystophyte *Cyanophora paradoxa*. *Biochimica et Biophysica Acta*, *1817*, 1428–1435.
Wei, L., Xin, Y., Wang, D., Jing, X., Zhou, Q., Su, X., et al. (2013). Nannochloropsis plastid and mitochondrial phylogenomes reveal organelle diversification mechanism and intragenus phylotyping strategy in microalgae. *BMC Genomics*, *14*, 534.
Yabuki, A., Kamikawa, R., Ishikawa, S. A., Kolisko, M., Kim, E., Tanabe, A. S., et al. (2014). *Palpitomonas bilix* represents a basal cryptist lineage: Insight into the character evolution in Cryptista. *Scientific Reports*, *4*, 4641.
Yang, E. C., Boo, S. M., Bhattacharya, D., Saunders, G. W., Knoll, A. H., Fredericq, S., et al. (2016). Divergence time estimates and the evolution of major lineages in the florideophyte red algae. *Scientific Reports*, *6*, 21361.
Yoon, H. S., Hackett, J. D., Ciniglia, C., Pinto, G., & Bhattacharya, D. (2004). A molecular timeline for the origin of photosynthetic eukaryotes. *Molecular Biology and Evolution*, *21*, 809–818.
Yoon, H. S., Nakayama, T., Reyes-Prieto, A., Andersen, R. A., Boo, S. M., Ishida, K., et al. (2009). A single origin of the photosynthetic organelle in different *Paulinella* lineages. *BMC Evolutionary Biology*, *9*, 98.
Yoon, H. S., Reyes-Prieto, A., Melkonian, M., & Bhattacharya, D. (2006). Minimal plastid genome evolution in the *Paulinella* endosymbiont. *Current Biology*, *16*, R670–R672.
Yurchenko, T., Ševčíková, T., Strnad, H., Butenko, A., & Eliáš, M. (2016). The plastid genome of some eustigmatophyte algae harbours a bacteria-derived six-gene cluster for biosynthesis of a novel secondary metabolite. *Open Biology*, *6*, 160249.
Zhang, J., Ma, J., Liu, D., Qin, S., Sun, S., Zhao, J., et al. (2017). Structure of phycobilisome from the red alga *Griffithsia pacifica*. *Nature*, *551*, 57–63.
Zhao, S., Burki, F., Brate, J., Keeling, P. J., Klaveness, D., & Shalchian-Tabrizi, K. (2012). Collodictyon—An ancient lineage in the tree of eukaryotes. *Molecular Biology and Evolution*, *29*, 1557–1568.

CHAPTER FIVE

Evolution of the Plastid Genomes in Diatoms

Mengjie Yu*[,1], Matt P. Ashworth*, Nahid H. Hajrah[†], Mohammad A. Khiyami[‡], Mumdooh J. Sabir[§], Alawiah M. Alhebshi[†], Abdulrahman L. Al-Malki[¶], Jamal S.M. Sabir[†], Edward C. Theriot*, Robert K. Jansen*[,||]

*Department of Integrative Biology, University of Texas at Austin, Austin, TX, United States
[†]Genomics and Biotechnology Research Group, Faculty of Science, King Abdulaziz University, Jeddah, 21589, Saudi Arabia
[‡]King Abdulaziz City for Science and Technology, Riyadh, 11442, Saudi Arabia
[§]Faculty of Computing and Information Technology, King Abdulaziz University, Jeddah, 21589, Saudi Arabia
[¶]Department of Biochemistry, Faculty of Science, King Abdulaziz University, Jeddah, 21589, Saudi Arabia
[||]Genomics and Biotechnology Research Group, Faculty of Science, King Abdulaziz University, Jeddah, Saudi Arabia
[1]Corresponding author: e-mail address: annaymj_2010@utexas.edu

Contents

Advances in Botanical Research, Volume 85
ISSN 0065-2296
https://doi.org/10.1016/bs.abr.2017.11.009

Abstract

Diatoms are a monophyletic group of eukaryotic, single-celled heterokont algae. Despite years of phylogenetic research, relationships among major groups of diatoms remain uncertain. Here we assess diatom phylogenetic relationships using the plastid genome (plastome). The 22 previously published diatom plastomes showed variable genome size, gene content and extensive rearrangement. We report another 18 diatom plastome sequences ranging in size from 119,120 to 201,816 bp. *Plagiogramma staurophorum* had the largest plastome sequenced so far due to large inverted repeats and a 2971 bp group II intron insertion in *petD*. The previously reported loss of *psaE*, *psaI* and *psaM* genes in *Rhizosolenia imbricata* also occurred in the closely related species *Rhizosolenia fallax*. In the largest genome-scale phylogeny yet published for diatoms based on 103 shared plastid-coding genes from 40 diatoms and *Triparma laevis* as the outgroup, *Leptocylindrus* was recovered as sister to the remaining diatoms and the clade of *Attheya* plus *Biddulphia* was recovered as sister to pennate diatoms, strongly rejecting monophyly of two of the three proposed classes of diatoms. Our study also revealed extensive gene loss and a strong positive correlation between sequence divergence and gene order change in diatom plastomes.

1. INTRODUCTION

Diatoms are photoautotrophic eukaryotic, single-celled heterokont algae and play an important role in the global geological cycle, being responsible for one quarter of primary production, as well as being the primary biological mediators of the silica cycle in the oceans (Nelson, Treguer, Brzezinski, Leynaert, & Queguiner, 1995). They have delicate siliceous cell walls, which can be used to identify structural groups of convenience, which may or may not be reflective of phylogeny. The two major groups are centrics and pennates, with the former typically with structures more or less radially arranged around a central point, and the latter with structures arranged more or less perpendicularly to a longitudinal rib or bar. Each can be further subdivided. The so-called radial centrics lack any significant polarity to shape and/or lack structures conferring any obvious degree of polarity. The bi- or multipolar centrics often have elongate outlines, and/or two or more structures (e.g., setae, fields of pores) that superimpose bilateral symmetry over the radial symmetry. The pennates can be further divided into two groups, those with slits in the rib or bar (the raphe bearing or raphid pennates) and those without raphe slits (the araphid pennates).

Traditional classification schemes can be discussed in terms of these structural groups. The following are not meant to be exhaustive, but to indicate that traditional classifications represent strikingly different hypotheses.

Steinecke (1931) proposed that centrics and pennates were each monophyletic sister taxa, and raphid pennates were monophyletic and nested within araphid pennates. In stark contrast, Simonsen (1979) concluded that centrics were paraphyletic, and araphids were monophyletic and nested within paraphyletic raphids. In disagreement with the previous two classifications, Round and Crawford (1981, 1984) later argued that the three major lineages (centrics, araphid pennates and raphid pennates) were derived independently and were thus each monophyletic.

Molecular phylogenies were similar to traditional phylogenies in that relationships varied from study to study, without a clear consensus as to arrangement of radial and (bi- or multi-) polar centrics (Theriot, Ashworth, Ruck, Nakov, & Jansen, 2010; Theriot, Cannone, Gutell, & Alverson, 2009). Again, a few studies have produced radically different topologies, and relationships among diatoms are still a matter of debate (Chesnick, Kooistra, Wellbrock, & Medlin, 1997). Here, we cite only a range of results to illustrate our point. Araphid monophyly, as proposed by Round and Crawford (1981, 1984), was supported by analysis of the *coxI* gene dataset with limited taxon sampling (Ehara, Inagaki, Watanabe, & Ohama, 2000). Centric monophyly was recovered using the nuclear-encoded small subunit ribosomal RNA (SSU) dataset (Van de Peer, Van der Auwera, & De Wachter, 1996). These studies led to a reclassification of diatoms with Medlin and Kaczmarska (2004) naming the bulk of radial centrics as the Coscinodiscophyceae, the bi- and multipolar centrics plus the order Thalassiosirales as the Mediophyceae and the pennates as the Bacillariophyceae. Each was argued to be monophyletic based on analysis of nuclear-encoded SSU. This classification, referred as the CMB hypothesis, has been under debate because different taxon sampling, alignments and optimality criteria can yield different results with radials being either monophyletic or not and polars (plus Thalassiosirales) being monophyletic or not (Alverson, Jansen, & Theriot, 2009; Chesnick et al., 1997; Theriot, Ashworth, Ruck, Nakov, & Jansen, 2015). Incongruence in phylogeny was also reported using diatom plastid protein-encoded genes vs nuclear-encoded SSU (Theriot et al., 2010).

The variations in results have led to inclusion of more sources of molecular data for resolving diatom relationships. The focus has been primarily on plastid genes due to the challenges of using nuclear data. The nuclear genome of eukaryotes is composed largely of multiple copy genes, making it difficult to reliably determine orthology. A more complex issue is that the diatom nuclear genome may be a chimeric assemblage due to multiple horizontal gene transfer events through their evolutionary history (Bowler et al.,

2008). In contrast, the plastome is largely composed of single copy genes, with limited horizontal gene transfer events (Ruck, Nakov, Jansen, Theriot, & Alverson, 2014). Plastid protein-coding genes are also easily aligned across a wide range of diatoms (Theriot et al., 2015). A recent study testing the phylogenetic informativeness using a broader suite of diatom plastid genes showed that the addition of plastid data adds signal instead of noise, and these same authors suggested that a phylogenomic study of plastid genes would provide valuable information for resolving the diatom phylogeny (Theriot et al., 2015).

Advances in sequencing technology have opened the door for generating genomic sequences more cheaply and quickly to better understand diatom evolution. The plastome organization potentially provides insights into diatom evolution. The first two diatom plastomes were sequenced in 2007 (Oudot-Le Secq et al., 2007), since then the number of sequenced diatom plastid genomes has increased 10-fold. Although the overall organization of these plastomes is conserved, all have a quadripartite organization with a large single copy (LSC) region, small single copy (SSC) region and two inverted repeats (IR). Sequencing of phylogenetically diverse diatoms showed remarkable variation in genome size, gene content and gene order (Ruck et al., 2014), with expansion of the IR and intergenic regions being the primary cause of plastome size variation (Ruck et al., 2014; Sabir et al., 2014). Extensive plastome sequencing in Thalassiosirales, an order with a moderately well-resolved multigene phylogeny, showed a high level of conservation of genome organization among closely related species (Sabir et al., 2014). One environmentally driven gene transfer event was reported in *T. oceanica*, where the *petF* gene encoding ferredoxin was transferred from the plastid to the nucleus, contributing to the ecological success of *T. oceanica* in iron limited environment by replacing the iron–sulfur protein with iron-free flavodoxin (Lommer et al., 2010). A plastid to nuclear gene transfer event of the acyl carrier protein gene *acpP* was also reported in all Thalassiosirales (Sabir et al., 2014).

Owing to the limited number of plastome sequences previously available, phylogenomics has previously not been an option for resolving questions about diatom systematics. In addition to the paucity of diatom plastome data, the lack of genomes from potential outgroups meant early attempts at phylogenomics were unrooted. Thus monophyly of the Coscinodiscophyceae, which previous single and multigene phylogenies recover as either monophyletic or a basal grade, could not be tested. The sister group to pennate diatoms, which recovered as the bipolar diatom *Attheya* in a phylogeny with nine nuclear and plastid genes (Sorhannus & Fox, 2011), could also not be tested as the genome was not available. A phylogenetic framework with more complete taxonomic sampling is necessary to identify and understand patterns and processes of diatom plastome evolution.

In this study, we nearly doubled the number of sequenced plastomes and added critical taxa such as *Attheya*. We also included the recently sequenced genome of *Triparma*, a close relative of diatoms (Tajima et al., 2016), to provide a more in-depth examination of diatom plastome evolution and to resolve phylogenetic relationships among major clades.

2. MATERIALS AND METHODS

2.1 Diatom Strains and DNA Extraction

Eighteen diatom strains were collected from different sources described in Table 1. Taxon sampling was based on the phylogeny in Theriot et al. (2015). All DNAs were extracted from cultured materials. Diatom cells were pelleted in a Sorvall RC-5B refrigerated superspeed centrifuge (DuPont Company, Newton, CT, USA) for 20 min at 7649 × *g* from a culture in the late logarithmic phase of growth. Cells were lysed using a PARR Cell Disruption Bomb (Parr Instrument Company, Moline, IL, USA) filled with nitrogen gas at 1500 psi. Isolation of DNA was performed following Doyle and Doyle (1987) with modifications. Cetyl trimethylammonium bromide (CTAB) buffer was augmented with 3% PVP and 3% beta-mercaptoethanol (Sigma, St. Louis MO, USA). Organic phase separation was repeated until the aqueous fraction was clear. DNA pellets were resuspended in ~200 μL DNase-free water. Following treatment with RNase A (ThermoScientific, Lafayette, CO, USA) samples were again subjected to phase separation with chloroform and DNA was recovered by ethanol precipitation. Samples were resuspended in DNase-free water, evaluated for concentration by NanoDrop and stored at −20°C.

2.2 DNA Sequencing and Genome Assembly

Paired-end (PE) libraries with insert sizes of 400 bp were prepared at the Genome Sequence and Analysis Facility (GSAF) at the University of Texas at Austin. Illumina HiSeq 2000 platform (Illumina, San Diego, CA, USA) was used to sequence total genomic DNA. The 100 bp PE Illumina reads were assembled with Velvet v.1.2.08 (Zerbino & Birney, 2008; Zerbino, McEwen, Marguiles, & Birney, 2009) using multiple odd number *k*-mers ranging from 71 to 83 and 100–600 × coverage on the Stampede supercomputer at the Texas Advanced Computing Center (TACC). Plastid contigs

Table 1 Taxa Used for Plastid Genome Sequencing With Source and GenBank Accession Numbers

Taxon	Source/Locality	GenBank Accession
Acanthoceras zachariasii	Lake Okoboji, Iowa, USA	MG755808
Actinocyclus subtilis	University of Guam Marine Lab outflows, Guam, USA	MG755799
Astrosyne radiata	Gab Gab Beach, Guam, USA	MG755807
Attheya longicornis	CCMP 214	MG755798
Biddulphia biddulphiana	Gab Gab Beach, Guam, USA	MG755805
Biddulphia tridens	Long Beach, California, USA	MG755806
Discostella pseudostelligera	Upper Bull Shoals Lake, Missouri, USA	MG755804
Entomoneis sp.	Jeddah, Saudi Arabia	MG755800
Eunotogramma sp.	Atlantic Coast, South Florida, USA	MG755797
Guinardia striata	Port O'Connor, Texas, USA	MG755796
Licmophora sp.	Duba, Saudi Arabia	MG755795
Plagiogramma staurophorum	Taelayag Beach, Guam, USA	MG755792
Plagiogrammopsis van heurckii	Moss Landing, California, USA	MG755794
Proboscia sp.	Duba, Saudi Arabia	MG755791
Psammoneis obaidii	Markaz Al Shoaibah, Saudi Arabia	MG755803
Rhizosolenia fallax	Duba, Saudi Arabia	MG755802
Rhizosolenia setigera	Lady's Island, South Carolina, USA	MG755793
Triceratium dubium	Al-Wajh, Saudi Arabia	MG755801

were identified by BLAST analyses of the assembled contigs against publicly available diatom plastid genomes from NCBI. The boundaries between IR and single copy regions were confirmed using Motif search in Geneious R6 v6.1.6 (http://www.geneious.com). Bowtie2 mapping (Langmead & Salzberg, 2012) was utilized to fill gaps in the plastid genome sequences.

2.3 Genome Annotation and Analysis

Plastid genomes were annotated using Dual Organellar GenoMe Annotator (DOGMA) (Wyman, Jansen, & Boore, 2004), followed by manual corrections for start codons using Geneious R6 v.6.1.6. tRNA genes were predicted using DOGMA (Wyman et al., 2004) and tRNAscan-SE 1.21 (Schattner, Brooks, & Lowe, 2005). Boundaries of rRNA genes, tmRNA *ssra* gene and signal recognition particle RNA *ffs* gene were delimited by direct comparison to sequenced diatom orthologs with Geneious R6 v.6.1.6. Annotated plastid genomes are available from GenBank (Table 1). The length of total genome, IR, SSC and LSC are shown in Supplementary Information A in the online version at https://doi.org/10.5061/dryad.vb44k (Dryad). Genome length variation was analysed using APE library in R (Paradis, Claude, & Strimmer, 2004).

2.4 Phylogenetic Analysis

Sequences of 103 shared plastid protein-encoding genes from 40 diatom taxa and the outgroup *Triparma laevis* were aligned with MAFFT (Katoh, Kuma, Toh, & Miyata, 2005) based on translated protein sequences. This included 22 published diatom plastid genomes, 1 outgroup species *T. laevis* and the 18 plastid genomes newly sequenced in this study. Three different partitioning schemes were analysed including no partitioning (1 partition), partition by codon position (3 partitions), and partition by codon position and gene functional group (21 partitions). Genes in each functional group were listed in Table 2. A maximum likelihood tree for each partition was computed on TACC Stampede supercomputer using RAxML 8.2.9 (Stamatakis, 2014) with the substitution model GTR + G and "-f a" option. One thousand bootstrap replicates were performed. The probabilities conferred upon the molecular data by trees in which Araphids, Mediophyceae, Coscinodiscophyceae and Coscinodiscophyceae plus Mediophyceae were each constrained as monophyletic were tested using the AU (approximately unbiased) and SH (Shimodara–Hasegawa) tests (Shimodaira, 2002).

To test the possibility of recombination in diatom plastid genomes, 11 conserved gene order blocks occurring in most diatoms were identified (Dryad: Supplementary Information B in the online version at https://doi.org/10.5061/dryad.vb44k). Gene blocks 1–4 and 6–10 were concatenated due to short sequence length. Four resulting concatenated sequence alignments (gene blocks 1–4, gene block 5, gene blocks 6–10 and gene block 11) were used to construct phylogenetic trees using RAxML with codon

Table 2 103 Shared Protein-Coding Genes Partitioned by Functional Groups

Category	Genes
Photosystem	*psaA, psaB, psaD, psaF, psaJ, psaL, psbA, psbB, psbC, psbD, psbE, psbF, psbH, psbI, psbJ, psbK, psbL, psbN, psbT, psbV, psbX, psbY, psbZ*
Cytochrome b/f complex	*petA, petB, petD, petG, petL, petM, petN*
ATP synthase	*atpA, atpB, atpD, atpE, atpF, atpG, atpH, atpI*
RubisCo subunit	*rbcL, rbcS, rbcR*
RNA polymerase	*rpoA, rpoB, rpoC1, rpoC2*
Ribosomal proteins	*rpl1, rpl2, rpl3, rpl4, rpl5, rpl6, rpl11, rpl12, rpl13, rpl14, rpl16, rpl18, rpl19, rpl20, rpl21, rpl22, rpl23, rpl24, rpl27, rpl29, rpl31, rpl32, rpl33, rpl34, rpl35, rps2, rps3, rps4, rps5, rps7, rps9, rps10, rps11, rps12, rps13, rps14, rps16, rps17, rps18, rps19, rps20*
Other genes	*cbbX, ccs1, ccsA, chlI, clpC, dnaB, ftsH, groEL, secA, secG, secY, sufB, sufC, tatC, ycf3, ycf12, ycf46*

partition. SH tests (Shimodaira, 2002) were run among the four resulting trees to test the congruence with the concatenated tree using 103 protein-coding genes.

2.5 Gene Order Analysis

Genome rearrangements were estimated with MAUVE after eliminating one copy of the inverted repeat (IR_B copy) (Darling, Mau, Blattner, & Perna, 2004). The rearrangement distances between gene orders were measured by Genome Rearrangements in Man and Mouse (GRIMM) and visualized using d3heatmap library in R (Tesler, 2002). Correlation between substitution rates (estimated from branch lengths on the ML tree) and genome rearrangement distances were analysed using Pearson correlation coefficient and Pearson test with Bonferroni multiple testing correction. The gene order tree with varying branch lengths to best fit the constrained ML sequence tree was constructed using PAUP v 4.0b10 (Swofford, 2003) not allowing negative branch lengths.

2.6 Gene Content Analysis

Gene loss and gain events were mapped to the ML cladogram using Dollo parsimony in MacClade v4.08 (Maddison & Maddison, 2000) based on the gene content comparison table (Dryad: Supplementary Information C in the online version at https://doi.org/10.5061/dryad.vb44k). The presence and absence of genes were encoded as 1 and 0, respectively. Gene pseudogenization events were encoded as 2, and the states (absent, present and pseudogenized) were treated as ordered. Dollo parsimony was used as an approximation of the assumption that genes were more likely to be lost from the plastome than gained, and that functioning genes are more likely to become pseudogenes than the reverse.

3. RESULTS

3.1 Phylogenomic Analysis

All partition schemes yielded trees with identical topologies and very similar branch lengths and bootstrap (BS) support values (Fig. 1). We present the results of the dataset partitioned by functional category and codon position. The maximum likelihood tree has 100% BS support values on most nodes (Fig. 1). Raphid pennate diatoms (labelled "Raphid") were recovered as a monophyletic group sister to a clade of araphid pennate diatoms (Araphid 2) with 100% BS support. Within raphid diatoms, *Eunotia naegelii* was sister to the rest of the raphid diatoms with 100% BS support. The model diatom *Phaeodactylum tricornutum* was recovered as sister to *Didymosphenia geminata*, but with only 52% BS support. Within araphid 2, *Astrosyne radiata* was recovered on an extremely long branch. Araphid 1 was sister to araphid 2 plus the raphid group with 100% BS.

Mediophyceae (bi- and multipolar diatoms plus the Thalassiosirales) were contained in three clades ('polar 1', 'polar 2' and 'polar 3') and were paraphyletic. *Attheya longicornis* formed polar clade 3 with the two *Biddulphia* species, and together were sister to the pennate diatoms (araphid 1 and 2, plus raphid) with 100% BS support. The clade polar 2 was sister to the polar 1 clade with 94% BS support. The Thalassiosirales (including the euryhaline model diatom *Cyclotella nana* Hustedt, which was sister to two undescribed freshwater species of *Cyclotella*), were in polar 1 clade and were monophyletic with 100% BS support. *Eunotogramma* sp. and *Lithodesmium undulatum* were sequentially related to the Thalassiosirales with 100% BS support.

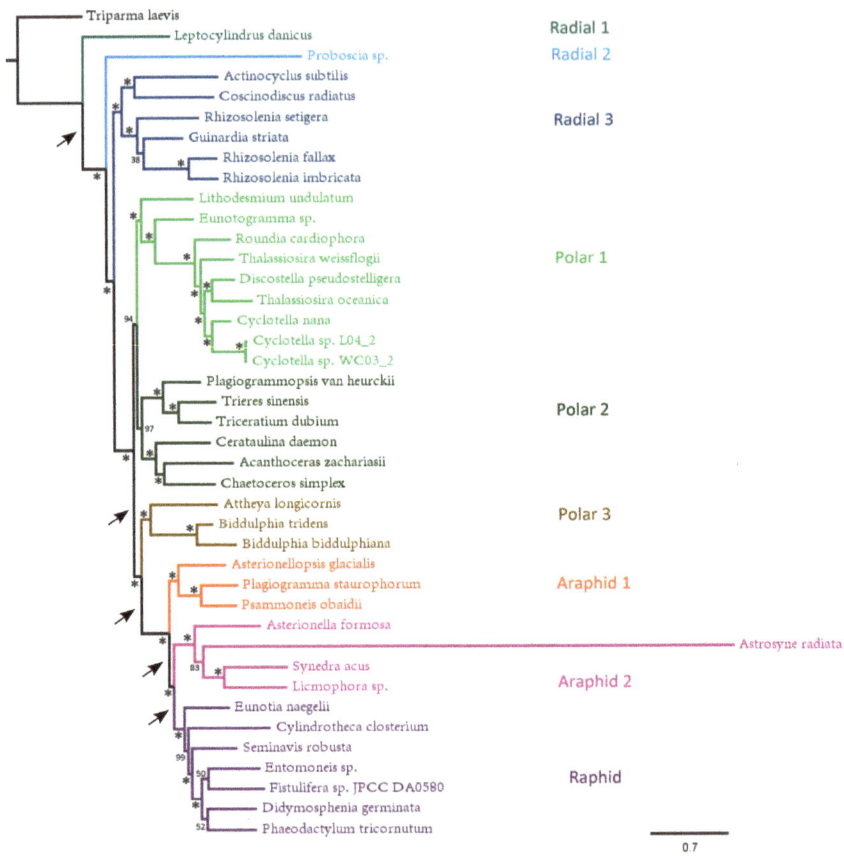

Fig. 1 Maximum likelihood tree inferred from 103 shared plastid genes of 40 diatom species and the outgroup *Triparma laevis*. Branch lengths are proportional to the number of nucleotide changes as indicated by the scale bar (0.7 substitutions per site). *Asterisks* at nodes indicate 100% bootstrap support; *numbers* indicate bootstrap support values. *Different colours* indicate different diatom groups based on Theriot et al. (2015). The *arrows* indicate consistent branches separating different clades in phylogenetic analyses of gene blocks.

Biddulphia plus *Attheya* formed a clade with 100% BS support, and that clade was sister to pennates with 100% BS support.

The radial centrics of the Coscinodiscophyceae (Radials 1, 2 and 3) formed a basal grade. Within radial 3 *Guinardia striata* was nested within *Rhizosolenia* spp. with low BS support. The two remaining radial centric groups, *Proboscia* sp. (Radial 2) and *Leptocylindrus danicus* (Radial 1) formed a grade at the base of the tree with each node having 100% BS support.

Monophyly of araphids, Mediophyceae, Coscinodiscophyceae and Mediophyceae plus Coscinodiscophyceae were each strongly rejected in favour of the best unconstrained tree by AU and SH tests (*P*-values < 0.005).

Comparison of the maximum likelihood tree constructed by four different gene order blocks revealed the conservation of five internal branches separating major clades as indicated by arrows in Fig. 1 and red lines in Fig. 2. All trees showed the following relationships: *L. danicus* sister to the rest of diatoms; polar diatoms paraphyletic with *Biddulphia* plus *Attheya* sister to pennates; raphids monophyletic within the monophyletic pennates. These relationships were consistent with the tree constructed using 103 concatenated genes in Fig. 1. The SH tests also showed none of those trees was significantly worse than the concatenated tree.

3.2 Genome Size

Plastome length varied across clades (Fig. 3) with *Plagiogramma staurophorum* exhibiting the largest size of 201,816 bp among all sequenced diatoms (Dryad: Supplementary Information A in the online version at https://doi.org/10.5061/dryad.vb44k). The araphid 1 group (indicated in red), where *P. staurophorum* was recovered, showed relatively larger genome size compared to other groups (Fig. 3). Large variation in IR length was found in araphid 2 (violet) and raphid (purple) groups, where the longest IR was almost 2–3 times longer than the shortest (Fig. 3). Sister to araphid and raphid groups, the polar 3 clade (brown) displayed a relatively conserved genome length, with little variation within the LSC, SSC and IR.

Polar 1 (light green) and polar 2 (dark green) groups also showed relatively conserved genome lengths, with *Eunotogramma* sp. and *Plagiogrammopsis van heurckii* showing the largest genome size in the polar 1 and polar 2 clades, respectively. The radial 3 group (dark blue) had relatively conserved genome length ranging from 118,120 bp to 125,283 bp (Dryad: Supplementary Information A in the online version at https://doi.org/10.5061/dryad.vb44k). *T. laevis*, the outgroup species, showed the longest LSC and the shortest IR in the dataset (Fig. 3; Supplementary Information A in the online version at https://doi.org/10.5061/dryad.vb44k).

IR length showed more variation across the groups than the length of LSC and SSC (Fig. 3). Phylogenetic independent contrast analysis showed that IR length contributed to the majority of the plastome size variation with $R^2 = 0.6875$. In comparison, the LSC and SSC contributed a relatively smaller portion, with $R^2 = 0.2959$ and 0.1036, respectively (Fig. 4).

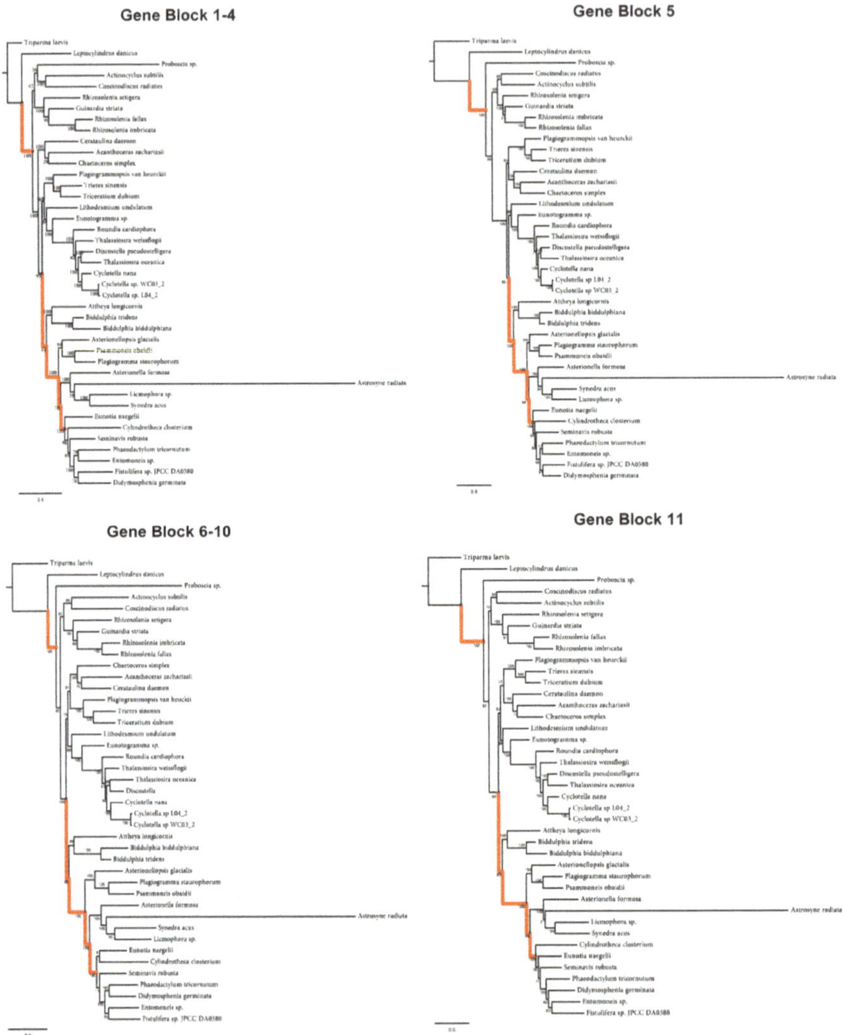

Fig. 2 Comparison of maximum likelihood tree constructed from four different gene blocks with codon partition. The *five branches in red* represent the consistent branches separating Radial 1 from the rest of clades, separating Polar 2 from Polar 3 and the Pennate, separating Polar 3 from the Pennate, separating Araphid 1 from Araphid 2 and Raphid, separating Araphid 2 from Raphid, respectively. The *branches in red* are consistent with the corresponding *branches with arrow* in Fig. 1.

3.3 Gene Content

Dollo parsimony was used to optimize gene losses and gains on the diatom phylogeny as an approximation of the higher likelihood that genes are lost from the plastome rather than gained (Fig. 5). Three genes involved in

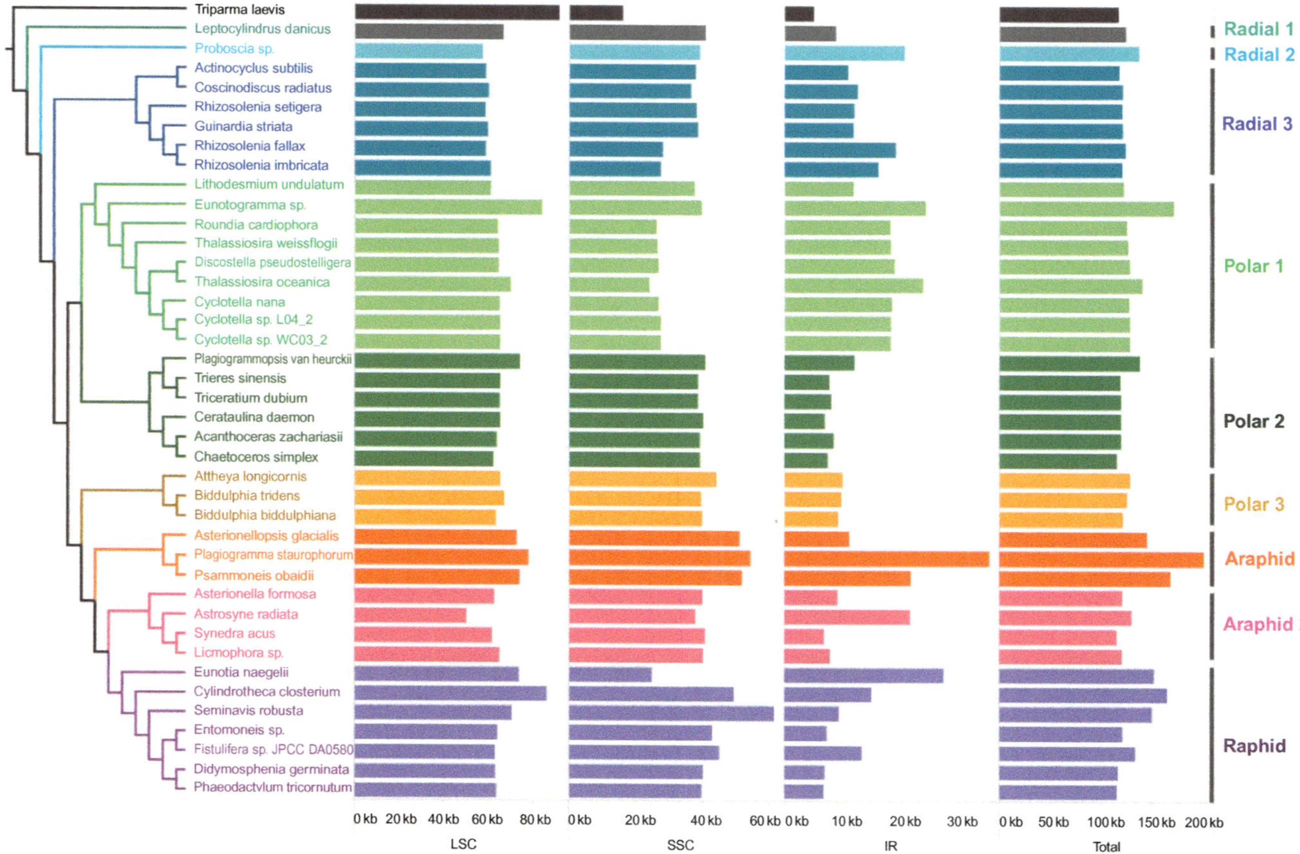

Fig. 3 Genome length variation across 40 diatom species and the outgroup *Triparma laevis*. *Colours* indicate different diatom groups as Fig. 1. *LSC*, large single copy; *SSC*, small single copy, *IR*, inverted repeats. The length of LSC, SSC and IR was scaled differently. Scale on *x*-axis in kilobase (kb).

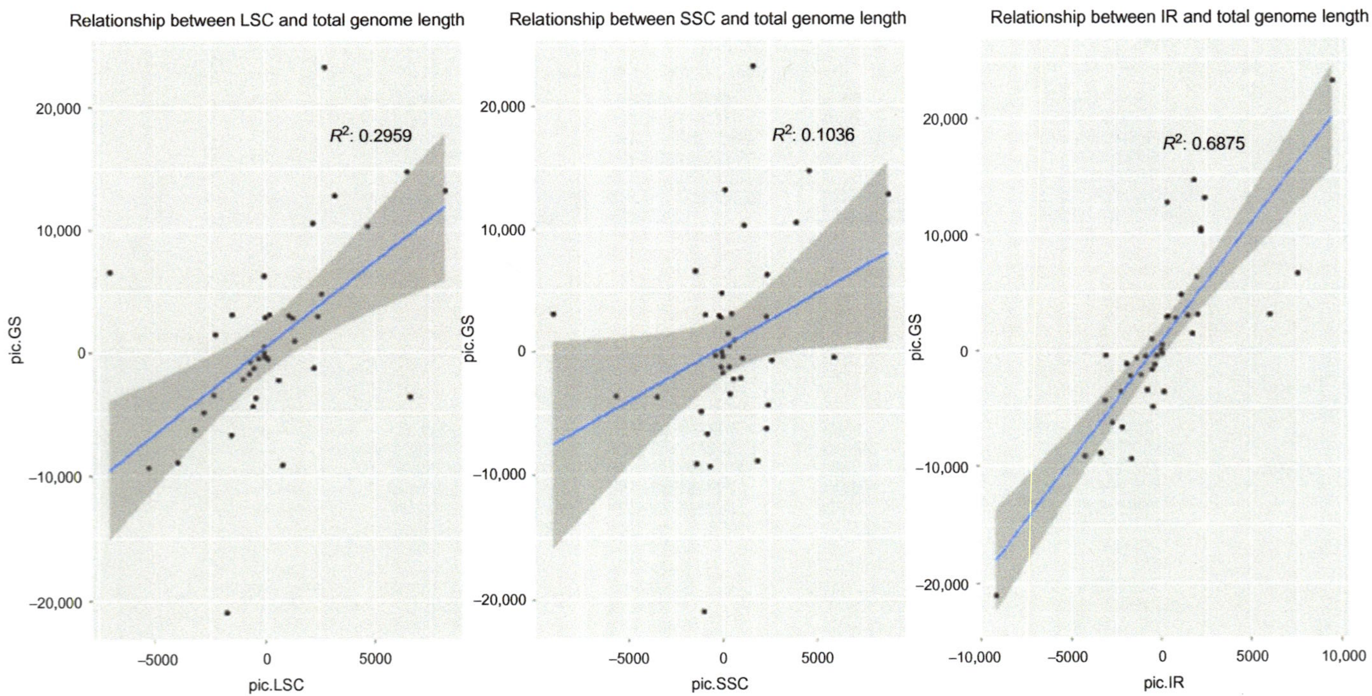

Fig. 4 Relationship between total genome size and LSC, SSC and IR, respectively, after applying phylogenetic independent contrast analysis. The *blue line* indicates the regression line. The *shaded area* indicates 95% of confidence interval. The coefficient of determination is indicated by R^2.

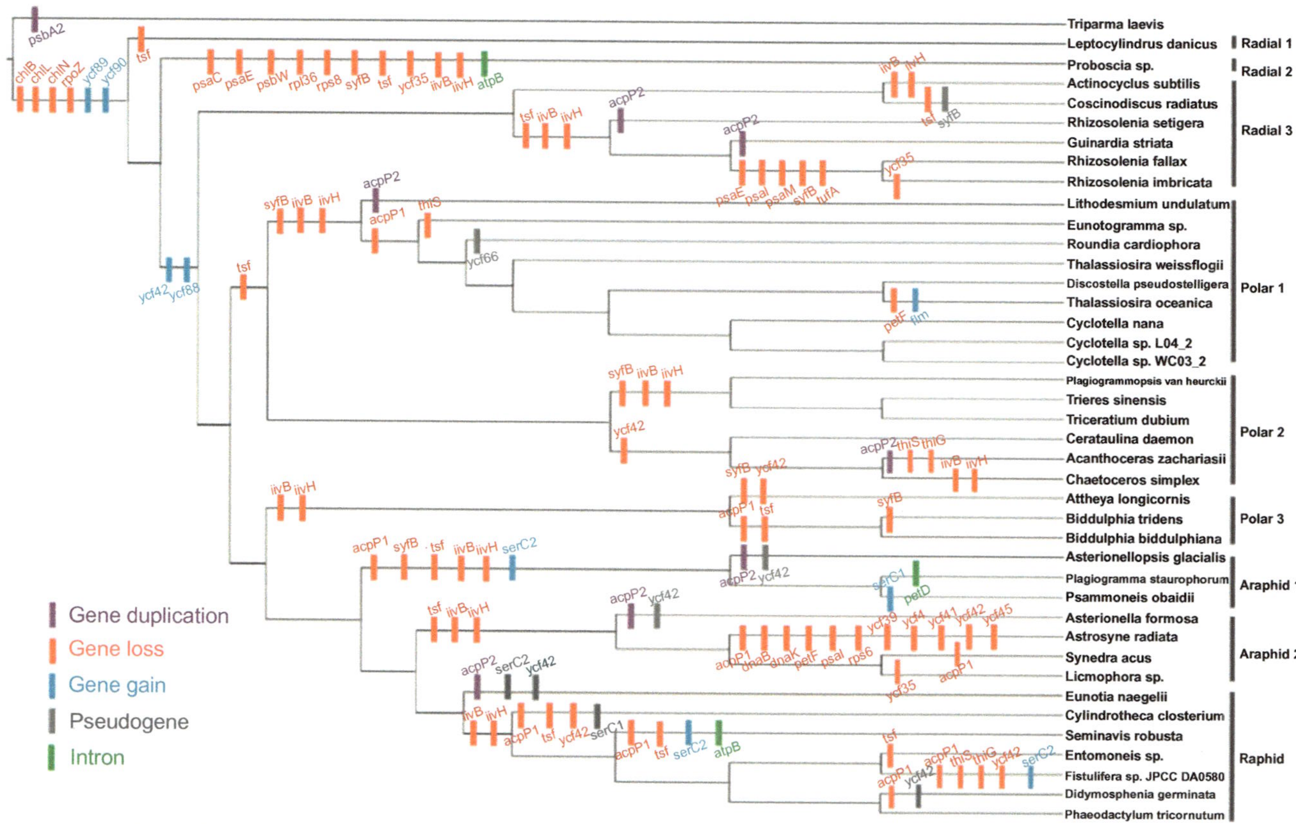

Fig. 5 Gene and intron loss and gain events mapped on the cladogram of the ML plastid gene tree using Dollo parsimony.

light-independent chlorophyll *a* biosynthesis, *chlB*, *chlL* and *chlN*, together with RNA polymerase omega subunit *rpoZ*, were entirely absent in the 40 sequenced diatom plastid genomes. In contrast, two hypothetical plastid ORFs with unknown functions (*ycf89* and *ycf90*) were absent in the out-group species *T. laevis* but present in all 40 diatom plastomes (Fig. 5).

Other genes appear to have experienced multiple losses, such as elongation factor Ts *tsf*, which was lost 11 times, and the acetolactate synthase large and small subunits *IlvB* and *IlvH*, which were lost 10 times.

Pseudogenes were relatively uncommon. The phenylalanyl–tRNA synthetase beta chain gene *syfB* showed seven losses and one pseudogenization event. The gene *ycf66* underwent one pseudogenization event but no losses. The gene *ycf42* was an exception with four pseudogenization events.

The branches with the largest number of gene losses (*Proboscia* sp. and *A. radiata*, 11 each) were also those with the greatest amount of inferred nucleotide substitution based on branch lengths (cf. Figs 1 and 5).

Finally, introns were detected in *atpB* in radial 2 species *Proboscia* sp. and in *petD* in araphid 1 species in *P. staurophorum*. A Conserved Domain Database (Marchler-Bauer & Bryant, 2004) search of these introns revealed a reverse transcriptase with group II intron origin with *E*-values of 5.24×10^{-44} and 7.89×10^{-40} for *atpB* and *petD*. BLAST comparisons of the intron-encoded proteins against NCBI revealed that the top hits were green algae reverse transcriptase with 50% and 54% nucleotide sequence identity, respectively.

3.4 Gene Order

The 40 diatom plastomes exhibit varying degrees of gene order rearrangement (Fig. 6; Dryad: Supplementary Information D in the online version at https://doi.org/10.5061/dryad.vb44k). The MAUVE alignment identified 42 locally collinear blocks (LCBs) shared by the plastid genomes examined (Dryad: Supplementary Information E in the online version at https://doi.org/10.5061/dryad.vb44k). Closely related species share more similar gene orders. Identical gene orders were found in radial 3, polar 1, polar 3 and raphid groups. The most extensive sampled polar 1 clade showed six very similar gene orders, with four Thalassiosirales (*Roundia cardiophora*, *Thalassiosira weissflogii*, *Discostella pseudostelligera* and *C. nana*) having exactly the same gene order, and the two closely related *Cyclotella* taxa differ by one inversion (Dryad: Supplementary Information E in the online version at https://doi.org/10.5061/dryad.vb44k).

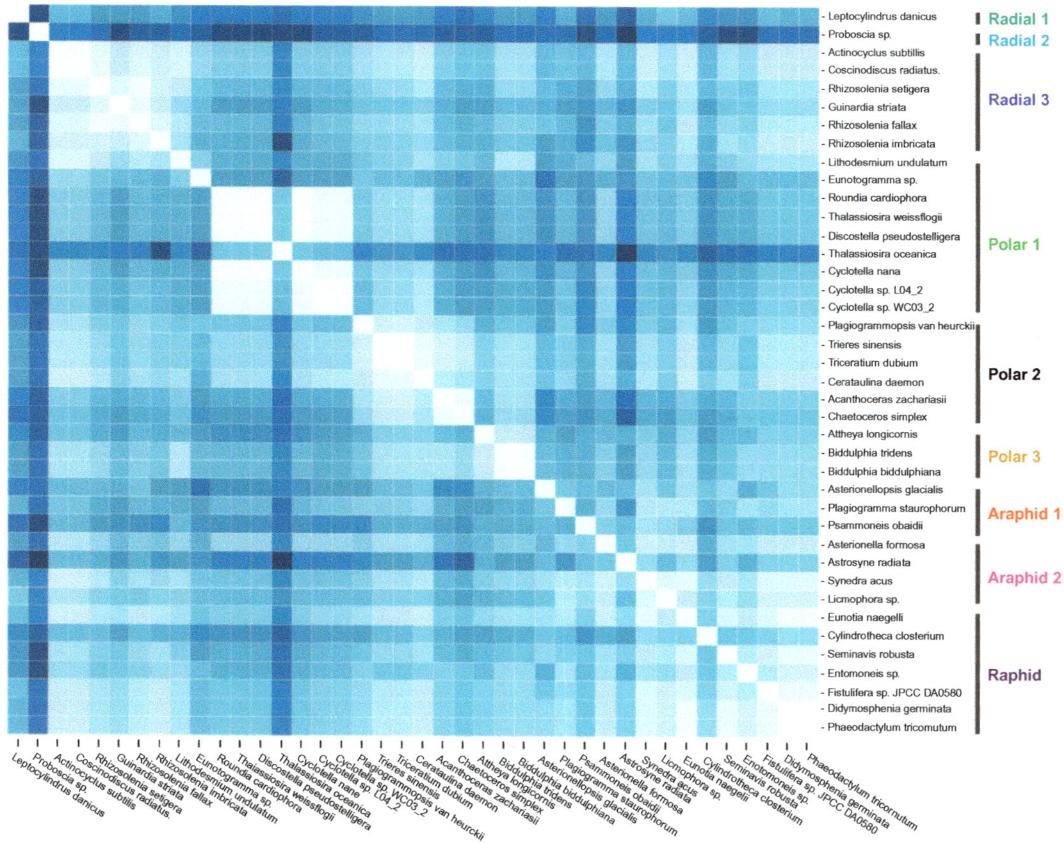

Fig. 6 Heatmap of pairwise genomic rearrangement distance estimated by GRIMM. The intensity of the colour is proportional to the degree of genome rearrangement. *Dark blue* indicates higher degree of genome rearrangement and *light colour* indicates lower degree of genome rearrangement.

Gene order and sequence divergence were strongly positively correlated in some regions of the tree. Approximately 40% of the Bonferroni corrected *P*-values of the Pearson correlation between pairwise branch length and gene order rearrangement distances were significant (Dryad: Supplementary Information F in the online version at https://doi.org/10.5061/dryad.vb44k). For example, *A. radiata*, which had the longest branch in the sequence tree (Fig. 1), also exhibited a high level of gene order rearrangement and had a high correlation value of 0.71 (Dryad: Supplementary Information F in the online version at https://doi.org/10.5061/dryad.vb44k). Similarly, *Proboscia* sp. had the next longest branch and also exhibited high levels of gene order rearrangement (Figs 1 and 6).

4. DISCUSSION

The advent of sequencing technology and powerful computers made it possible to sequence whole plastomes in a short amount of time at a reasonable cost. Given the phylogenetic diversity of diatoms, it is critical that more diversity be studied for their genomic properties to better understand their evolutionary history. In this study, we sampled extensively across the diatom phylogeny, especially taxa whose phylogenetic placement remains controversial. Our results provide deeper insights into diatom phylogeny and the dynamics of their plastome evolution.

4.1 Phylogeny of Diatoms

Medlin (2017) and Medlin and Kaczmarska (2004) proposed a classification with three monophyletic classes based primarily on SSU rDNA sequence analysis, Coscinodiscophyceae (radial centrics), Mediophyceae (polar centrics) and Bacillariophyceae (araphid and raphid pennates) or the CMB hypothesis. This classification has been adopted by several authors (e.g., Adl et al., 2005; Cox, 2015), but there is considerable disagreement as to whether the classification is natural. In fact, Adl et al. (2005) explicitly considered the Coscinodiscophyceae and Mediophyceae each to be paraphyletic. Frequently, studies recover the radials as paraphyletic, the polars as paraphyletic, or both as paraphyletic or grade groups (Medlin, 2016; Theriot et al., 2010; Theriot, Ruck, Ashworth, Nakov, & Jansen, 2011). The foundational problem is that the taxon sampling and molecular sampling to date have simply not generated a robust result. For example, the CMB hypothesis is only seven steps longer than the grade hypothesis (the most parsimonious hypothesis, $L = 14{,}094$ steps) using SSU data alone

(Theriot et al., 2009). Theriot et al. (2010) analysed SSU, *rbcL* and *psbC* for 136 diatoms under ML; the optimal solution was again the grade hypothesis, but it was not statistically significantly different than the CMB hypothesis. In short, for most data and taxon sets in the diatom literature, it takes little to turn the CMB hypothesis into the grade hypothesis and vice versa.

In a search for more genes that might provide information about the diatom phylogeny, Theriot et al. (2015) found that individual plastid genes return results that disagree with traditional views, the CMB hypothesis, the grade hypothesis and indeed even with one another. In instances where plastids are biparentally inherited, there is the possibility that species hybridization could lead to recombination in the plastome, and to conflict between gene trees (D'Alelio & Ruggiero, 2015; Sullivan, Schiffthaler, Thompson, Street, & Wang, 2017). Such instances might result in different plastid genes yielding different but strongly supported trees. The individual gene trees recovered by Theriot et al. (2015), however, were not robustly supported. After studying the potential for saturation, and analysing signal/noise ratios, they argued that individual plastid genes could be concatenated. When this was done the grade hypothesis was recovered with strong support. Their conclusion was that the signal in the individual genes was low, but that it was additive. While the noise levels were high, they were not correlated and did not sum to a positively misleading signal. Thus, incongruence among plastid genes seemed to be best explained simply by noise.

We examined the potential for plastome recombination as a source of misleading signal by analysing four subsets of the plastome genome: two large blocks of genes that each seem to be inherited as a single locus and two concatenated subsets of smaller blocks of genes with each smaller block acting as a single locus. All trees from these analyses reject the CMB topology in the same manner (*Leptocylindrus* sister to all other diatoms; *Attheya* plus *Biddulphia* sister to pennates). We cannot reject the hypothesis that (relatively minor) examples of plastome recombination are occurring and may affect some parts of the tree. But it seems certain there are not two or more different strong signals for different relationships, and it seems certain that signal for the tree in Fig. 1 comes from across the plastome.

We also tested the 103 combined plastid genes with three different partitions. All phylogenetic analyses showed the same tree topology with slightly different bootstrap support. The resulting ML tree partitioned by codon and gene functional group showed the Coscinodicophyceae (radial centrics) and Mediophyceae (bi- and multipolar centrics) were not monophyletic, while the Bacillariophyceae (raphid diatoms) were monophyletic with high bootstrap support (Fig. 1). The AU tests of araphid pennate monophyly suggested by Simonsen (1979) and the CMB monophyly suggested by Medlin and Kaczmarska (2004) were both strongly rejected with *P* values less than 0.05.

Our results are congruent with the conclusion of Alverson, Beszteri, and Theriot (2011) that the model diatom *C. nana* (now usually referred to as *Thalassiosira pseudonana*) is more closely related to the euryhaline genus *Cyclotella* (Fig. 1). Another model diatom *P. tricornutum* is sister to *D. geminata* in the raphid clade with low bootstrap support (Fig. 1). Presently, this diatom is classified in its own suborder and family, reflecting its unique morphology (lack of a full siliceous frustule). However, it was once argued that it bore some similarity to the genus *Cymbella* (Lewin, 1958), in the same family as *Didymosphenia*. We hesitate to advocate moving *Phaeodactylum* to the Cymbellaceae on the basis of our results, given the low BS support and the fact that raphid diatoms are such a diverse clade. More extensive taxon sampling in this group may further elucidate the phylogenetic position of this model organism.

4.2 Plastome Evolution

Plastome size varies considerably within diatoms, ranging from 116,251 to 201,816 bp (Fig. 3, Dryad: Supplementary Information A in the online version at https://doi.org/10.5061/dryad.vb44k). Several factors such as expansion or contraction of the IR, loss and duplication of genes, gain of introns and expansion of intergenic spacer regions are responsible for variation in plastome sizes (Jansen & Ruhlman, 2012). It has been previously reported that the larger plastid genome size in *Thalassiosirales* was mainly due to expansion of the IR (Sabir et al., 2014). Our study reports the largest diatom plastome at 201,816 bp in *P. staurophorum* (Dryad: Supplementary Information A in the online version at https://doi.org/10.5061/dryad.vb44k). This species also has the largest IR among diatoms at 34,888 bp (Fig. 3, Dryad: Supplementary Information A in the online version at https://doi.org/10.5061/dryad.vb44k). The large size of the genome is mainly due to the IR expansion. An introduction of a 2971 bp group II intron in *petD* also contributed to the larger size of *P. staurophorum*. This is consistent with our phylogenetic independent contrast analysis that IR length contributed to the majority of the plastome size variation (Fig. 4).

Our extensive sampling across diatom phylogeny also showed the similarity of genome sizes across most clades (Fig. 3), which is consistent with previous finding within Thalassiosirales (Sabir et al., 2014). Ruck et al. (2014) reported that larger intergenic spacer regions and the introduction of foreign genes played an important role in the expansion of plastome size. Within the araphid 1 clade, the introduction of *SerC*1 gene probably contributed to the relative larger size of *Psammoneis obaidii*.

Massive numbers of gene losses occur across diatom plastomes (Fig. 5). The four gene losses (*chlB*, *chlL*, *chlN* and *rpoZ*) together with two hypothetical protein gains (*ycf89* and *ycf90*) appear to be synapomorphies for diatoms. Gene loss in plastomes is often associated with a functional gene transfer to the nucleus. Acyl carrier protein *acpP1*, the gene involved in the lipid metabolism pathway, was reported missing in all *Thalassiosirales* and a hypothetical transfer from plastid to nucleus transfer was proposed (Sabir et al., 2014). In this study, expanded taxon sampling in the polar 1 group again confirmed the order-wide loss of *acpP1* in all *Thalassiosirales* and *Eunotogramma* (Fig. 5), and we found the gene loss event occurred at the split between *Lithodesmium* and *Thalassiosirales*. Ferredoxin gene *petF*, an ecologically driven plastid to nucleus transfer in *T. oceanica* (Lommer et al., 2012), is also absent from the *A. radiata* plastome. *A. radiata* has not only undergone extensive gene order rearrangement and sequence divergence (Fig. 1, Dryad: Supplementary Information G in the online version at https://doi.org/10.5061/dryad.vb44k), it has also experienced extreme morphological divergence, having entirely lost the symmetry of pennate morphological structure (Ashworth, Ruck, Lobban, Romanovicz, & Theroit, 2012). Gene loss was suggested as a pervasive source of genetic change that potentially causes adaptive phenotype diversity (Albalat & Canestro, 2016). Our gene content comparison showed massive gene loss (11 losses) in the *A. radiata* plastome. The connection between plastid evolution and morphological evolution suggests that perhaps the nuclear genome of *A. radiata* also experienced radical change.

Another long branch bearing species, *Proboscia* sp., has experienced massive gene loss (Fig. 5, 10 losses) and a rare instance of an intron gain in *atpB*. However, in this case gene losses seem only weakly correlated with gene order rearrangement. *Actinocyclus* and *Coscinodiscus* are morphologically similar, identical in gene order and exhibit two losses each of functional genes (one due to pseudogenenization in *Coscinodiscus*). In contrast, the extensively sampled diatom order Thalassiosirales showed a pattern of stasis in gene content and gene order except for *T. oceanica*, which has a high degree of reorganization but only one gene loss and one gene gain (Sabir et al., 2014). The branch leading to *Rhizosolenia fallax* and *Rhizosolenia imbricata* exhibits the next highest level of gene loss (five losses), but very few gene order changes (Fig. 6).

Photosynthetic gene loss is rare in diatom plastomes. Three noteworthy gene losses reported in diatom plastomes were the photosynthetic genes *psaE*, *psaI* and *psaM* missing from *R. imbricata* (Sabir et al., 2014). Our study also documented the loss of *psaE*, *psaI* and *psaM* in *R. fallax*, a species sister to *R. imbricata* but these genes are present in *R. setigera*, an earlier diverging *Rhizosolenia* in the Radial 3 clade (Fig. 5). This indicates that the loss of these three photosynthetic genes occurred at the split between *Guinardia* and the more recently derived *Rhizosolenia* species.

There has been a history of repeated loss of the acetolactate synthase large and small subunits, *ilvB* and *ilvH* among diatom plastomes (Ruck et al., 2014; Sabir et al., 2014). The tRNA synthetase gene, *syfB*, has a similar history of repeated loss in several diatom plastid genomes (Fig. 5). A pseudogene is retained in *Coscinodiscus radiatus* indicating that losses are ongoing. The translation factor gene *tsf* shows a similar pattern (Fig. 5). Ruck et al. (2014) proposed a single deep plastid-to-nuclear transfer of *tsf*. In our study, we also found repeated losses of *tsf*, but data are not available at this time to determine if there have been multiple transfers to the nucleus.

Group II introns are mostly found in plants, fungi, eubacteria and archaea. The first group II intron-encoding intronic maturase was found in tRNA-Met in the red alga *Gracilaria* (Janouškovec et al., 2013). There were reports of a group II intron in the *atpB* gene of the diatoms *Seminavis robusta* and *psaA* gene of *Toxarium undulatum* (Brembu et al., 2013; Ruck, Linard, Nakov, Theriot, & Alverson, 2016). We found two additional group II introns, one in *petD* gene in *P. staurophorum*, and another in *atpB* gene in *Proboscia* sp. Reverse transcriptases within the introns are most similar to those in green algae. There have been studies reporting genes of green algal origin in diatom nuclear genomes (Bowler et al., 2008), and an endosymbiotic gene transfer from green algae was proposed (Moustafa et al., 2009). More intensive molecular investigation across diatoms would likely reveal evidence for the origin and evolution of those introns.

Highly conserved gene order within clades and extensively rearranged gene orders across groups have been reported in previous diatom plastome studies (Ruck et al., 2014; Sabir et al., 2014). Our extended sampling further confirmed the conservation of gene order in closely related species and extensive rearrangement in distantly related species (Fig. 6). Correlations between rates of nucleotide substitution and genomic rearrangements were detected in angiosperms (Jansen et al., 2007; Weng, Blazier, Govindu, & Jansen, 2014). A significant positive correlation between nucleotide substitution and gene order rearrangement is present on the long-branch leading to *A. radiata* (Supplementary Information F in the online version at https://doi.org/10.5061/dryad.vb44k). The longest branch in polar 1 group, *T. oceanica*, also showed a significant correlation between sequence divergence and genome rearrangement (Supplementary Information F in the online version at https://doi.org/10.5061/dryad.vb44k).

Doubling the size of available plastome data of diatoms has greatly expanded our understanding of plastome evolution across this large and diverse photosynthetic clade. With the inclusion of *T. laevis* as the outgroup, we strongly rejected the CMB hypothesis of diatom classification. Our data suggest that radial diatoms evolved as a grade, polar diatoms and araphid diatoms are paraphyletic, and raphid diatoms are monophyletic and nested within the pennates. The 103 plastid gene dataset also strongly suggests that *Attheya* together with the *Biddulphia* group is the sister to the pennate diatoms. Our expanded sampling again confirmed that expansion of IR played the major role of plastome size variation. Gene content and order of closely related species are much more conserved than distantly related species. Extensive gene loss events were also observed. Although recombination of parts of the chloroplast genome may occur in some diatoms (D'Alelio & Ruggiero, 2015), we found no evidence that wholesale recombination was occurring across the diatoms. We found multiple gene blocks that appear to have been inherited as single loci, and each of those carried the same phylogenetic signal. Our study also shows a strong positive correlation between sequence divergence and genome rearrangement in diatoms, a phenomenon that has been documented in flowering plants (Jansen et al., 2007; Schwarz et al., 2017; Weng et al., 2014). Expanded studies of the sequence divergence in terms of substitution rates will provide more insights into the driving force for diatom plastome evolution.

Supplementary data to this article can be found online at https://doi.org/10.5061/dryad.vb44k.

ACKNOWLEDGEMENTS

The authors acknowledge the President of King Abdulaziz University, Prof. Abdulrahman O. Alyoubi, for funding support, the Genome Sequencing and Analysis Facility (GSAF) at the University of Texas at Austin for performing Illumina sequencing, the Texas

Advanced Computing Center (TACC) at the University of Texas at Austin for access to supercomputers, Tracey Ruhlman, Jin Zhang and Mao-lun Weng for their assistance with DNA isolations and genome assembly and annotation and three anonymous reviewers for valuable comments on an earlier draft. M.Y. was also supported by NSF DEB-1208256.

AUTHOR CONTRIBUTIONS

Conceived and designed experiments: M.Y., M.P.A., M.A.K., M.J.S., A.M.A., A.L.A., E.C.T. and R.K.J. Performed experiments: M.Y. and M.P.A. Analysed data: M.Y. Wrote paper: M.Y., E.C.T. and R.K.J. Secured funding: N.H.H., M.J.S., J.S.M.S. and R.K.J.

REFERENCES

Adl, S. M., Simpson, A. G., Farmer, M. A., Andersen, R. A., Anderson, O. R., Barta, J. R., et al. (2005). The new higher level classification of eukaryotes with emphasis on the taxonomy of protists. *The Journal of Eukaryotic Microbiology*, *52*(5), 399–451.

Albalat, R., & Canestro, C. (2016). Evolution by gene loss. *Nature Reviews Genetics*, *17*(7), 379–391.

Alverson, A. J., Beszteri, B., & Theriot, E. C. (2011). The model marine diatom *Thalassiosira pseudonana* likely descended from a freshwater ancestor in the genus *Cyclotella*. *BMC Evolutionary Biology*, *11*, 125.

Alverson, A. J., Jansen, R. K., & Theriot, E. C. (2009). Response to Medlin and Kaczmarska (2008). *Molecular Phylogenetics and Evolution*, *50*(2), 409–410.

Ashworth, M. P., Ruck, E. C., Lobban, C. S., Romanovicz, D. K., & Theroit, E. C. (2012). A revision of the genus *Cyclophora* and description of *Astrosyne* gen. nov. (Bacillariophyta), two genera with the pyrenoids contained within pseudosepta. *Phycologia*, *51*(6), 684–699.

Bowler, C., Allen, A. E., Badger, J. H., Grimwood, J., Jabbari, K., Kuo, A., et al. (2008). The *Phaeodactylum* genome reveals the evolutionary history of diatom genomes. *Nature*, *456*(7219), 239–244.

Brembu, T., Winge, P., Tooming-Klunderud, A., Nederbragt, A. J., Jakobsen, K. S., & Bones, A. M. (2013). The chloroplast genome of the diatom *Seminavis robusta*: New features introduced through multiple mechanisms of horizontal gene transfer. *Marine Genomics*, *16*, 17–27.

Chesnick, J. M., Kooistra, W. H., Wellbrock, U., & Medlin, L. K. (1997). Ribosomal RNA analysis indicates a benthic pennate diatom ancestry for the endosymbionts of the dinoflagellates *Peridinium foliaceum* and *Peridinium balticum* (Pyrrhophyta). *Journal of Eurkaryotic Microbiology*, *44*(4), 314–320.

Cox, E. J. (2015). Coscinodiscophyceae, Mediophyceae, Fragilariophyceae, Bacillariophyceae (diatoms). In W. Frey (Ed.), *Adolf Engler's Syllabus der Pflanzenfamilien: Vol. 2. Syllabus of plant families* (13th ed., pp. 64–103). Stuttgart: Borntraeger Science Publishers. 1. *Photoautotrophic eukaryotic algae Glaucocystophyta, Cryptophyta, Dinophyta/Dinozoa, Heterokontophyta/Ochrophyta, Chlorarachniophyta/Cercozoa, Euglenophyta/Euglenozoa, Chlorophyta, Streptophyta*. 324 pp.

D'Alelio, D., & Ruggiero, M. V. (2015). Interspecific plastidial recombination in the diatom genus *Pseudo-nitzschia*. *Journal of Phycology*, *51*(6), 1024–1028.

Darling, A. C., Mau, B., Blattner, F. R., & Perna, N. T. (2004). Mauve: Multiple alignment of conserved genomic sequence with rearrangements. *Genome Research*, *14*(7), 1394–1403.

Doyle, J. J., & Doyle, J. L. (1987). A rapid DNA isolation procedure for small quantities of fresh leaf tissue. *Phytochemical Bulletin*, *19*, 11–15.

Ehara, M., Inagaki, Y., Watanabe, K. I., & Ohama, T. (2000). Phylogenetic analysis of diatom *coxI* genes and implications of a fluctuating GC content on mitochondrial genetic code evolution. *Current Genetics*, *37*(1), 29–33.

Janouškovec, J., Liu, S. L., Martone, P. T., Carré, W., Leblanc, C., Collén, J., et al. (2013). Evolution of red algal plastid genomes: Ancient architectures, introns, horizontal gene transfer, and taxonomic utility of plastid markers. *PLoS One*, *8*(3), e59001.

Jansen, R. K., Cai, Z., Raubeson, L. A., Daniell, H., Depamphilis, C. W., Leebens-Mack, J., et al. (2007). Analysis of 81 genes from 64 plastid genomes resolves relationships in angiosperms and identifies genome-scale evolutionary patterns. *Proceedings of the National Academy of Sciences of the United States of America*, *104*(49), 19369–19374.

Jansen, R. K., & Ruhlman, T. A. (2012). Plastid genomes of seed plants. In R. Bock & V. Knoop (Eds.), *Genomics of chloroplasts and mitochondria* (pp. 103–126). Netherlands: Springer.

Katoh, K., Kuma, K., Toh, H., & Miyata, T. (2005). MAFFT version 5: Improvement in accuracy of multiple sequence alignment. *Nucleic Acids Research*, *33*(2), 511–518.

Langmead, B., & Salzberg, S. (2012). Fast gapped-read alignment with bowtie 2. *Nature Methods*, *9*(4), 357–359.

Lewin, J. C. (1958). The taxonomic position of *Phaeodactylum tricornutum*. *Journal of General Microbiology*, *18*, 427–432.

Lommer, M., Roy, A. S., Schilhabel, M., Schreiber, S., Rosenstiel, P., & LaRoche, J. (2010). Recent transfer of an iron-regulated gene from the plastid to the nuclear genome in an oceanic diatom adapted to chronic iron limitation. *BMC Genomics*, *11*, 718.

Lommer, M., Specht, M., Roy, A. S., Kraemer, L., Andreson, R., Gutowska, M. A., et al. (2012). Genome and low-iron response of an oceanic diatom adapted to chronic iron limitation. *Genome Biology*, *13*(7), R66.

Maddison, W., & Maddison, D. (2000). *MacClade 4: Analysis of phylogeny and character evolution*. Sunderland, MA: Sinauer Associates Inc.

Marchler-Bauer, A., & Bryant, S. H. (2004). CD-search: Protein domain annotations on the fly. *Nucleic Acids Research*, *32*(Suppl. 2), W327–W331.

Medlin, L. K. (2016). Evolution of the diatoms: Major steps in their evolution and a review of the supporting molecular and morphological evidence. *Phycologia*, *55*, 79–103.

Medlin, L. K. (2017). Evolution of the diatoms: IX. Two datasets resolving monophyletic classes of diatoms are used to explore the validity of adding short clone library sequences to the analysis. *European Journal of Phycology*, *52*(1), 90–103.

Medlin, L. K., & Kaczmarska, I. (2004). Evolution of the diatoms V: Morphological and cytological support for the major clades and a taxonomic revision. *Phycologia*, *43*(3), 245–270.

Moustafa, A., Beszteri, B., Maier, U. G., Bowler, C., Valentin, K., & Bhattacharya, D. (2009). Genomic footprints of a cryptic plastid endosymbiosis in diatoms. *Science*, *324*(5935), 1724–1726.

Nelson, D. M., Treguer, P., Brzezinski, M. A., Leynaert, A., & Queguiner, B. (1995). Production and dissolution of biogenic silica in the ocean—revised global estimates, comparison with regional data and relationship to biogenic sedimentation. *Global Biogeochemical Cycles*, *9*(3), 359–372.

Oudot-Le Secq, M.-P., Grimwood, J., Shapiro, H., Armbrust, E. V., Bowler, C., & Green, B. R. (2007). Chloroplast genomes of the diatoms *Phaeodactylum tricornutum* and *Thalassiosira pseudonana* and comparison with other plastid genomes of the red lineage. *Molecular Genetics and Genomics*, *277*(4), 427–439.

Paradis, E., Claude, J., & Strimmer, K. (2004). APE: Analyses of phylogenetics and evolution in R language. *Bioinformatics*, *20*(2), 289–290.

Round, F. E., & Crawford, R. M. (1981). The lines of evolution of the Bacillariophyta. 1. Origin. *Proceedings of the Royal Society of London Series B: Biological Sciences*, *211*(1183), 237–260.

Round, F. E., & Crawford, R. M. (1984). The lines of evolution of the Bacillariophyta. 2. The centric series. *Proceedings of the Royal Society of London Series B: Biological Sciences*, *221*(1223), 169–188.

Ruck, E. C., Linard, S. R., Nakov, T., Theriot, E. C., & Alverson, A. J. (2016). Hoarding and horizontal transfer led to an expanded gene and intron repertoire in the plastid genome of the diatom, *Toxarium undulatum* (Bacillariophyta). *Current Genetics*, *63*(3), 1–9.

Ruck, E. C., Nakov, T., Jansen, R. K., Theriot, E. C., & Alverson, A. J. (2014). Serial gene losses and foreign DNA underlie size and sequence variation in the plastid genomes of diatoms. *Genome Biology and Evolution*, *6*(3), 644–654.

Sabir, J. S. M., Yu, M., Ashworth, M. P., Baeshen, N. A., Baeshen, M. N., Bahieldin, A., et al. (2014). Conserved gene order and expanded inverted repeats characterize plastid genomes of Thalassiosirales. *PLoS One*, *9*(9), e107854.

Schattner, P., Brooks, A. N., & Lowe, T. M. (2005). The tRNAscan-SE, snoscan and snoGPS web servers for the detection of tRNAs and snoRNAs. *Nucleic Acids Research*, *33*, W686–689.

Schwarz, E. N., Ruhlman, T. A., Weng, M. L., Khiyami, M. A., Sabir, J. S. M., Hajarah, N. H., et al. (2017). Plastome-wide nucleotide substitution rates reveal accelerated rates in Papilionoideae and correlations with genome features across legume subfamilies. *Journal of Molecular Evolution*, *84*(4), 187–203.

Shimodaira, H. (2002). An approximately unbiased test of phylogenetic tree selection. *Systematic Biology*, *51*, 492–508.

Simonsen, R. (1979). The diatom system: Ideas on phylogeny. *Bacillaria*, *2*, 9–71.

Sorhannus, U., & Fox, M. G. (2011). Phylogenetic analyses of a combined data set suggest that the *Attheya* lineage is the closest living relative of the pennate diatoms (Bacillariophyceae). *Protist*, *163*(2), 252–262.

Stamatakis, A. (2014). RAxML version 8: A tool for phylogenetic analysis and post-analysis of large phylogenies. *Bioinformatics*, *30*(9), 1312–1313.

Steinecke, F. (1931). Die Phylogenie der Algophyten: Versuch einer morphologischen Begründung des natürlichen Systems der Algen. *Schriften der Königsberger Gelehrten Gesellschaft*, *8*(5), 127–298.

Sullivan, A. R., Schiffthaler, B., Thompson, S. L., Street, N. R., & Wang, X. R. (2017). Interspecific plastome recombination reflects ancient reticulate evolution in *Picea* (Pinaceae). *Molecular Biology and Evolution*, *34*(7), 1689–1701.

Swofford, D. (2003). *PAUP**: *Phylogenetic analysis using parsimony. (* and other methods)*. Ver. 4.0b10. Sunderland, MA: Sinauer Associates.

Tajima, N., Saitoh, K., Sato, S., Maruyama, F., Ichinomiya, M., Yoshikawa, S., et al. (2016). Sequencing and analysis of the complete organellar genomes of Parmales, a closely related group to Bacillariophyta (diatoms). *Current Genetics*, *62*(4), 887–896.

Tesler, G. (2002). GRIMM: Genome rearrangements web server. *Bioinformatics*, *18*, 492–493.

Theriot, E. C., Ashworth, M., Ruck, E., Nakov, T., & Jansen, R. K. (2010). A preliminary multigene phylogeny of the diatoms (Bacillariophyta): Challenges for future research. *Plant Ecology and Evolution*, *143*(3), 278–296.

Theriot, E. C., Ashworth, M., Ruck, E., Nakov, T., & Jansen, R. K. (2015). Dissecting signal and noise in diatom chloroplast protein encoding genes with phylogenetic information profiling. *Molecular Phylogenetics and Evolution*, *89*, 28–36.

Theriot, E. C., Cannone, J. J., Gutell, R. R., & Alverson, A. J. (2009). The limits of nuclear-encoded SSU rDNA for resolving the diatom phylogeny. *European Journal of Phycology*, *44*(3), 277–290.

Theriot, E. C., Ruck, E., Ashworth, M., Nakov, T., & Jansen, R. K. (2011). Status of the pursuit of the diatom phylogeny: Are traditional views and new molecular paradigms really that different? In J. Seckbach & P. Kociolek (Eds.), *The diatom world. Cellular origin, life in extreme habitats and astrobiology: Vol. 19* (pp. 119–142). Netherlands: Springer.

Van de Peer, Y., Van der Auwera, G., & De Wachter, R. (1996). The evolution of stramenopiles and alveolates as derived by "substitution rate calibration" of small ribosomal subunit RNA. *Journal of Molecular Evolution*, *42*(2), 201–210.

Weng, M.-L., Blazier, J. C., Govindu, M., & Jansen, R. K. (2014). Reconstruction of the ancestral plastid genome in Geraniaceae reveals a correlation between genome rearrangements, repeats and nucleotide substitution rates. *Molecular Biology and Evolution*, *31*, 645–659.

Wyman, S., Jansen, R. K., & Boore, J. L. (2004). Automatic annotation of organellar genomes with DOGMA. *Bioinformatics*, *20*(17), 3252–3255.

Zerbino, D. R., & Birney, E. (2008). Velvet: Algorithms for de novo short read assembly using de Bruijn graphs. *Genome Research*, *18*(5), 821–829.

Zerbino, D. R., McEwen, G. K., Marguiles, E. H., & Birney, E. (2009). Pebble and rock band: Heuristic resolution of repeats and scaffolding in the velvet short-read *de novo* assembler. *PLoS One*, *4*(12), e8407.

CHAPTER SIX

Evolution of the Plastid Genome in Green Algae

Monique Turmel[1], **Claude Lemieux**
Université Laval, Québec, QC, Canada
[1]Corresponding author: e-mail address: monique.turmel@bcm.ulaval.ca

Contents

Abstract

Comparative analyses of green algal plastid genomes (plastomes) have flourished in the past decade. In addition to improving our understanding of the phylogenetic relationships among green algal lineages, the expanded collection of plastome sequences has provided new insights into the ancestral architecture of this genome in the common ancestor of all green algae and into the changes that it underwent during lineage diversification in the two major divisions of the Viridiplantae (Chlorophyta and Streptophyta). The level of plastome diversity is much greater in the Chlorophyta than in the Streptophyta, with important variations seen at several levels—including genome size, presence/absence and size of the large inverted repeat encoding the rRNA operon, pattern of gene partitioning among single-copy regions, gene content, gene order, intron content, and amount of repetitive sequences—both within and across the main

Advances in Botanical Research, Volume 85
ISSN 0065-2296
https://doi.org/10.1016/bs.abr.2017.11.010

lineages of these two divisions. Here, we present an overview of the structural changes that sustained the plastome during the evolution of both chlorophyte and streptophyte algae. We begin by examining the range of variations observed at the above-mentioned levels in 112 chlorophyte taxa and then summarize what we learned for the Streptophyta based on the plastomes of 17 taxa. The chapter ends with a presentation of issues that need to be resolved in future studies.

1. INTRODUCTION

The green algae represent one of the most successful groups of photosynthetic eukaryotes, but surprisingly little is known about their evolutionary history. Their plastids (chloroplasts), like those of other members of the Archaeplastida or Plantae sensu *lato* (green plants, red algae, and glaucophytes), can be traced back to a single endosymbiotic event between a freshwater cyanobacterium and a heterotrophic eukaryotic host (Adl et al., 2005; Archibald, 2009; Ponce-Toledo et al., 2017; Reyes-Prieto, Weber, & Bhattacharya, 2007). Together with the land plants, the green algae constitute the Viridiplantae (the so-called green plant lineage) or Chloroplastida. The Viridiplantae split early (~1200–750 Mya) into two divisions (Becker, 2013; Leliaert et al., 2016; Lemieux, Otis, & Turmel, 2007): the Chlorophyta, containing the majority of the green algae, and the Streptophyta, containing all land plants and their closest green algal relatives, also known as charophytes (Leliaert et al., 2012; Lewis & McCourt, 2004).

With the recent revolution of DNA sequencing technologies, the number of complete or near-complete plastome (plastid genome) sequences that have become available for green algae has increased considerably since Lang and Nedelcu (2012) published their review on the coding capacity and organization of algal plastomes. Early studies of green algal plastomes, which were mostly sampled from the Chlorophyta, uncovered tremendous structural differences at several levels between green algal lineages as well as little similarity with their land plant counterparts (Bélanger et al., 2006; de Cambiaire, Otis, Lemieux, & Turmel, 2006; de Cambiaire, Otis, Turmel, & Lemieux, 2007; Maul et al., 2002; Pombert, Lemieux, & Turmel, 2006; Pombert, Otis, Lemieux, & Turmel, 2005; Turmel, Otis, & Lemieux, 1999; Wakasugi et al., 1997). As a matter of fact, a major theme of green algal plastome evolution is the extraordinary diversity of their architecture, which is in sharp contrast with the remarkable conservation observed for most land plants. As reviewed in other chapters of this volume, land plant plastomes typically consist of 120–150 kb circular-mapping molecules that contain approximately 120 genes and feature two copies of a rRNA operon-encoding inverted repeat (IR) separated by small and large single-copy (SC) regions (Jansen & Ruhlman, 2012; Wicke, Schneeweiss, dePamphilis, Muller, & Quandt, 2011). Intramolecular as well as intermolecular recombination between the two IR copies produce isomers that differ in the relative orientations of the SC regions (Palmer, 1983). But plastid DNA molecules, at least in maize, do not occur predominantly as circles in vivo, but rather as multiple genomic, linear-branched structures, which are thought to result from recombination-dependent replication (Bendich, 2004). Whether this observation can be generalized to all land plants is not clear nor do we know whether it applies to algae. Despite the different conformations/configurations of the plastome, both the gene content and gene order of each genomic region have sustained relatively minor changes during land plant evolution (Jansen & Ruhlman, 2012).

This chapter summarizes our current understanding of how the plastome changed in the course of green algal evolution. It is based primarily on the 129 complete and near-complete plastome sequences of photosynthetic green algae that were deposited in public databases as of January 2017 (see Table 1 for the accession numbers and names of taxa): 112 are from chlorophytes and 17 from streptophytes. We will begin by examining what these genomes revealed about the characteristics of the ancestral green algal plastome and then we will review the various types of changes that the plastome sustained in the Chlorophyta and Streptophyta. Considering that plastome evolution was much more dynamic in the Chlorophyta than in the Streptophyta, these two divisions will be treated in separate sections. But before getting into the heart of these topics, it is essential to provide basic knowledge of the interrelationships between the major lineages of green algae.

2. CLASSIFICATION AND PHYLOGENY OF GREEN ALGAE

2.1 Chlorophyta

Four groups of chlorophyte green algae traditionally recognized as classes have been distinguished on the basis of ultrastructural data derived from the mitotic, cytokinetic, and flagellar apparatus: the predominantly marine, unicellular, Prasinophyceae; the predominantly marine and morphologically diverse Ulvophyceae; and the freshwater or terrestrial, morphologically diverse Trebouxiophyceae and Chlorophyceae (Leliaert et al., 2012;

Table 1 GenBank Accession Numbers of Complete or Near-Complete Green Algal Plastomes Available as of January 2017

Taxon	Accession
Streptophytes	
Mesostigma viride	NC_002186
Chlorokybus atmophyticus	NC_008822
Klebsormidium flaccidum	NC_024167
Entransia fimbriata	NC_030313
Chara vulgaris	NC_008097
Chaetosphaeridium globosum	NC_004115
Coleochaete scutata	NC_030358
Mesotaenium endlicherianum	NC_024169
Zygnema circumcarinatum	NC_008117
Cylindrocystis brebissonii	NC_030359
Spirogyra maxima	NC_030355
Netrium digitus	NC_030356
Roya anglica	NC_024168
Roya obtusa	NC_030315
Closterium baillyanum	NC_030314
Cosmarium botrytis	NC_030357
Staurastrum punctulatum	NC_008116
Prasinophytes	
Prasinophyceae sp. MBIC10622	KJ746602
Prasinoderma coloniale	NC_024817
Verdigellas peltata	NC_030220
Prasinococcus sp. CCMP 1194	KJ746597
Cymbomonas tetramitiformis	NC_030169
Pyramimonas parkeae	FJ493499
Monomastix sp. OKE-1	NC_012101
Ostreococcus tauri	NC_008289
Micromonas commoda	NC_012575
Nephroselmis astigmatica	NC_024829
Nephroselmis olivacea	NC_000927
Pycnococcus provasolii	NC_012097
Picocystis salinarum	NC_024828
Prasinophyceae sp. CCMP 1205	KJ746601
Chlorodendrophyceae	
Scherffelia dubia	NC_029807
Tetraselmis sp. CCMP 881	KU167097
Pedinophyceae	
Marsupiomonas sp. NIES 1824	KM462870
Pedinomonas tuberculata	NC_025530
Pedinomonas minor	NC_016733
Trebouxiophyceae, Chlorellales	
Auxenochlorella protothecoides	NC_023775
Parachlorella kessleri	NC_012978
Dicloster acuatus	NC_025546
Marvania geminata	NC_025549
Pseudochloris wilhelmii	NC_025547
Chlorella sorokiniana	NC_023835
Chlorella sp. ArM0029B	KF554427
Chlorella vulgaris	NC_008097
Chlorella variabilis	NC_015359

Taxon	Accession
Core Trebouxiophyceae	
Oocystis solitaria	FJ968739
Planctonema lauterbornii	NC_025541
Koliella corcontica	NC_025536
Geminella terricola	NC_025542
Geminella minor	NC_025544
Gloeotilopsis sterilis	NC_025538
Pleurastrosarcina brevispinosa	KM462875
Neocystis brevis	NC_025535
Stichococcus bacillaris	NC_025527
Prasiolopsis sp. SAG 84.81	KM462862
Chlorella mirabilis	NC_025528
Koliella longiseta	NC_025531
Pabia signiensis	NC_025529
Parietochloris pseudoalveolaris	NC_025532
Leptospira terrestris	NC_009681
Xylochloris irregularis	NC_025534
Microthamnion kuetzingianum	NC_025537
Fusochloris perforatum	NC_025543
Myrmecia israelensis	NC_025525
Lobosphaera incisa	NC_025533
Dictyochloropsis reticulata	NC_025524
Watanabea reniformis	NC_025526
Botryococcus braunii	NC_025545
Choricystis minor	NC_025539
Elliptochloris bilobata	NC_025548
Trebouxiophyceae sp. MX-AZ01	NC_018569
Coccomyxa subellipsoidea	NC_015084
Paradoxia multiseta	NC_025540
Ulvophyceae, Bryopsidales	
Bryopsis hypnoides	NC_013359
Bryopsis plumosa	NC_026795
Caulerpa cliftonii	NC_031368
Caulerpa racemosa	NC_032042
Codium decorticatum	NC_032043
Derbesia sp. WEST4838	NC_031367
Lambia antarctica	NC_032284
Tydemania expeditionis	NC_026796
Core Ulvophyceae	
Ignatius tetrasporus	NC_034712
Pseudocharacium americanum	NC_034711
Oltmannsiellopsis viridis	NC_008099
Dangemannia microcystis	NC_034713
Pseudoneochloris marina	NC_034710
Ulva fasciata	NC_029040
Ulva sp. UNA00071828	KP720616
Ulva linza	NC_030312
Chamaetrichon capsulatum	NC_034714
Pseudendoclonium akinetum	NC_008114
Trichosarcina mucosa	NC_034709
Gloeotilopsis planctonica	KX306824
Gloeotilopsis sarcinoidea	KX306821

Taxon	Accession
Chlorophyceae, OCC clade	
Oedogonium cardiacum	NC_011031
Oedocladium carolinianum	NC_031510
Floydiella terrestris	NC_014346
Stigeoclonium helveticum	NC_008372
Schizomeris leibleinii	NC_015645
Chlorophyceae, Sphaeropleales	
Treubaria triappendiculata	NC_028578
Jenufa perforata	NC_028581
Jenufa minuta	NC_028582
Ankyra judayi	NC_029735
Bracteacoccus aerius	NC_029675
Bracteacoccus giganteus	NC_028586
Bracteacoccus minor	NC_029674
Pseudomuriella schumacherensis	NC_029669
Chromochloris zofingiensis	NC_029672
Mychonastes jurisii	NC_028579
Mychonastes homosphaera	NC_029671
Kirchneriella aperta	NC_029676
Neochloris aquatica	NC_029670
Chlorotetraedron incus	NC_029673
Acutodesmus obliquus	NC_008101
Monoraphidium neglectum	NW_014013626
Chlorophyceae, Chlamydomonadales	
Carteria cerasiformis	NC_028585
Hafniomonas laevis	NC_028583
Carteria sp. SAG 8-5	KT625419
Oogamochlamys gigantea	NC_028580
Chlamydomonas reinhardtii	NC_005353
Gonium pectorale	NC_020438
Pleodorina starrii	NC_021109
Volvox carteri f. *nagariensis*	GU084820
Phacotus lenticularis	NC_028587
Characiochloris acuminata	NC_028584
Dunaliella salina	NC_016732
Chlamydomonas applanata	KT625417
Chlamydomonas leiostraca	NC_032109
Chloromonas perforata	KT625416

Lewis & McCourt, 2004). The interpretation of these ultrastructural data led to the proposal that the Prasinophyceae—which displays predominantly unicellular algae bearing scales on their cell body/flagella or having lost them—evolved first, followed by the Ulvophyceae, and then the Trebouxiophyceae and Chlorophyceae. Phylogenetic studies based on molecular data, in particular the nuclear-encoded 18S rRNA gene and multiple plastid genes, have enabled the evaluation of this early hypothesis, yielding several alternative evolutionary scenarios and many taxonomic changes.

The consensus tree shown in Fig. 1 summarizes our current understanding of the phylogenetic relationships among the major lineages of green algae. There is now general agreement that the prasinophytes form several independent lineages at the base of the Chlorophyta, with the Palmophyllophyceae (Prasinococcales + Palmophyllales) representing the deepest branch (Leliaert et al., 2016). However, the branching order among the other major chlorophyte lineages, collectively designated as the core Chlorophyta (Leliaert et al., 2012), remains contentious, as variable topologies have been reported depending upon taxon and character sampling and method of phylogenetic inference (Fang, Leliaert, Zhang, Penny, & Zhong, 2017; Fucikova et al., 2014; Leliaert & Lopez-Bautista, 2015; Lemieux, Otis, & Turmel, 2014a; Melton, Leliaert, Tronholm, & Lopez-Bautista, 2015; Sun et al., 2016; Turmel, de Cambiaire, Otis, & Lemieux, 2016; Turmel, Otis, & Lemieux, 2017). Moreover, it is uncertain whether the traditional classes Trebouxiophyceae and Ulvophyceae are monophyletic. It has been hypothesized that the phycoplast—a microtubule structure mediating cell division—evolved early during the radiation of core chlorophytes. Like prasinophytes, the Pedinophyceae lack a phycoplast, and it is considered that the Ulvophyceae secondarily lost it (Leliaert et al., 2012). Consistent with this view, phylogenies based on nuclear and plastid rDNA operons recovered the Pedinophyceae as the earliest-diverging lineage of the core Chlorophyta, followed by the Chlorodendrophyceae and the traditionally recognized classes of green algae (Marin, 2012). More recently, maximum likelihood trees inferred from concatenated plastid genes coding for proteins and RNAs were shown to be largely congruent with the latter topology (Turmel, de Cambiaire, et al., 2016; Turmel et al., 2017).

The lack of resolution of the interrelationships between the major clades of the core Chlorophyta is most probably the result of poor taxon sampling and inappropriate methods of phylogenetic analysis in which the applied models poorly fit the nucleotide or amino acid data (Fang et al., 2017; Turmel, de Cambiaire, et al., 2016). To infer more reliable and robust plastid

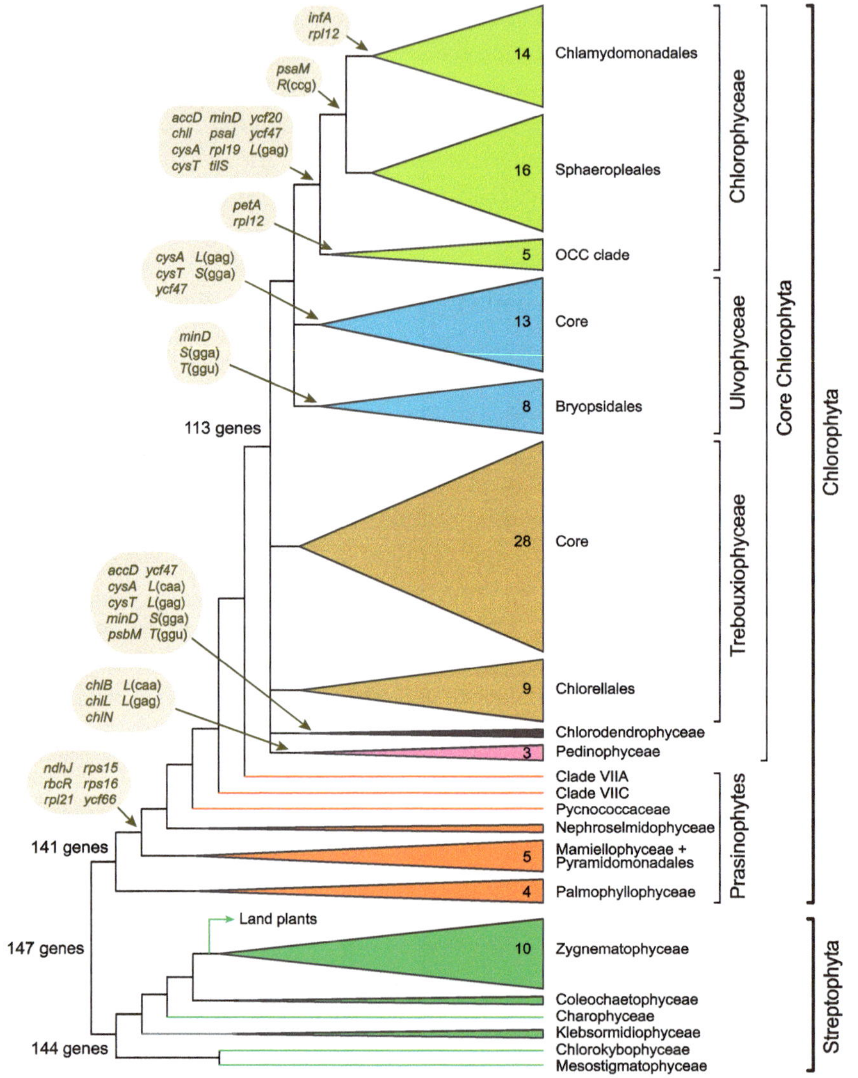

Fig. 1 Consensus relationships among major green algal lineages inferred in recent plastid phylogenomic studies. The *triangles* represent highly supported clades; the number of taxa in each clade (≥3 taxa) is indicated inside the corresponding *triangle*. Poorly resolved nodes are denoted by polytomies. The plastid genes that experienced losses at various nodes of the chlorophyte topology are indicated. As indicated by the *green arrow*, land plants are sister to the Zygnematophyceae. The plastomes of the last common ancestors of green plants, streptophytes, chlorophytes, and core chlorophytes contained 147, 144, 141, and 113 genes, respectively.

phylogenomic trees, it will be necessary to increase taxon sampling by adding representatives of previously unsampled or poorly sampled lineages and to use more realistic models of sequence evolution that account for among-site and among-branch compositional heterogeneities as well as lineage-specific codon-usage biases.

2.2 Streptophyta

The Streptophyta includes six main lineages of freshwater green algae (Fig. 1) that display a variety of cellular organizations, ranging from unicellular (e.g. *Mesostigma viride*, the only species of the Mesostigmatophyceae and some species of the Zygnematophyceae), to packets of cells (*Chlorokybus atmophyticus*, Chlorokybophyceae) or filaments (Klebsormidiophyceae and Zygnematophyceae), and to multicellular organization (Coleochaetophyceae and Charophyceae) (Graham, Cook, & Busse, 2000; McCourt, Delwiche, & Karol, 2004; Umen, 2014). The relationships among the streptophyte algal lineages have been resolved using concatenated plastid (Civan, Foster, Embley, Seneca, & Cox, 2014; Lemieux et al., 2007; Lemieux, Otis, & Turmel, 2016; Turmel, Otis, & Lemieux, 2006; Turmel, Pombert, Charlebois, Otis, & Lemieux, 2007) and nuclear genes (Laurin-Lemay, Brinkmann, & Philippe, 2012; Timme, Bachvaroff, & Delwiche, 2012; Wodniok et al., 2011). The earliest-diverging lineage is occupied by a clade comprising both the Mesostigmatophyceae and Chlorokybophyceae, and is followed by the Klebsormidiophyceae (Fig. 1). Cell division in the latter lineages occurs by furrowing, but as in land plants, the morphologically more complex Charophyceae and Coleochaetophyceae, which emerged following the divergence of the Klebsormidiophyceae, use a mechanism of cell division involving a phragmoplast and possess cell walls with plasmodesmata (Graham et al., 2000; McCourt et al., 2004; Umen, 2014). Sister to all plants is the Zygnematophyceae, a morphologically diverse group of green algae that reproduce sexually by conjugation. This is the only streptophyte algal class that displays substantial diversity (at least 4000 species). Based on the structure of the cell wall, members of the Zygnematophyceae were divided into two orders: the Zygnematales feature a smooth cell wall (the ancestral trait) and the Desmidiales an ornamented and segmented cell wall. However, molecular phylogenies are not in agreement with this traditional taxonomic structure: the Zygnematales were shown to be paraphyletic as the root of the Zygnematophyceae was positioned within this order (Gontcharov, 2008; Lemieux et al., 2016).

It is worth noting that recently published phylogenetic analyses of proteins from cyanobacteria and plastids of photosynthetic eukaryotes failed to support the notion that the *Mesostigma* + *Chlorokybus* clade represents the earliest branch of the Streptophyta (Sanchez-Baracaldo, Raven, Pisani, & Knoll, 2017). This clade was instead resolved before the divergence of the Streptophyta and Chlorophyta, a topology that was also recovered by Lemieux, Otis, and Turmel (2000) in their plastome-based phylogenomic study that included *Mesostigma* and only a few chlorophytes and streptophytes. Using increased taxon sampling, Lemieux et al. (2007) later showed that the position inferred for the *Mesostigma* + *Chlorokybus* clade is greatly influenced by the nature of the data set and sampling of characters, and concluded that the basal placement of this clade within the Streptophyta likely reflects the true organismal relationships. The more recent analyses of Leliaert et al. (2016) corroborated this conclusion, revealing that not only the nature of the data set but also the methods of phylogenetic inference have an impact on the statistical support observed for the *Mesostigma* + *Chlorokybus* clade. In short, the position of this clade has become again a matter of controversy, and additional phylogenomic analyses will be required to resolve with certainty the deep-branching relationships of the Viridiplantae.

3. PLASTOME ARCHITECTURE OF THE COMMON ANCESTOR OF ALL GREEN ALGAE

Although prasinophyte plastomes are extremely variable in structure and gene content, their comparative analyses with those of the earliest-diverging streptophytes have provided insights into the architecture of the ancestral plastome of green plants (Leliaert et al., 2016; Lemieux, Otis, & Turmel, 2014b; Turmel, Gagnon, O'Kelly, Otis, & Lemieux, 2009). It has been predicted that this ancestral plastome possessed two copies of a large rRNA operon-encoding IR separated by large and small SC regions and that the gene contents of the SC regions closely matched those observed in most extant streptophyte algae and land plants. A minimum of 147 genes (encoding 107 proteins and 40 RNAs) were present, all likely devoid of introns (Lemieux et al., 2014b). Many of them formed operons or clusters that have been maintained in red algae and glaucophytes in addition to several lineages of the Chlorophyta and Streptophyta. Six genes (*odpB*, *rpl33*, *ycf61*, *trnA*(ggc), *bioY*, and *ycf22*) have been retained exclusively in the Streptophyta, whereas only three (*ycf47*, *rnpB*, and *rne*) are specific to the Chlorophyta.

Among the prasinophyte plastomes analysed to date, that of the macroscopic *Verdigellas peltata* (Palmophyllales, Palmophyllophyceae) is remarkably similar to the plastome of the streptophyte *M. viride* at both the gene content and gene order levels (Leliaert et al., 2016). The degree of synteny it displays with this streptophyte is even greater compared to those observed with more closely related algae belonging to the Prasinococcales (Palmophyllophyceae). Note that the recently sequenced plastome of *Palmophyllum crassum* (Palmophyllales) is entirely collinear with its *Verdigellas* counterpart (Furukawa, Kunugi, Ihara, Takabayashi, & Tanaka, 2017). While both palmophyllalean plastomes lack a large IR, prasinophytes from the Pyramimonadales and Nephroselmidophyceae have retained this feature and an ancestral partitioning of genes among the SC regions.

4. PLASTOME EVOLUTION IN THE CHLOROPHYTA

In this section, we will first consider the changes that the chlorophyte plastome underwent at the levels of size and nucleotide composition. Then, we will examine the variety of changes associated with the IR and look at the variations in gene content, intron content, and gene organization.

4.1 Plastome Size

Plastome size ranges from 64 kb (for Prasinophyceae sp. CCMP 1205, a member of the lineage sister to all core chlorophytes) to 521 kb (for *Floydiella terrestris*, a member of the Chaetopeltidales, Chlorophyceae) (Brouard, Otis, Lemieux, & Turmel, 2010; Lemieux et al., 2014b) in the Chlorophyta. There is important size variation not only across major lineages but also within individual classes and orders (Fig. 2A). The prasinophyte plastomes are found at the lowest end of the size range, with 10 of the 14 genomes surveyed being less 100 kb. At the other extreme are found the Chlamydomonadales (Chlorophyceae), also known as Volvocales, with 11 of the 14 plastomes exceeding 200 kb. Size variability as revealed by the interquartile range was found to be maximal in the core Ulvophyceae, a group comprising the Oltmannsiellopsidales, Ignatiales, Ulvales, and Ulotrichales.

Plastome size variations within major lineages are mainly attributable to a combination of three factors: differences in length of intergenic regions, changes in intron content, and contractions/expansions of the large IR (Brouard et al., 2010; Lemieux et al., 2014b; Marcelino, Cremen, Jackson, Larkum, & Verbruggen, 2016; Smith et al., 2013; Turmel, de

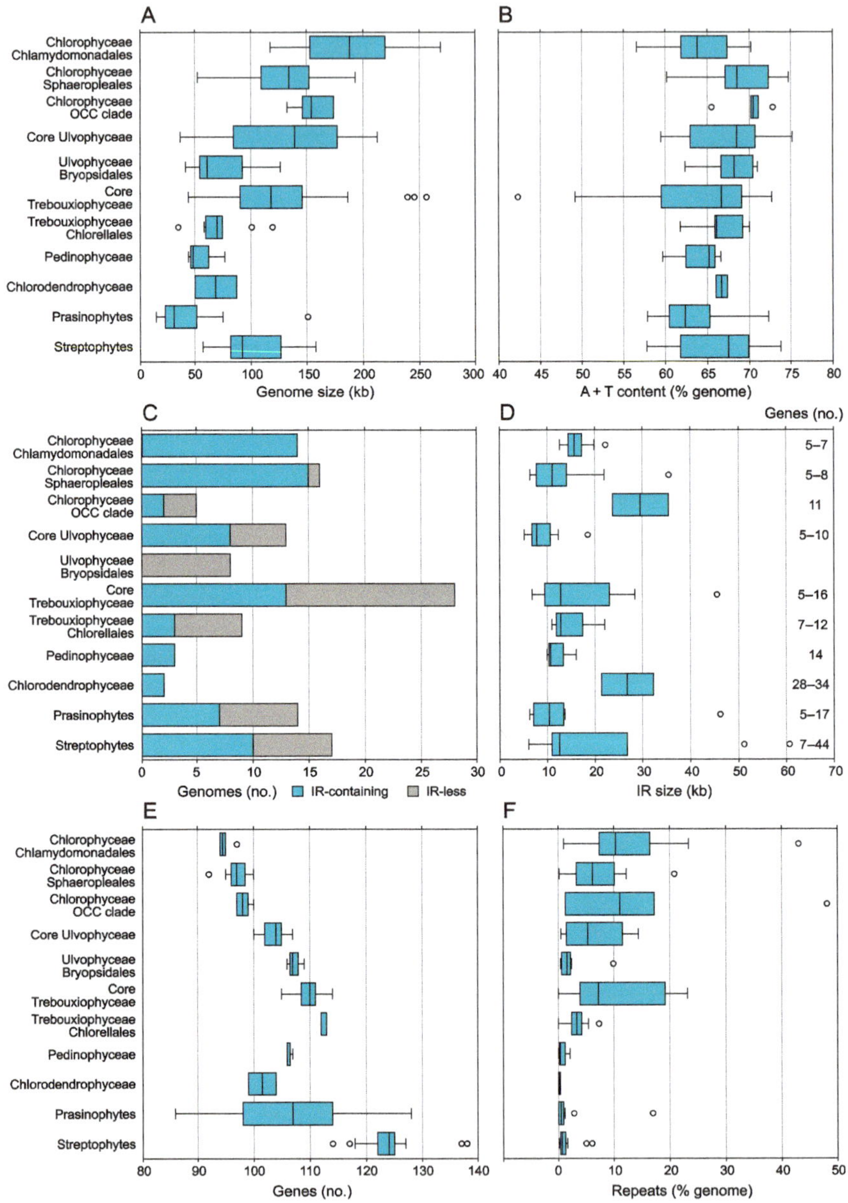

Fig. 2 Variations of plastome features within and among the major clades of green algae: size (A), AT content (B), IR presence/absence (C), IR size and gene content (D), total number of canonical genes (E), and proportion of dispersed repeats (F). These data are based on complete or near-complete plastome sequences. Each box plot encloses 50% of the data with the median value displayed as a line, while the lines extending on each side of the boxes mark the minimum and maximum values, with the outliers displayed as individual points. Note that the large plastomes of *Floydiella terrestris* (521 kb) and *Volvox carteri* (461 kb) are not represented in (A) and that all data derived from streptophytes exclude the land plants.

Cambiaire, et al., 2016; Turmel et al., 2017; Turmel, Otis, & Lemieux, 2015). The amounts of intron and intergenic sequences, in particular, show extensive fluctuations but no obvious patterns can be discerned in a comparative phylogenetic context. In a large-scale study comparing three members of the Pedinophyceae and 35 taxa representing the major clades recognized in the Trebouxiophyceae (Turmel et al., 2015), the intergenic regions proved to be by far the noncoding sequences that contribute the most to plastome size variation. Wide changes in the proportion of intergenic sequences were even seen in closely related species, as exemplified by the 117-kb and 306-kb IR-less genomes of *Prasiolopsis* sp. and *Stichococcus bacillaris* (*Prasiola* clade, Trebouxiophyceae), where 68.0% and 16.3% of the sequences, respectively, correspond to intergenic regions. The plastomes with the largest sizes also featured the greatest abundance of dispersed repeats. Similarly, an important variation in noncoding content and gene density has been observed among the plastomes examined in the core Ulvophyceae (Turmel et al., 2017), Bryopsidales (Ulvophyceae) (Marcelino et al., 2016), and Chlorophyceae (Brouard et al., 2010; Smith et al., 2013).

The large size variations characterizing major chlorophyte lineages and their subclades are most likely the consequences of nonadaptive processes. In addition to random genetic drift, mutation rates and DNA maintenance pathways (DNA replication and DNA repair mechanisms) play a central role in shaping the plastome architecture (Smith, 2016). In recent years, the mutational hazard hypothesis, which is a general theory for explaining architectural diversity of genomes (Lynch, 2007), has been assessed using organelle genomes of diverse organisms, including picoplanktonic prasinophytes and chlamydomonadalean green algae (Smith, 2016). This hypothesis predicts that lineages with expanded genomes will tend to have lower mutation rates and smaller effective population sizes than those with more compact or streamlined genomes. While it gained support from studies of green algal plastomes, it was refuted by data derived from some land plant organelle genomes, which in turn led to a new model postulating that differences in double-strand break (DSB) repair systems are responsible for the expansion/contraction of organelle genomes (Christensen, 2014). According to this model, DSB in coding sequences are repaired accurately, while error-prone systems causing additions/deletions are used for intergenic regions.

Analyses of green algal plastomes have pointed to additional factors that may influence genome size. All pico- and nanoplanktonic chlorophytes sampled so far are at the low end of the observed plastome size variation, thus strengthening the notion that small cells tend to have small and compact genomes (Lemieux et al., 2014b; Turmel et al., 2015). Given their shorter replication times, small genomes in bloom-forming species could confer a selective advantage (Cavalier-Smith, 2005). On the other hand, the 82.0-kb plastome of the endolithic (limestone-boring) seaweed *Ostreobium quekettii* (Bryopsidales), which is the smallest and most gene-dense plastome among the Ulvophyceae, is thought to have been shaped primarily by adaptation to low light conditions (Marcelino et al., 2016): both its higher degree of compaction and significantly slower rate of molecular evolution compared to plastomes from other families of the same order are in agreement with the expected effects of low light (e.g. energy limitation). Similarly, the 79.4-kb plastome of *V. peltata* (Palmophyllales), which occurs in deep waters, has been suggested to be the consequence of low light-driven genome reduction (Marcelino et al., 2016).

In the volvocine lineage of the Chlamydomonadales, which is a model lineage for studying the origins and evolution of multicellularity, the plastome architectures of *Chlamydomonas reinhardtii*, *Gonium pectorale*, *Pleodorina starrii*, and *Volvox carteri* revealed that plastome complexity correlates positively with organismal complexity (Smith et al., 2013). Therefore, in line with the mutational hazard hypothesis, it has been suggested that this trend is the result of lower mutation rates and/or smaller effective population sizes in multicellular vs unicellular volvocines. More recently, however, the plastome of the basal four-celled *Tetrabaena socialis* (>405 kb) has been found to be more expanded than those of some volvocines with more complex cellular organizations (including *Gonium* and *Pleodorina*) (Featherston, Arakaki, Nozaki, Durand, & Smith, 2016), raising the hypothesis that shift from a unicellular to a colonial organization coincided with plastome expansion, possibly due to increased random genetic drift.

4.2 Nucleotide Composition

The overall content of A and T nucleotides (AT content) in chlorophyte plastomes ranges from 42.3% (for Trebouxiophyceae sp. MX-AZ01) to 75.2% (for *Ulva linza*, core Ulvophyceae), with only nine taxa displaying values less than 58.0%: the prasinophyte *Nephroselmis olivacea* (Nephroselmidophyceae), six trebouxiophyceans, all from the *Elliptochloris* + *Choricystis* clade (a late-emerging clade of the core Trebouxiophyceae), as well as the chlamydomonadaleans *Chlamydomonas leiostraca* and *V. carteri* (Fig. 2B). Chlorophyte plastomes are typically AT-biased possibly due to selection for translational efficiency (i.e. selection for preferred codons matching the anticodons of plastid tRNAs) and/or to AT mutation pressure coupled with inefficient DNA repair systems (Smith, 2012). In contrast to what has been observed for the majority of examined chlorophyte plastomes, variation in GC content of protein-coding genes in the *Elliptochloris/Choricystis* clade, including the polar alga *Coccomyxa subellipsoidea* (Smith et al., 2011), is higher at third codon positions (29%–64%) than at the more functionally constrained first and second codon positions (Smith et al., 2011; Turmel et al., 2015). Moreover, the GC bias of these trebouxiophycean plastomes is also evident at noncoding regions by the occurrence of small repeats rich in G and C (Turmel et al., 2015). Unusually high levels of these nucleotides have also been documented for the mitogenome and nuclear genome of *C. subellipsoidea* (Smith et al., 2011) as well as the mitogenome of Trebouxiophyceae sp. MX-AZ01 (Servin-Garciduenas & Martinez-Romero, 2012). Thus, it has been argued that the forces driving the nucleotide composition towards G and C in both organelles of *C. subellipsoidea* are neutral and linked to a nuclear mutation affecting GC-biased gene conversion or cell-wide features such as life history-related traits, environment, and/or metabolic features (Smith et al., 2011). It remains to be seen whether the mitogenomes and nuclear genomes of the other members of the *Elliptochloris/Choricystis* clade have GC-biased nucleotide compositions.

4.3 Presence/Absence of the IR

4.3.1 Multiple IR Losses

About two-thirds (76/112) of the chlorophyte plastomes compared in this review contain a large IR sequence encoding the rRNA genes. As shown in Fig. 2C, the proportion of IR-containing plastomes is variable across major lineages; all taxa from the Chlorodendrophyceae, Pedinophyceae, and Chlamydomonadales (Chlorophyceae), but none from the Bryopsidales (Ulvophyceae), display an IR. Given the currently known relationships among green algae and assuming that de novo creation of an IR from an IR-less plastome is very unlikely (Turmel et al., 2015), it was inferred that the IR was lost at least four times in prasinophytes (Lemieux et al., 2014b), seven times in the Trebouxiophyceae (Turmel et al., 2015), three times in the Ulvophyceae (Leliaert & Lopez-Bautista, 2015; Turmel et al., 2017), and twice in the Chlorophyceae (Brouard et al., 2010) (for the second IR loss observed in this class, see GenBank accession NC_028581 of the *Jenufa perforata* plastome). As discussed later, the mechanisms leading to IR loss are still largely unknown.

4.3.2 IR Size Variation

The size of the IR sequence varies almost 10-fold across chlorophyte lineages, from 5.2 kb in the core ulvophycean *Chamaetrichon capsulatum* to 46.1 kb in the prasinophyte *N. olivacea* (Fig. 2D). There is important IR size variation even within individual lineages; for example, in the *Prasiola* clade of the core Trebouxiophyceae, the IRs of '*Chlorella*' *mirabilis* (6.8 kb) and *Pabia signiensis* (27.3 kb) show a fourfold size difference (Turmel et al., 2015). The number of canonical genes (i.e. genes commonly found in plastomes) encoded in the IR is also variable (Fig. 2D), with the IRs of 13 chlorophytes from diverse lineages (Pyramimonadales, core Trebouxiophyceae, core Ulvophyceae, Sphaeropleales and Chlamydomonadales orders of the Chlorophyceae) containing only the five genes that make up the rRNA operon (*rrs*, *trnI*(gau), *trnA*(ugc), *rrl*, and *rrf*) and the large IRs of the two investigated chlorodendrophycean taxa encoding up to 34 genes (Turmel, de Cambiaire, et al., 2016). The variability in IR gene content is often due to the integration of sequences found in the neighbouring SC regions or to sequence excision from the IR termini. Small changes at the endpoints of the IR are very common in land plants (Raubeson & Jansen, 2005), and this phenomenon, also known as the ebb and flow of the IR, has been explained by homologous recombination and gene conversion (Goulding, Olmstead, Morden, & Wolfe, 1996). On the other hand, major IR expansion events leading to the incorporation of multiple genes have been attributed to double reciprocal recombination or DSB repair combined with gene conversion (Raubeson & Jansen, 2005). But variations in the number of canonical genes are not the only cause of IR size fluctuations, as exemplified by the *P. signiensis* IR which encodes a single extra gene compared to the fourfold shorter IR of '*Chlorella*' *mirabilis*. IR size variations also occur via shrinkage or growth of internal intergenic regions, including the gains of genes putatively acquired by horizontal transfers (Brouard, Otis, Lemieux, & Turmel, 2008; Brouard, Turmel, Otis, & Lemieux, 2016; Turmel et al., 2009, 1999). Notably, the IR has been suggested to be a hot spot for the integration of foreign sequences in the Chlorophyta (Brouard et al., 2016).

4.3.3 Nonidentical IR Copies in the Ulvophyceae

Remarkably, the copies of the short IRs (<7.8 kb) found in *Ignatius tetrasporus* and *Pseudocharacium americanum* (Ignatiales), *Pseudoneochloris marina* (Ulvales), and *C. capsulatum* (Ulotrichales) are not identical in sequence (Turmel et al., 2017). The *trnI*(gau) and *trnA*(ugc) genes of the rRNA operon are missing in one of the IR copies of these core ulvophyceans;

moreover, intergenic regions and intron sequences of the *rrs* and *rrl* genes exhibit large indels. Although these observations are unprecedented for the Viridiplantae, nonidentical IR copies with indels mapping to the tRNA genes have also been documented for two haptophyte lineages: the Phaeocystales (Smith, Arrigo, Alderkamp, & Allen, 2014) and Prymnesiales (Hovde et al., 2014). The situation for the IRs of the two analysed *Phaeocystis* species (Phaeocystales) is identical to that prevailing in the Ulvophyceae (i.e. a standard rRNA operon in one IR copy and only the rRNA genes in the other), whereas each IR copy of *Chrysochromulina tobin* (Prymnesiales) is lacking a single tRNA gene (*trnI*(gau) or *trnA*(ugc)).

The *Chamaetrichon* plastome is exceptional in containing three nonidentical copies of the IR sequence. This finding lends credit to the hypothesis that an IR can be created de novo from an IR-lacking plastome (Turmel et al., 2017). In a previous study, the discovery of a 8.3-kb IR lacking any rRNA genes in the plastome of the trebouxiophycean *S. bacillaris* had led Turmel et al. (2015) to speculate that this unusual IR originated de novo, but the hypothesis that it represents a remnant of a conventional rDNA-encoding IR could not be ruled out.

The two IR copies of *Ignatius* can undergo intramolecular recombination, but no isomers were detected for the *Pseudoneochloris* plastome (Turmel et al., 2017). Similar to the *Chrysochromulina* plastome, the absence of flip-flop recombination in the latter plastome was correlated with the accumulation of nucleotide polymorphisms in coding sequences of the rRNA genes, supporting the notion that pairing of the two IR copies for recombination provides a copy correction mechanism. Nucleotide polymorphisms were also uncovered in the *Chamaetrichon* rRNA genes, but in this case, it remains to be seen whether the IR copies participate in flip-flop recombination.

4.3.4 Mechanisms of IR Loss

At least three models have been proposed to explain how the IR is lost (Turmel et al., 2017). First, IR loss may be the ultimate consequence of repeated events of IR contraction by the ebb and flow mechanism; however, no convincing evidence supports this model. Indeed, the IRs of all photosynthetic green plants investigated so far, with the exception of the angiosperm *Monsonia speciosa* (Geraniaceae) (Guisinger, Kuehl, Boore, & Jansen, 2011), contain all five genes making up the rRNA operon, suggesting that erosion of the IR is impeded when the IR/SC boundaries reach this operon. Alternatively, excision of one of the IR copy may occur in a single step through intramolecular recombination between short direct repeats at the endpoints of the IR sequence. Comparative gene order analyses of IR-lacking and IR-containing plastomes from three independent trebouxiophycean lineages (Turmel et al., 2015), the streptophyte class Coleochaetophyceae (Lemieux et al., 2016), and land plant lineages (Raubeson & Jansen, 2005) are consistent with this mechanism of IR loss. Third, a model of IR loss has recently been proposed for the Ulotrichales based on a comparison of gene order between the IR-containing and IR-lacking plastomes of the distantly related *Pseudendoclonium akinetum* and *Gloeotilopsis planctonica*, respectively (Turmel, Otis, & Lemieux, 2016). This model entails the differential elimination of sequences within the rDNA operon from the two IR copies, and it is consistent with the finding that the IR copies in the plastomes from three distinct ulvophycean lineages differ in both gene/intron contents (see Section 4.3.3).

4.3.5 Diversity of Gene Partitioning Patterns in IR-Containing Plastomes

In the IR-containing plastomes of certain prasinophytes such as the members of the Nephroselmidophyceae and Pyramimonadales, the pattern of gene distribution among the three genomic regions is similar to that observed for the earliest-diverging charophytes and most other streptophytes (Lemieux et al., 2014b; Turmel et al., 2009, 1999). Indeed, despite differences in IR gene content due to shifts of the IR/SC boundaries, the genes typically found in the small SC (SSC) region have not been relocated to the large SC (LSC) region and vice versa.

This ancestral gene partitioning pattern has undergone various degrees of changes in the course of chlorophyte evolution. Most green algae that were sampled from the Pedinophyceae and core Trebouxiophyceae revealed minor deviations from the ancestral pattern, which are largely accounted for by the relocalization of a few genes (*psbM*, *trnS*(uga), *trnD*(guc), *trnMe*(cau), and *trnG*(gcc)) ancestrally present in the LSC region to the IR and/or to the immediately adjacent SSC sequence (Turmel et al., 2015). In the case of the Chlorellales (Trebouxiophyceae), the set of reshuffled genes (*trnI*(cau), *rps14*, *rbcL*, and *psbA*) was found to be different. These gene relocalizations likely occurred through IR/SSC and IR/LSC boundary shifts, presumably caused by inversions or DSB repairs (Goulding et al., 1996) as well as inversions of internal IR sequences.

Although the plastomes of the chlorodendrophyceans *Scherffelia dubia* and *Tetraselmis* sp. CCMP 881 are unique in exhibiting unusually large gene-rich IRs and very short SSC regions devoid of any gene, their IRs share with pedinophycean and most trebouxiophycean plastomes several genes that are encoded by the LSC region in ancestral-type plastomes (Turmel, de Cambiaire, et al., 2016). Based on these observations, it was proposed that the acquisition of a set of seven genes from the LSC region led to the expansion of the IR in the common ancestor of all core chlorophytes.

Later during chlorophyte evolution, extensive transfers of genes from the LSC to the SSC region coincided with the emergence and diversification of the Ulvophyceae (Turmel et al., 2017). Distinct waves of gene transfers were inferred for the plastomes of the Ignatiales, Oltmannsiellopsidales, and Ulvales/Ulotrichales, and the gene partitioning pattern observed for the latter lineage was found to be the most similar to that predicted for the common ancestor of all chlorophytes.

More radical departures from the ancestral gene partitioning pattern took place independently in the Chlorophyceae (Brouard et al., 2008; de Cambiaire et al., 2006), the prasinophyte clades II (Mamiellophyceae) and VIIC (Lemieux et al., 2014b), and the lineage occupied by the trebouxiophycean *Xylochloris irregularis* (Turmel et al., 2015). Restructuring of these plastomes involved multiple transfers from both LSC to SSC and SSC to LSC. As in the Ulvophyceae, the highly dynamic nature of the plastome architecture in the Chlorophyceae is reflected by the extremely different gene partitioning patterns observed for the Oedogoniales, Sphaeropleales, and Chlamydomonadales (Brouard et al., 2008; de Cambiaire et al., 2006).

4.4 Gene Content

Free-standing genes, either canonical genes or coding sequences putatively acquired by horizontal transfers, are dealt with in this section. They are translated using the bacterial, archaeal, and plant plastid genetic code (code 11, National Center for Biotechnology Information) in all chlorophytes examined so far, with the exception of *Jenufa minuta* (Sphaeropleales, Chlorophyceae) where the codon UGA, which is normally a stop codon, is used as a sense codon by a subset of canonical genes and translated as tryptophan. This sphaeroplealean plastome (GenBank NC_028582) is currently the only known green plant plastome with a deviant genetic code.

4.4.1 Repertoire of Canonical Genes and Gene Losses

Chlorophyte plastomes contain between 86 and 128 distinct canonical genes (in the prasinophytes *Micromonas commoda* (Mamiellophyceae) and *N. olivacea*, respectively) (Fig. 2E). The latter form a repertoire of 141 genes (not counting the tRNA genes that arose from duplications), of which only 71 have been retained in all chlorophytes (Fig. 3). Based on the data

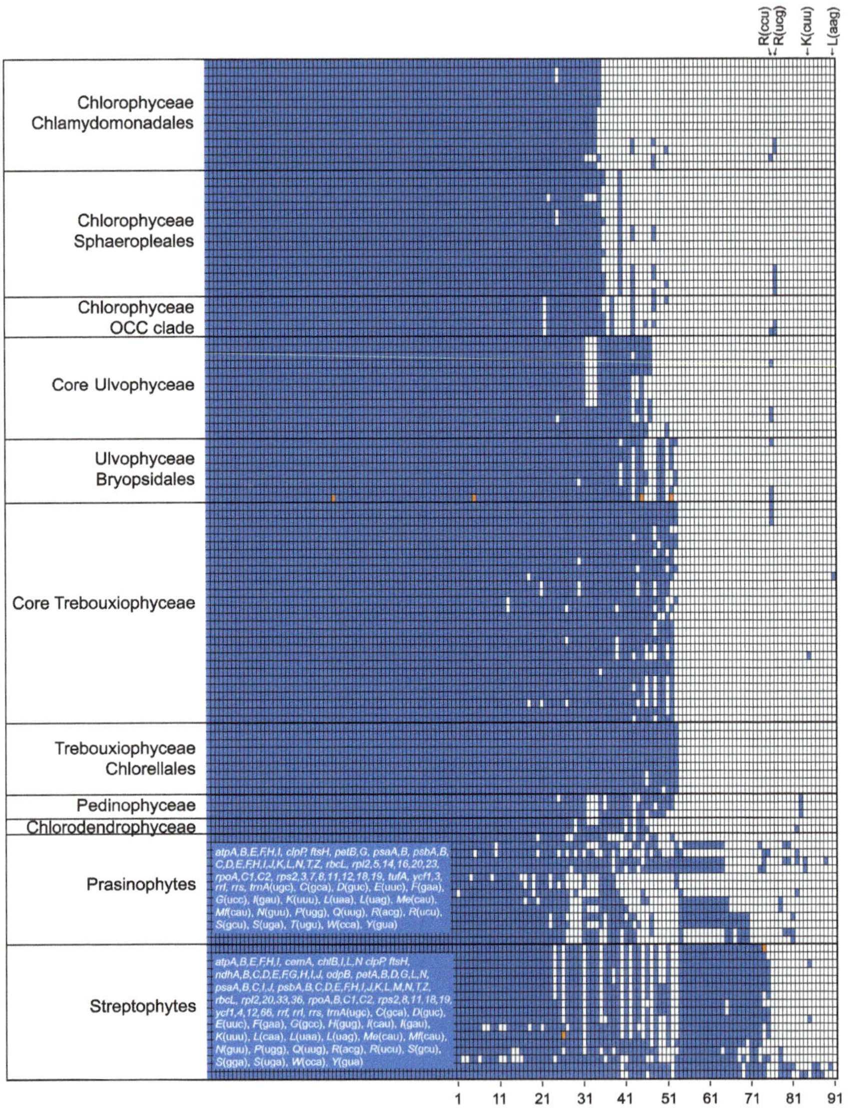

Fig. 3 Distribution of canonical genes in green algal plastomes. The 151 genes found in these plastomes (*vertical columns*) are sorted by decreasing level of conservation from *left* to *right*. *Blue boxes* denote the presence of genes and *orange boxes* the presence of pseudogenes in *Bryopsis hypnoides*, *Chaetosphaeridium globosum*, and *Staurastrum punctulatum*. Taxa in each major lineage are displayed in reverse order relative to Table 1. The 71 genes common to all chlorophytes and the 96 common to all streptophytes are listed in the prasinophyte and streptophyte panels, respectively. Genes that underwent losses are numbered consecutively at the bottom of the figure: (1–10), *psaC*, *rpl14,16,36*, *rpoB*, *rps3,7,12*, *trnH*(gug), *psaJ*; (11–20), *rpl23*, *ycf3,12,*

(Continued)

compiled in this review, 113 genes are predicted to have been present in the common ancestor of all core chlorophytes, implying that 28 of the 55 genes that sustained losses or migrated to the nucleus during prasinophyte evolution were not vertically transmitted to this ancestor (Leliaert et al., 2016; Lemieux et al., 2014b; Turmel et al., 2009). Included in these genes are *ndhJzrbcR*, *rpl21*, *rps15*, *rps16*, and *ycf66*, six genes that were identified solely in the Palmophyllophyceae (Leliaert et al., 2016).

Many genes display recurring losses in the Chlorophyta (Fig. 3), with some events coinciding with the emergence of different classes or orders (Fig. 1). For instance, of the 11 genes missing in all 35 examined chlorophyceans, 4 (*cysA*, *cysT*, *tilS*, and *trnL*(gag)) are also absent in the 13 members of the core ulvophyceans and 5 (*cysA*, *cysT*, *minD*, *ycf47*, and *trnL*(gag)) in the 2 members of the Chlorodendrophyceae. More resolved chlorophyte phylogenies are needed to determine whether these gene losses occurred independently or in a common ancestor of these monophyletic groups.

4.4.2 Gains of Unusual Genes via Horizontal DNA Transfers

Diverse coding sequences with similarity to proteins of known functions and/or recognized protein domains, but no affinity to canonical plastid genes, have been identified as freestanding open reading frames (ORFs) in plastomes from various chlorophyte lineages (Brouard et al., 2008, 2016; Leliaert et al., 2016; Leliaert & Lopez-Bautista, 2015; Turmel et al., 2009, 2015). All potentially code for products acting on DNA or RNA, such as DNA breaking-rejoining enzymes, DNA primases, DNA methyltransferases, reverse transcriptases, endonucleases, and maturases. These unusual sequences often represent remnants of genes. Several appear to be of bacterial or viral origin (Leliaert & Lopez-Bautista, 2015), some show strong similarity with organelle DNA sequences, possibly mobile

Fig. 3—Cont'd trnG*(ucc)*,T*(ugu)*,V*(uac)*, petD, rps4, ycf4, rrf; *(21–30)*, rps14, petA, trnG*(-gcc)*, rpl5, trnI*(cau)*, tufA, ccsA, cemA, psbM, petL; *(31–40)*, rps9, chlL,N,lB, rpl32, infA, psaI, M, chlI, rpl12; *(41–50)*, accD, rpl19, trnL*(caa)*, tilS, ycf20, trnR*(ccg)*, minD, trnS*(gga)*, cysT,A; *(51–60)*, trnT*(ggu)*, ycf47, trnL*(gag)*, petN, ndhA,B,C,D,E,F; *(61–70)*, ndhG,H,I,K, rpl22, ycf66, ndhJ, rps16, rpl21, rps15; *(71–80)*, odpB, rpl33, trnV*(gac)*,S*(cga)*,P*(ggg)*,R*(ccu)*, R*(ucg)*, ycf65, ftsI,W; *(81–90)*, rbcR, rnpB, ssrA, ycf81, trnK*(cuu)*, ycf61, trnA*(ggc)*, bioY, rne, ycf27; *(91)* trnL*(aag)*. *Note that* trnK*(cuu)*, trnL*(aag)*, *and* trnR*(ccu)* *probably arose from duplication and subsequent sequence divergence of preexisting tRNA genes (Lemieux et al., 2016; Turmel et al., 2015), and that* trnR*(ucg)* *was likely acquired through horizontal transfer from a mitochondrial or bacterial donor (Brouard et al., 2010). The data presented for streptophytes exclude the land plants.*

elements, from organisms other than green algae (Brouard et al., 2008), while those encoding putative homing endonucleases and reverse transcriptases may be vestiges of introns that were originally present in canonical plastid genes. The sporadic phylogenetic distribution of the seemingly foreign sequences suggests that they were gained independently through intercellular gene transfers. In the case of the Bryopsidales, putative horizontal DNA transfers might have been facilitated by the bacteria residing inside these giant siphonous green algae (Leliaert & Lopez-Bautista, 2015). In other lineages, they might have been promoted by closely associated bacteria (Leliaert et al., 2016).

The discovery of sequences possibly originating from a mitochondrial donor (*int* and *dpoB*) within a 10-kb region of the 35.5-kb IR of *Oedogonium cardiacum* (Oedogoniales, Chlorophyceae) provided the first case of horizontal transfer in which coding sequences of known function, not carried by introns, were gained by the plastome in the Viridiplantae (Brouard et al., 2008). The equivalent region of the 23.7-kb IR of the closely related *Oedocladium carolinianum* is missing the *int* and *dpoB* sequences, but houses instead two ORFs showing similarities to putative phage/bacterial DNA primases and to a previously reported hypothetical protein (Brouard et al., 2016). Considering that homologues of the *Oedocladium* ORFs have also been localized within or very near the IR in distantly related chlorophytes (*Pyramimonas*, *Nephroselmis*, and *Pleodorina*), it has been suggested that the IR could be a hot spot for the integration of foreign sequences (Brouard et al., 2016).

More recently, it has been shown that unusual sequences were also gained through intracellular interorganellar transfers (Turmel, Otis, et al., 2016). Comparisons of the plastomes and mitogenomes of *G. planctonica* and *Gloeotilopsis sarcinoidea* (Ulotrichales) unveiled short sequences of mitochondrial origin at two distinct loci of the *G. sarcinoidea* plastome, yielding the first evidence for the intracellular transfer of gene sequences from the mitochondria to the plastid in green algae.

4.5 Introns and Their Encoded Proteins

Chlorophyte plastomes contain a plethora of introns. The 643 group I introns and 442 group II introns identified in the plastomes examined in this review represent 92 and 143 distinct sites of insertion, respectively (Fig. 4). With the exception of two group II introns of the ulotrichalean *G. planctonica* (Turmel, Otis, et al., 2016), they are all located within coding

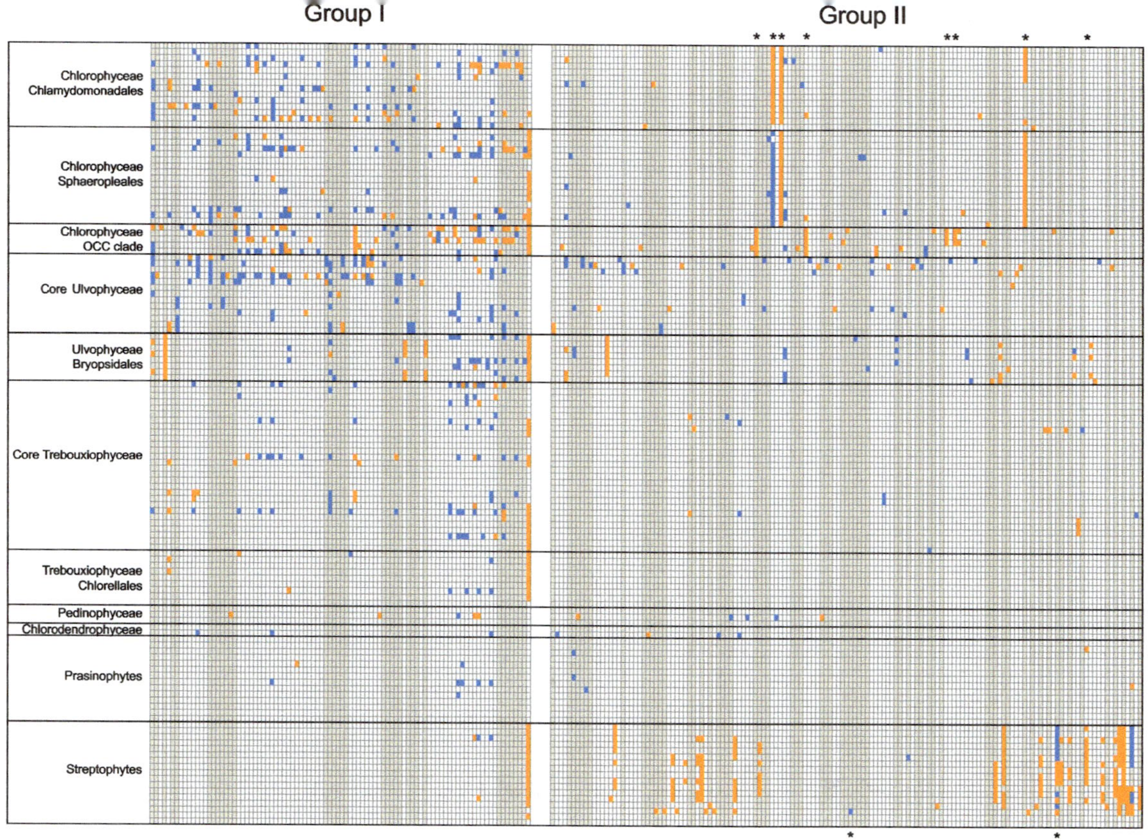

Fig. 4 Distribution of group I and group II introns in green algal plastomes. *Filled boxes* indicate the presence of introns, with *orange boxes* denoting introns with no ORF and *blue boxes* denoting introns that encode proteins with homing endonuclease (group I introns) or reverse transcriptase (group II introns) activities. Each column represents a distinct insertion site, with contiguous columns of identical shade denoting introns within the same gene. Genes are presented in the same order as in Fig. 5A, while taxa in each major lineage are displayed in reverse order relative to Table 1. *Asterisks* at the *top* and *bottom* of the figure indicate the insertion sites occupied by *trans*-spliced group II introns in chlorophyte and streptophyte plastomes, respectively. The data presented for streptophytes exclude the land plants.

regions of canonical genes. Nearly half of the canonical genes (66/141) are interrupted by introns, and while group II introns occur in most of these genes (63/66), group I introns are restricted to 18 genes (Fig. 5A). Genes encoding components of photosystems I and II are the most intron rich, with 28 insertion sites identified in *psbA* alone. The rRNA genes also contain introns at many sites, but these are exclusively of the group I family. While all chlorophyte group I introns are *cis*-spliced, *trans*-spliced group II introns occur in five genes of the Chlorophyceae—*psaA* (de Cambiaire et al., 2006; Fucikova, Lewis, & Lewis, 2016; Goldschmidt-Clermont et al., 1991) and *rpl32* (see GenBank accessions listed in Lemieux, Vincent, Labarre, Otis, & Turmel, 2015) in the Chlamydomonadales and Sphaeropleales, and *petD, psaC,* and *rbcL* in the OCC clade (Bélanger et al., 2006; Brouard et al., 2010)—as well as in the *ycf3* gene of the clade-VIIC prasinophyte *Picocystis salinarum* (Lemieux et al., 2014b).

Group I introns are particularly abundant in the Chlorophyceae and core Ulvophyceae (Figs 4 and 5B), whereas group II introns are most prevalent in the OCC clade of the Chlorophyceae and in the core Ulvophyceae (Fig. 5C), where they have been recently shown to multiply by intragenomic proliferation of existing introns via retrohoming (Brouard et al., 2016; Turmel, Otis, et al., 2016; Turmel et al., 2017)—the mobility mechanism used to maintain group II introns at cognate sites (Lambowitz & Belfort, 2015). Retrohoming of these introns at noncognate target sites has been shown to require mutations in the exon-binding sequences (Brouard et al., 2016; Turmel, Otis, et al., 2016). Owing to their mobility and also their limited lifetime following insertion into new sites, group I and group II introns generally show highly variable distribution patterns (Fig. 4). Contrary to most group II introns, group I introns are frequently found at the same sites in distantly related taxa, reflecting a higher frequency of insertions and/or a more restricted number of target sites for the group I introns. It is notable that chlorophycean *trans*-spliced group II introns differ from *cis*-spliced introns by their maintenance over long evolutionary time; this is likely the result of a lower frequency of intron loss events caused by recombination of reverse-transcribed mRNAs (Brouard et al., 2010).

The majority of the group I intron insertion sites (64/92) are occupied by putative mobile introns encoding a homing endonuclease (Fig. 4). Of the three recognized families of homing endonucleases, the LAGLIDADG family is the most prevalent, with mobile introns found at 31 insertion sites (compared to 18 and 15 sites for the HNH and GIY-YIG families, respectively). All introns sharing a given site carry the same type of homing

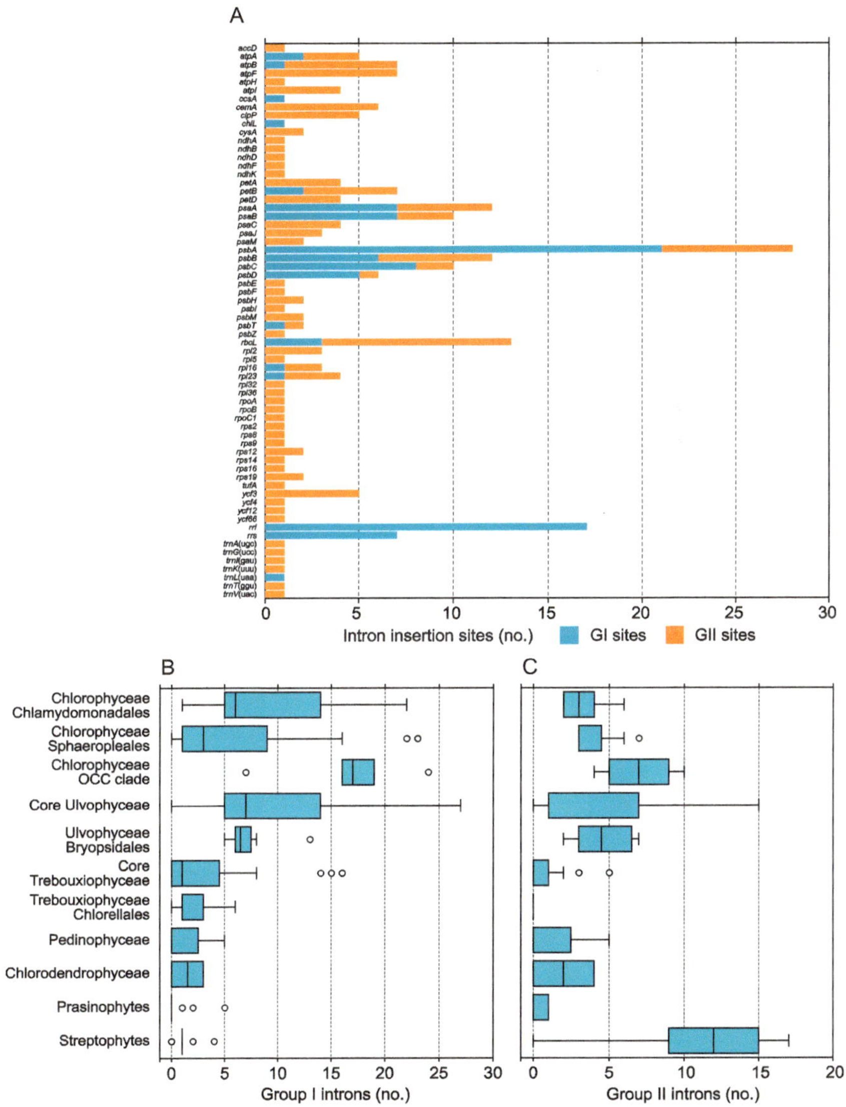

Fig. 5 Statistics for introns in green algal plastomes: number of group I and group II intron sites per canonical gene in the set of 129 plastomes listed in Table 1 (A), and variations of group I (B) and group II (C) intron contents within and among the major clades of *green* algae. The *box plots* in (B) and (C) enclose 50% of the data with the median value displayed as a line, while the lines extending on each side of the boxes mark the minimum and maximum values, with the outliers displayed as individual points. The data presented for streptophytes exclude the land plants.

endonuclease. Group II introns also encode proteins conferring mobility (i.e. proteins with reverse transcriptase, intron maturase, and HNH-endonuclease domains), but only 40% of the occupied sites (57/143) display at least one putative mobile intron.

4.6 Gene Rearrangements

4.6.1 Extent and Impact of Gene Rearrangements

Gene order was reshuffled to various degrees among and within major groups of chlorophytes (Brouard et al., 2010; Leliaert et al., 2016; Turmel, de Cambiaire, et al., 2016; Turmel et al., 2015, 2017). These events often caused the disruption of ancestral clusters (i.e. clusters conserved between streptophytes and chlorophytes), several of which were inherited from the cyanobacterial ancestor of the plastid. Moreover, as mentioned earlier, reconfiguration of gene order in IR-containing plastomes was sporadically associated with major alterations in the pattern of gene partitioning among SC regions (Brouard et al., 2010; Turmel et al., 2015, 2017).

In the Chlorophyta, the highest level of conservation of ancestral gene clusters has been observed in certain prasinophyte lineages (Leliaert et al., 2016; Lemieux et al., 2014b; Turmel et al., 2009, 1999). Within the core Chlorophyta, erosion of ancestral gene clusters was minimal in the Chlorodendrophyceae and reached maximal level in the Chlorophyceae (Brouard et al., 2010; Turmel, de Cambiaire, et al., 2016). Some ancestral clusters were broken only once during chlorophyte evolution, while others were fragmented independently multiple times. The highest frequency of breakage has been documented for the rDNA operon (Leliaert & Lopez-Bautista, 2015; Lemieux et al., 2014a; Turmel et al., 2009, 2015). For instance, this operon was disrupted at a minimum of four sites in late-diverging lineages of core trebouxiophyceans, and given the complexity of the associated rearrangements, the series of events that lead to the various configurations of genes in the operon could not be reconstructed (Turmel et al., 2015).

Plastome rearrangements also led to the breakup of coding regions within a few genes (such as *rpoB*, *rpoC2*, and *tilS*) in the core Trebouxiophyceae (de Cambiaire et al., 2007; Turmel et al., 2015), Ulvophyceae (Leliaert & Lopez-Bautista, 2015; Turmel et al., 2017), and Chlorophyceae (Brouard et al., 2010), as well as to the formation of *trans*-spliced group II introns in the Chlorophyceae (Brouard et al., 2010) and the prasinophyte *P. salinarum* (Lemieux et al., 2014b). The highest frequency of gene fragmentation has been documented for *rpoB*, and in most cases, the two resulting ORFs have remained contiguous on the plastome.

4.6.2 Utility of Gene Rearrangements in Assessing Phylogenetic Hypotheses

Analyses of breakpoints within ancestral and derived gene clusters have proven useful to assess conflicting phylogenetic hypotheses (Brouard et al., 2008, 2010; Leliaert et al., 2016; Turmel, de Cambiaire, et al., 2016; Turmel et al., 2015). For example, Turmel, de Cambiaire, et al. (2016) compared derived gene pairs to evaluate the relationships among the major lineages of the core Chlorophyta: six synapomorphic gene pairs were found to unite the Chlorellales with core trebouxiophyceans, supporting the monophyly of the Trebouxiophyceae and providing evidence against the affiliation between the Pedinophyceae and Chlorellales.

4.6.3 Factors Influencing Gene Rearrangements

The rate of gene rearrangements varies among lineages of the Chlorophyceae and core Trebouxiophyceae. Twice as many inversions were estimated for the Chlamydomonadales compared to the Chaetophorales (Brouard, Otis, Lemieux, & Turmel, 2011), and within the OCC clade, the pace of gene rearrangements in the Oedogoniales is even slower compared to the Chaetophorales (Brouard et al., 2016). The IR-less plastomes of late-diverging taxa in the core Trebouxiophyceae were reported to rearrange at a faster rate than their IR-containing homologues, and an accelerated rate of sequence evolution was also noted in these lineages (Turmel et al., 2015). A similar correlation between increased rearrangements and accelerated substitution rates has been documented for the plastomes of angiosperms belonging to the Geraniaceae (Weng, Blazier, Govindu, & Jansen, 2014).

Inversions caused by nonhomologous recombination between repeated sequences are thought to be the main mechanism for gene shuffling in plastomes (Jansen & Ruhlman, 2012; Palmer, 1991). Consistent with this mechanism, the extent of plastome rearrangements in the Geraniaceae and other land plant lineages has been correlated with the proportion and numbers of repeated sequences (Weng et al., 2014). For the Chlorophyta, however, no strict correlation has been established between the proportion/sizes of dispersed repeats and the degree of gene rearrangements nor have repeats been reported to be prevalent at inversion endpoints. Dispersed repeats in chlorophyte plastomes show tremendous variations in both abundance and sequence, and the plastomes packed with such sequences are generally found in lineages displaying extensive plastome rearrangements (i.e. core Trebouxiophyceae, core Ulvophyceae, and Chlorophyceae) (Fig. 2F). When dispersed repeats are bountiful, they are typically evenly dispersed throughout the plastome. The extremely dynamic evolution of dispersed repeats in the Chlorophyta has likely obscured or eliminated signals of past rearrangement events in which these sequences participated (Turmel et al., 2015). In this context, it is worth pointing out that breakpoints of gene rearrangements have been associated with tRNA genes in some green plant plastomes, raising the possibility that homologous recombination between tRNA genes was responsible for inversions in these plastomes (Turmel et al., 2015).

In principle, when dispersed repeats in plastomes are present in direct orientation, they confer the risk of generating two or more plastid DNA molecules differing in gene content. However, among all green plant plastomes that have been fully sequenced to date, only that recently reported for the epiphytic green alga *Koshicola spirodelophila* (Chaetopeltidales, Chlorophyceae) was assembled as more than one circle (Watanabe, Fucikova, Lewis, & Lewis, 2016). Genes in this 384.9-kb IR-less plastome are distributed on three distinct circular chromosomes.

5. PLASTOME EVOLUTION IN STREPTOPHYTE ALGAE

The 17 plastomes currently available for streptophyte algae represent all six charophyte lineages, with the Klebsormidiophyceae and Coleochaetophyceae being each represented by two taxa and the Zygnematophyceae by 10 taxa (Lemieux et al., 2016). These plastomes are more conserved than those of chlorophytes at the levels of size (Fig. 2A), gene order, and gene partitioning among SC regions. As in the Chlorophyta, IR losses (Fig. 2C), IR expansions/contractions (Fig. 2D), gene losses (Fig. 3), and intron gains/losses (Figs 4 and 5) contributed in a major way to their variations. Of the 144 canonical genes predicted in the common ancestor of all streptophytes, 54 were lost during streptophyte evolution, with about half of these genes associated with unique events (Lemieux et al., 2016). Introns in charophyte plastomes, most of which belong to the group II family, are located at 38 different insertion sites (Fig. 4) and include the 21 group II introns shared with land plants. The latter were gained during four distinct evolutionary periods—before the emergence of the Klebsormidiophyceae, Charophyceae, Coleochaetophyceae, and Zygnematophyceae (Lemieux et al., 2016).

The *M. viride* and *C. atmophyticus* plastomes are the richest in ancestral traits: they feature the largest gene repertoire, are almost entirely devoid of introns, and share a very similar quadripartite architecture with some of their prasinophyte homologues (Lemieux et al., 2000, 2007, 2016; Turmel et al., 2007). The *Chara vulgaris* (Charophyceae) plastome is the next featuring the most ancestral gene organization, suggesting a very slow rate of evolution in the Charophyceae (Lemieux et al., 2016; Turmel et al., 2006) that would mirror the evolutionary stasis observed for the mitogenome in the same class (Turmel, Otis, & Lemieux, 2013). Remarkably, the *Chara* and *Chaetosphaeridium globosum* (Coleochaetophyceae) IRs are almost identical to their bryophyte counterparts in both gene content and gene order; as in liverworts and mosses, their IR/LSC boundaries are located between *trnV*(gac) and the second exon of *rps12* (Lemieux et al., 2016). Although the Zygnematophyceae displays the highest levels of plastome diversity (Civan et al., 2014; Lemieux et al., 2016; Turmel, Otis, & Lemieux, 2005), only a few structural modifications took place during the transition from green algae to land plants (Civan et al., 2014; Lemieux et al., 2016; Turmel et al., 2006, 2007).

Below, we focus on the main evolutionary changes that account for the specific features and diversity of the streptophyte plastome in the Klebsormidiophyceae, Coleochaetophyceae, and Zygnematophyceae.

5.1 IR Expansions, Gene Losses, and Intron Gains in the Klebsormidiophyceae

The plastomes of *Klebsormidium flaccidum* and *Entransia fimbriata* (Klebsormidiophyceae), which are among the largest among streptophyte algae, are characterized by greatly reduced gene contents and vastly expanded IRs (Civan et al., 2014; Lemieux et al., 2016). Eight canonical genes, including four tRNA genes, were lost before the split of the lineages leading to *Klebsormidium* and *Entransia*, and 22 genes losses occurred subsequently in these lineages (Lemieux et al., 2016). Given that the tRNAs encoded in these algal plastomes are not sufficient to decode all codons, it has been suggested that the missing tRNAs are imported from the cytosol into the plastid (Lemieux et al., 2016). With their large sizes (51.1 and 60.6 kb) and abundance of genes (32 and 44 canonical genes with only 19 shared), the IRs of these plastomes resemble those of chlorodendrophycean chlorophytes (Fig. 2D). Considerable expansion of the IR took place towards the SSC region before the split of the two klebsormidiophycean lineages and further expansion, predominantly towards the LSC region, occurred following this divergence (Lemieux et al., 2016).

In addition to numerous lineage-specific group II introns, klebsormidiophycean plastomes contain group II introns that are shared with land plants, including the *cis*-spliced *trnK*(uuu) and *trans*-spliced *rps12*_114 introns. In land plants, the *trnK*(uuu) intron houses an ORF encoding a maturase (MatK) that assists splicing of other plastid group II introns (Zoschke et al., 2010), but this intron ORF is missing in *Klebsormidium* and a number of other streptophyte algae. Acquisition of the *rps12*_114 intron by the Klebsormidiophyceae led to breakage of the ancestral *str* operon (5′-*rps12*-*rps7*-*tufA*-3′) and ultimately to the transfer of the *tufA* gene encoding the elongation factor EF-Tu to the SSC region in *Klebsormidium* and to complete loss of this gene in *Entransia* (Lemieux et al., 2016).

5.2 Plastome Streamlining in the Coleochaetophyceae

At 107.2 kb, the *Coleochaete scutata* (Coleochaetophyceae) plastome is the smallest among the charophytes examined to date (Lemieux et al., 2016). Reductive evolution of this plastome relative to that of *C. globosum* entailed IR loss, shortening of intergenic regions, and deletions of eight canonical genes. Notably, losses of *rps4*, *rps7*, *rps12*, and *rps14* represent unique events in the evolutionary scenario inferred for streptophyte algae. Otherwise, plastome gene order is highly conserved in the two sampled coleochaetophyceans, with the organization of the *Coleochaete* genes formerly present in the IR, SSC, and LSC regions conforming to the ancestral partitioning pattern.

Considering the unusual divergence and rapid evolution of the *tufA* sequence in the plastome of the Coleochaetophyceae, it has been questioned whether it plays any functional role or represents a pseudogene (Baldauf, Manhart, & Palmer, 1990; Lemieux et al., 2016; Turmel et al., 2006). In the Zygnematophyceae and land plants, the elongation factor EF-Tu is encoded in the nucleus and there is no trace of this gene in the plastome. Remarkably, *tufA* has been maintained as an intact ORF in the coleochaetophycean plastome, despite the absence of conserved amino acids essential for the function of the putatively encoded product in protein synthesis and the documented evidence for the presence of nuclear-encoded copies (Baldauf et al., 1990). Recent analyses support the notion that the plastid *tufA* sequence is undergoing pseudogenization and that the functional coding sequence of this gene resides in the nucleus (Lemieux et al., 2016).

5.3 Highly Dynamic Evolution in the Zygnematophyceae

In the Zygnematophyceae, the plastome (130.0–207.9 kb) underwent multiple IR losses and more extensive rearrangements than in any other streptophyte algal classes (Lemieux et al., 2016). At the gene order level, all 10 examined zygnematophycean plastomes are united by a single synapomorphy, which corresponds to loss of linkage between the *trnI*(gau) and *trnA*(ugc) genes of the rRNA operon. Only four of them, all originating from late-diverging lineages, feature an IR. The IRs of these plastomes display a disrupted rRNA operon with two to four breakage sites depending on the species, and their size variation (12.6–26.8 kb) is mainly explained by varying amounts of noncoding sequences. Mapping of the IR presence/absence on the zygnematophycean phylogeny uncovered a minimum of five IR losses.

The zygnematophycean plastome shows astonishing variability at the intron level. It has been inferred that 17 of the 21 group II introns usually present in land plant plastomes underwent one to six independent losses during the diversification of the Zygnematophyceae (Lemieux et al., 2016). Just three introns, including the *trans*-spliced *rps12*_114 intron, have been retained in all investigated taxa. Conversely, only the introns in *trnI*(gau)—which is part of the rDNA operon—and *trnV*(uac)—which also maps to the IR in *Chara*, *Coleochaete*, and bryophytes—are lacking in all taxa, implying that losses of the latter introns from the IR and breakage of the rRNA operon were early events in the evolutionary history of the Zygnematophyceae.

Sequences encoding integrases/recombinases and DNA primases of phage/viral origin are present in some zygnematophycean plastomes (Civan et al., 2014; Lemieux et al., 2016). Given that putative integrase/recombinase genes have also been identified in the mitogenomes of various streptophytes (Turmel et al., 2013), including zygnematophyceans, interorganellar DNA transfers might account for the presence of these foreign sequences in both the plastid and mitochondria.

Why is the zygnematophycean plastome so prone to gene rearrangements, IR loss, and intron deletions? The underlying causes of this instability remain unclear, although various hypotheses have been proposed. The intron losses may be the result of retroposition events, but the source of the protein providing the required reverse transcriptase activity is not obvious (Lemieux et al., 2016). Some zygnematophycean plastomes feature a moderate proportion of dispersed repeats; however, no strict correlation could be established between the acquisition of these repetitive sequences and IR losses and/or increased gene rearrangements (Civan et al., 2014; Lemieux et al., 2016). Moreover,

IR losses could not be linked with increased rates of rearrangements. Early invasions of phages/viruses or retroviruses/retrotransposons in the plastid might have contributed to the IR instability and triggered massive plastome rearrangements (Civan et al., 2014; Lemieux et al., 2016). Alternatively, nuclear-encoded, plastid-targeted genes involved in DNA replication, recombination, and repair might have played a major role in reshuffling gene order.

6. FUTURE DIRECTIONS

Comparative analyses of green algal plastomes have uncovered an impressive range of variations, providing insights into the timing and variety of genetic changes that took place during the evolution of chlorophytes and streptophytes. These studies revealed that the green algal plastome followed its own evolutionary trajectory within each class and that extensive changes in overall architecture and gene organization generally coincided with the emergence of major lineages. Despite multiple independent losses, the IR has been remarkably preserved across green algal lineages, and among the plastomes that retained the IR, the gene partitioning pattern among the SC regions is markedly conserved within major lineages, suggesting that the IR plays a role in stabilizing the plastome architecture.

Although considerable progress has been accomplished during the past decade, there are still numerous gaps in our understanding of green algal plastome evolution. This situation stems from the facts that the relationships among the major monophyletic groups of the Chlorophyta remain uncertain and that an extremely small portion of the green algal diversity has been sampled for plastome analysis. For instance, in the Chlorophyta, three prasinophyte lineages (clade-VIIA, clade-VIIC, and the Pycnococcaceae) are each represented by a single species and no plastome sequence is available for several orders of the Ulvophyceae (Cladophorales, Dasycladales, Scotinosphaerales, and Trentepohliales). To gain deeper insights into the evolutionary history of the chlorophyte and streptophyte plastomes, poorly sampled and previously unexplored lineages will need to be investigated. Among the phylogenetic questions that must be settled are the precise positions of late-diverging prasinophyte lineages, the branching order of major lineages within the core Chlorophyta, the monophyletic vs polyphyletic status of the Trebouxiophyceae and Ulvophyceae, and the identities of the earliest-diverging lineages of the Chlamydomonadales and Sphaeropleales (Chlorophyceae). In addition to contributing important information on the evolution of green algae and their plastome, the new phylogenetically targeted plastomes are expected to unveil unusual features that have not previously been documented and perhaps open new avenues of research.

Delineating the functional role of the IR, if any, would certainly be a valuable goal, but achieving this objective is challenging, as this will probably require a better understanding of the interrelationships between the processes of flip-flop recombination, gene conversion, and DNA replication. Given the highly dynamic evolution of the IR and the exceptional presence of divergent IR copies in the Ulvophyceae (Ignatiales, Ulvales, and Ulotrichales), it would be important to undertake studies on the mechanisms of IR loss and expansion/contraction as well on the cause and impact of IR sequence divergence in this class. Investigating, for instance whether the level and distribution of mutations in the IR copies are tied to the frequency of flip-flop recombination would allow to determine if this mechanism plays a major role in the process of gene conversion. Aside from the Ulvophyceae, it would be worth examining additional taxa in the Klebsormidiophyceae and Chlorodendrophyceae to determine whether the impressive IR size differences documented for the two taxa representing separate lineages in each of these classes are due to an acceleration of the rate of IR expansion after the emergence of these lineages or to a significant IR expansion event coincident with the divergence of these lineages.

The unprecedented discovery that the *Koshicola* plastome consists of three separate chromosomes raises questions about the prevalence of a fragmented plastome architecture in the Chaetopeltidales, an order of the Chlorophyceae that appears to be characterized by enormous plastomes (Brouard et al., 2010; Watanabe et al., 2016). Similarly, it would be of interest to explore the evolution of plastome architecture in the Chlamydomonadales, another chlorophycean order displaying plastomes of exceptionally large size, because the plastome sequences of numerous chlamydomonadalean taxa were recently found to assemble as multiple linear contigs instead of unique circles (Del Vasto et al., 2015; Featherston et al., 2016; Lemieux et al., 2015). Although these sequence assemblies were certainly hampered by the presence of extremely abundant dispersed repeats, the existence of multipartite architectures cannot be entirely dismissed. Completion of these partially assembled chlamydomonadalean plastomes will probably require the use of Single-Molecule-Real-Time sequencing technologies that generate very long sequence reads.

Finally, considering that the high-throughput sequencing technologies that are currently available offer the possibility to analyse both the plastome and mitogenome in individual taxa, future studies should take advantage of this opportunity to identify whether the organelle genomes from the same species or lineage exhibit similar architectural or unusual features (e.g. foreign DNA and nonstandard genetic code). The shared characteristics that have been reported for the plastomes and mitogenomes of a few green algae (Pombert et al., 2005; Robbens et al., 2007; Smith et al., 2011, 2010; Smith & Lee, 2010; Turmel, Oits, et al., 2016) are not too surprising given that parallel evolution of these genomes can be mediated by nuclear-encoded, organelle-targeted proteins shared by the two organelles and/or by common forces influencing their genome architectures (Smith & Keeling, 2015). But how widespread among and within green algal lineages is the tendency of the two organelle genomes to evolve similar architectures needs to be explored further.

REFERENCES

Adl, S. M., Simpson, A. G., Farmer, M. A., Andersen, R. A., Anderson, O. R., Barta, J. R., et al. (2005). The new higher level classification of eukaryotes with emphasis on the taxonomy of protists. *The Journal of Eukaryotic Microbiology*, *52*(5), 399–451. https://doi.org/10.1111/j.1550-7408.2005.00053.x.

Archibald, J. M. (2009). The puzzle of plastid evolution. *Current Biology*, *19*(2), R81–88. https://doi.org/10.1016/j.cub.2008.11.067.

Baldauf, S. L., Manhart, J. R., & Palmer, J. D. (1990). Different fates of the chloroplast *tufA* gene following its transfer to the nucleus in green algae. *Proceedings of the National Academy of Sciences of the United States of America*, *87*(14), 5317–5321.

Becker, B. (2013). Snow ball earth and the split of Streptophyta and Chlorophyta. *Trends in Plant Science*, *18*(4), 180–183. https://doi.org/10.1016/j.tplants.2012.09.010.

Bélanger, A. S., Brouard, J. S., Charlebois, P., Otis, C., Lemieux, C., & Turmel, M. (2006). Distinctive architecture of the chloroplast genome in the chlorophycean green alga *Stigeoclonium helveticum*. *Molecular Genetics and Genomics*, *276*(5), 464–477. https://doi.org/10.1007/s00438-006-0156-2.

Bendich, A. J. (2004). Circular chloroplast chromosomes: The grand illusion. *Plant Cell*, *16*(7), 1661–1666. https://doi.org/10.1105/tpc.160771.

Brouard, J. S., Otis, C., Lemieux, C., & Turmel, M. (2008). Chloroplast DNA sequence of the green alga *Oedogonium cardiacum* (Chlorophyceae): Unique genome architecture, derived characters shared with the Chaetophorales and novel genes acquired through horizontal transfer. *BMC Genomics*, *9*, 290. https://doi.org/10.1186/1471-2164-9-290.

Brouard, J. S., Otis, C., Lemieux, C., & Turmel, M. (2010). The exceptionally large chloroplast genome of the green alga *Floydiella terrestris* illuminates the evolutionary history of the Chlorophyceae. *Genome Biology and Evolution*, *2*, 240–256. https://doi.org/10.1093/gbe/evq014.

Brouard, J. S., Otis, C., Lemieux, C., & Turmel, M. (2011). The chloroplast genome of the green alga *Schizomeris leibleinii* (Chlorophyceae) provides evidence for bidirectional DNA replication from a single origin in the Chaetophorales. *Genome Biology and Evolution*, *3*, 505–515. https://doi.org/10.1093/gbe/evr037.

Brouard, J. S., Turmel, M., Otis, C., & Lemieux, C. (2016). Proliferation of group II introns in the chloroplast genome of the green alga *Oedocladium carolinianum* (Chlorophyceae). *PeerJ*, *4*, e2627. https://doi.org/10.7717/peerj.2627.

Cavalier-Smith, T. (2005). Economy, speed and size matter: Evolutionary forces driving nuclear genome miniaturization and expansion. *Annals of Botany*, *95*(1), 147–175. https://doi.org/10.1093/aob/mci010.

Christensen, A. C. (2014). Genes and junk in plant mitochondria-repair mechanisms and selection. *Genome Biology and Evolution*, *6*(6), 1448–1453. https://doi.org/10.1093/gbe/evu115.

Civan, P., Foster, P. G., Embley, M. T., Seneca, A., & Cox, C. J. (2014). Analyses of charophyte chloroplast genomes help characterize the ancestral chloroplast genome of land plants. *Genome Biology and Evolution*, *6*(4), 897–911. https://doi.org/10.1093/gbe/evu061.

de Cambiaire, J.-C., Otis, C., Lemieux, C., & Turmel, M. (2006). The complete chloroplast genome sequence of the chlorophycean green alga *Scenedesmus obliquus* reveals a compact gene organization and a biased distribution of genes on the two DNA strands. *BMC Evolutionary Biology*, *6*, 37.

de Cambiaire, J. C., Otis, C., Turmel, M., & Lemieux, C. (2007). The chloroplast genome sequence of the green alga *Leptosira terrestris*: Multiple losses of the inverted repeat and extensive genome rearrangements within the Trebouxiophyceae. *BMC Genomics*, *8*, 213. https://doi.org/10.1186/1471-2164-8-213.

Del Vasto, M., Figueroa-Martinez, F., Featherston, J., Gonzalez, M. A., Reyes-Prieto, A., Durand, P. M., et al. (2015). Massive and widespread organelle genomic expansion in the green algal genus *Dunaliella*. *Genome Biology and Evolution*, 7(3), 656–663. https://doi.org/10.1093/gbe/evv027.

Fang, L., Leliaert, F., Zhang, Z.-H., Penny, D., & Zhong, B.-J. (2017). Evolution of the Chlorophyta: Insights from chloroplast phylogenomic analyses. *Journal of Systematics and Evolution*, *55*, 322–332. https://doi.org/10.1111/jse.12248.

Featherston, J., Arakaki, Y., Nozaki, H., Durand, P. M., & Smith, D. R. (2016). Inflated organelle genomes and a circular-mapping mtDNA probably existed at the origin of coloniality in volvocine green algae. *European Journal of Phycology*, *51*(4), 369–377. https://doi.org/10.1080/09670262.2016.1198830.

Fucikova, K., Leliaert, F., Cooper, E. D., Skaloud, P., D'hondt, S., De Clerck, O., et al. (2014). New phylogenetic hypotheses for the core Chlorophyta based on chloroplast sequence data. *Frontiers in Ecology and Evolution*, *2*, 63. https://doi.org/10.3389/fevo.2014.00063.

Fucikova, K., Lewis, L. A., & Lewis, P. O. (2016). Comparative analyses of chloroplast genome data representing nine green algae in Sphaeropleales (Chlorophyceae, Chlorophyta). *Data in Brief*, 7, 558–570. https://doi.org/10.1016/j.dib.2016.03.014.

Furukawa, R., Kunugi, M., Ihara, K., Takabayashi, A., & Tanaka, A. (2017). Complete chloroplast genome sequence of the early diverging green alga *Palmophyllum crassum*. *Genome Announcements*, *5*(10), e01745-16. https://doi.org/10.1128/genomeA.01745-16.

Goldschmidt-Clermont, M., Choquet, Y., Girard-Bascou, J., Michel, F., Schirmer-Rahire, M., & Rochaix, J. D. (1991). A small chloroplast RNA may be required for *trans*-splicing in Chlamydomonas reinhardtii. *Cell*, *65*(1), 135–143.

Gontcharov, A. (2008). Phylogeny and classification of Zygnematophyceae (Streptophyta): Current state of affairs. *Fottea*, *8*(2), 87–104. https://doi.org/10.5507/fot.2008.004.

Goulding, S. E., Olmstead, R. G., Morden, C. W., & Wolfe, K. H. (1996). Ebb and flow of the chloroplast inverted repeat. *Molecular & General Genetics*, *252*(1–2), 195–206.

Graham, L. E., Cook, M. E., & Busse, J. S. (2000). The origin of plants: Body plan changes contributing to a major evolutionary radiation. *Proceedings of the National Academy of Sciences of the United States of America*, *97*(9), 4535–4540.

Guisinger, M. M., Kuehl, J. V., Boore, J. L., & Jansen, R. K. (2011). Extreme reconfiguration of plastid genomes in the angiosperm family Geraniaceae: Rearrangements, repeats, and codon usage. *Molecular Biology and Evolution*, *28*(1), 583–600. https://doi.org/10.1093/molbev/msq229.

Hovde, B. T., Starkenburg, S. R., Hunsperger, H. M., Mercer, L. D., Deodato, C. R., Jha, R. K., et al. (2014). The mitochondrial and chloroplast genomes of the haptophyte *Chrysochromulina tobin* contain unique repeat structures and gene profiles. *BMC Genomics, 15*, 604. https://doi.org/10.1186/1471-2164-15-604.

Jansen, R. K., & Ruhlman, T. A. (2012). Plastid genomes of seed plants. In R. Bock & V. Knoop (Eds.), *Genomics of chloroplasts and mitochondria: Vol. 35* (pp. 103–126). Netherlands: Springer.

Lambowitz, A. M., & Belfort, M. (2015). Mobile bacterial group II introns at the crux of eukaryotic evolution. *Microbiology Spectrum, 3*(1), 1–26. MDNA3-0050-2014 https://doi.org/10.1128/microbiolspec.MDNA3-0050-2014.

Lang, B. F., & Nedelcu, A. M. (2012). Plastid genomes of algae. In R. Bock & V. Knoop (Eds.), *Genomics of chloroplasts and mitochondria: Vol. 35* (pp. 59–87). Netherlands: Springer.

Laurin-Lemay, S., Brinkmann, H., & Philippe, H. (2012). Origin of land plants revisited in the light of sequence contamination and missing data. *Current Biology, 22*(15), R593–R594. https://doi.org/10.1016/J.Cub.2012.06.013.

Leliaert, F., & Lopez-Bautista, J. M. (2015). The chloroplast genomes of *Bryopsis plumosa* and *Tydemania expeditiones* (Bryopsidales, Chlorophyta): Compact genomes and genes of bacterial origin. *BMC Genomics, 16*, 204.

Leliaert, F., Smith, D. R., Moreau, H., Herron, M. D., Verbruggen, H., Delwiche, C. F., et al. (2012). Phylogeny and molecular evolution of the green algae. *Critical Reviews in Plant Sciences, 31*, 1–46.

Leliaert, F., Tronholm, A., Lemieux, C., Turmel, M., DePriest, M. S., Bhattacharya, D., et al. (2016). Chloroplast phylogenomic analyses reveal the deepest-branching lineage of the Chlorophyta, Palmophyllophyceae class. nov. *Scientific Reports 6*, 25367. https://doi.org/10.1038/srep25367.

Lemieux, C., Otis, C., & Turmel, M. (2000). Ancestral chloroplast genome in *Mesostigma viride* reveals an early branch of green plant evolution. *Nature, 403*(6770), 649–652. https://doi.org/10.1038/35001059.

Lemieux, C., Otis, C., & Turmel, M. (2007). A clade uniting the green algae *Mesostigma viride* and *Chlorokybus atmophyticus* represents the deepest branch of the Streptophyta in chloroplast genome-based phylogenies. *BMC Biology, 5*, 2. https://doi.org/10.1186/1741-7007-5-2.

Lemieux, C., Otis, C., & Turmel, M. (2014a). Chloroplast phylogenomic analysis resolves deep-level relationships within the green algal class Trebouxiophyceae. *BMC Evolutionary Biology, 14*, 211. https://doi.org/10.1186/s12862-014-0211-2.

Lemieux, C., Otis, C., & Turmel, M. (2014b). Six newly sequenced chloroplast genomes from prasinophyte green algae provide insights into the relationships among prasinophyte lineages and the diversity of streamlined genome architecture in picoplanktonic species. *BMC Genomics, 15*, 857. https://doi.org/10.1186/1471-2164-15-857.

Lemieux, C., Otis, C., & Turmel, M. (2016). Comparative chloroplast genome analyses of streptophyte green algae uncover major structural alterations in the Klebsormidiophyceae, Coleochaetophyceae and Zygnematophyceae. *Frontiers in Plant Science, 7*, 697. https://doi.org/10.3389/fpls.2016.00697.

Lemieux, C., Vincent, A. T., Labarre, A., Otis, C., & Turmel, M. (2015). Chloroplast phylogenomic analysis of chlorophyte green algae identifies a novel lineage sister to the Sphaeropleales (Chlorophyceae). *BMC Evolutionary Biology, 15*, 264. https://doi.org/10.1186/s12862-015-0544-5.

Lewis, L. A., & McCourt, R. M. (2004). Green algae and the origin of land plants. *American Journal of Botany, 91*(10), 1535–1556. https://doi.org/10.3732/ajb.91.10.1535.

Lynch, M. (2007). *The origins of genome architecture*. Sunderland, MA: Sinauer Associates.

Marcelino, V. R., Cremen, M. C., Jackson, C. J., Larkum, A. A., & Verbruggen, H. (2016). Evolutionary dynamics of chloroplast genomes in low light: A case study of the endolithic green alga *Ostreobium quekettii*. *Genome Biology and Evolution*, *8*(9), 2939–2951. https://doi.org/10.1093/gbe/evw206.

Marin, B. (2012). Nested in the Chlorellales or independent class? Phylogeny and classification of the Pedinophyceae (Viridiplantae) revealed by molecular phylogenetic analyses of complete nuclear and plastid-encoded rRNA operons. *Protist*, *163*(5), 778–805. https://doi.org/10.1016/j.protis.2011.11.004.

Maul, J. E., Lilly, J. W., Cui, L., dePamphilis, C. W., Miller, W., Harris, E. H., et al. (2002). The *Chlamydomonas reinhardtii* plastid chromosome: Islands of genes in a sea of repeats. *The Plant Cell*, *14*(11), 2659–2679.

McCourt, R. M., Delwiche, C. F., & Karol, K. G. (2004). Charophyte algae and land plant origins. *Trends in Ecology & Evolution*, *19*(12), 661–666. https://doi.org/10.1016/j.tree.2004.09.013.

Melton, J. T., 3rd, Leliaert, F., Tronholm, A., & Lopez-Bautista, J. M. (2015). The complete chloroplast and mitochondrial genomes of the green macroalga *Ulva* sp. UNA00071828 (Ulvophyceae, Chlorophyta). *PLoS One*, *10*(4), e0121020. https://doi.org/10.1371/journal.pone.0121020.

Palmer, J. D. (1983). Chloroplast DNA exists in two orientations. *Nature*, *301*(5895), 92–93.

Palmer, J. D. (1991). Plastid chromosomes: structure and evolution. In I. K. Vasil (Ed.), *The molecular biology of plastids* (pp. 5–53). Academic Press.

Pombert, J. F., Lemieux, C., & Turmel, M. (2006). The complete chloroplast DNA sequence of the green alga *Oltmannsiellopsis viridis* reveals a distinctive quadripartite architecture in the chloroplast genome of early diverging ulvophytes. *BMC Biology*, *4*, 3. https://doi.org/10.1186/1741-7007-4-3.

Pombert, J. F., Otis, C., Lemieux, C., & Turmel, M. (2005). The chloroplast genome sequence of the green alga *Pseudendoclonium akinetum* (Ulvophyceae) reveals unusual structural features and new insights into the branching order of chlorophyte lineages. *Molecular Biology and Evolution*, *22*(9), 1903–1918.

Ponce-Toledo, R. I., Deschamps, P., Lopez-Garcia, P., Zivanovic, Y., Benzerara, K., & Moreira, D. (2017). An early-branching freshwater Cyanobacterium at the origin of plastids. *Current Biology*, *27*(3), 386–391. https://doi.org/10.1016/j.cub.2016.11.056.

Raubeson, L. A., & Jansen, R. K. (2005). Chloroplast genomes of plants. In R. J. Henry (Ed.), *Plant diversity and evolution. Genotypic and phenotypic variation in higher plants* (pp. 45–68). Cambridge University Press.

Reyes-Prieto, A., Weber, A. P., & Bhattacharya, D. (2007). The origin and establishment of the plastid in algae and plants. *Annual Review of Genetics*, *41*, 147–168. https://doi.org/10.1146/annurev.genet.41.110306.130134.

Robbens, S., Derelle, E., Ferraz, C., Wuyts, J., Moreau, H., & Van de Peer, Y. (2007). The complete chloroplast and mitochondrial DNA sequence of *Ostreococcus tauri*: Organelle genomes of the smallest eukaryote are examples of compaction. *Molecular Biology and Evolution*, *24*(4), 956–968. https://doi.org/10.1093/molbev/msm012.

Sanchez-Baracaldo, P., Raven, J. A., Pisani, D., & Knoll, A. H. (2017). Early photosynthetic eukaryotes inhabited low-salinity habitats. *Proceedings of the National Academy of Sciences of the United States of America*, *114*(37), E7737–E7745. https://doi.org/10.1073/pnas.1620089114.

Servin-Garciduenas, L. E., & Martinez-Romero, E. (2012). Complete mitochondrial and plastid genomes of the green microalga Trebouxiophyceae sp. strain MX-AZ01 isolated from a highly acidic geothermal lake. *Eukaryotic Cell*, *11*(11), 1417–1418. https://doi.org/10.1128/EC.00244-12.

Smith, D. R. (2012). Updating our view of organelle genome nucleotide landscape. *Frontiers in Genetics*, *3*, 175. https://doi.org/10.3389/fgene.2012.00175.

Smith, D. R. (2016). The mutational hazard hypothesis of organelle genome evolution: 10 years on. *Molecular Ecology*, *25*(16), 3769–3775. https://doi.org/10.1111/mec.13742.

Smith, D. R., Arrigo, K. R., Alderkamp, A. C., & Allen, A. E. (2014). Massive difference in synonymous substitution rates among mitochondrial, plastid, and nuclear genes of *Phaeocystis* algae. *Molecular Phylogenetics and Evolution*, *71*, 36–40. https://doi.org/10.1016/j.ympev.2013.10.018.

Smith, D. R., Burki, F., Yamada, T., Grimwood, J., Grigoriev, I. V., Van Etten, J. L., et al. (2011). The GC-rich mitochondrial and plastid genomes of the green alga *Coccomyxa* give insight into the evolution of organelle DNA nucleotide landscape. *PLoS One*, *6*(8), e23624. https://doi.org/10.1371/journal.pone.0023624.

Smith, D. R., Hamaji, T., Olson, B. J., Durand, P. M., Ferris, P., Michod, R. E., et al. (2013). Organelle genome complexity scales positively with organism size in volvocine green algae. *Molecular Biology and Evolution*, *30*(4), 793–797. https://doi.org/10.1093/molbev/mst002.

Smith, D. R., & Keeling, P. J. (2015). Mitochondrial and plastid genome architecture: Reoccurring themes, but significant differences at the extremes. *Proceedings of the National Academy of Sciences of the United States of America*, *112*(33), 10177–10184. https://doi.org/10.1073/pnas.1422049112.

Smith, D. R., & Lee, R. W. (2010). Low nucleotide diversity for the expanded organelle and nuclear genomes of Volvox carteri supports the mutational-hazard hypothesis. *Molecular Biology and Evolution*, *27*(10), 2244–2256. https://doi.org/10.1093/molbev/msq110.

Smith, D. R., Lee, R. W., Cushman, J. C., Magnuson, J. K., Tran, D., & Polle, J. E. (2010). The *Dunaliella salina* organelle genomes: Large sequences, inflated with intronic and intergenic DNA. *BMC Plant Biology*, *10*, 83. https://doi.org/10.1186/1471-2229-10-83.

Sun, L., Fang, L., Zhang, Z., Chang, X., Penny, D., & Zhong, B. (2016). Chloroplast phylogenomic inference of green algae relationships. *Scientific Reports*, *6*, 20528. https://doi.org/10.1038/srep20528.

Timme, R. E., Bachvaroff, T. R., & Delwiche, C. F. (2012). Broad phylogenomic sampling and the sister lineage of land plants. *PLoS One*, 7(1), e29696. https://doi.org/10.1371/journal.pone.0029696.

Turmel, M., de Cambiaire, J. C., Otis, C., & Lemieux, C. (2016). Distinctive architecture of the chloroplast genome in the chlorodendrophycean green algae *Scherffelia dubia* and *Tetraselmis* sp. CCMP 881. *PLoS One*, *11*(2), e0148934. https://doi.org/10.1371/journal.pone.0148934.

Turmel, M., Gagnon, M. C., O'Kelly, C. J., Otis, C., & Lemieux, C. (2009). The chloroplast genomes of the green algae *Pyramimonas, Monomastix*, and *Pycnococcus* shed new light on the evolutionary history of prasinophytes and the origin of the secondary chloroplasts of euglenids. *Molecular Biology and Evolution*, *26*(3), 631–648. https://doi.org/10.1093/molbev/msn285.

Turmel, M., Otis, C., & Lemieux, C. (1999). The complete chloroplast DNA sequence of the green alga *Nephroselmis olivacea*: Insights into the architecture of ancestral chloroplast genomes. *Proceedings of the National Academy of Sciences of the United States of America*, *96*, 10248–10253.

Turmel, M., Otis, C., & Lemieux, C. (2005). The complete chloroplast DNA sequences of the charophycean green algae *Staurastrum* and *Zygnema* reveal that the chloroplast genome underwent extensive changes during the evolution of the Zygnematales. *BMC Biology*, *3*, 22. https://doi.org/10.1186/1741-7007-3-22.

Turmel, M., Otis, C., & Lemieux, C. (2006). The chloroplast genome sequence of *Chara vulgaris* sheds new light into the closest green algal relatives of land plants. *Molecular Biology and Evolution*, *23*(6), 1324–1338. https://doi.org/10.1093/molbev/msk018.

Turmel, M., Otis, C., & Lemieux, C. (2013). Tracing the evolution of streptophyte algae and their mitochondrial genome. *Genome Biology and Evolution*, *5*(10), 1817–1835. https://doi.org/10.1093/gbe/evt135.

Turmel, M., Otis, C., & Lemieux, C. (2015). Dynamic evolution of the chloroplast genome in the green algal classes Pedinophyceae and Trebouxiophyceae. *Genome Biology and Evolution*, 7(7), 2062–2082. https://doi.org/10.1093/gbe/evv130.

Turmel, M., Otis, C., & Lemieux, C. (2016). Mitochondrion-to-chloroplast DNA transfers and intragenomic proliferation of chloroplast group II introns in *Gloeotilopsis* green algae (Ulotrichales, Ulvophyceae). *Genome Biology and Evolution*, *8*(9), 2789–2805. https://doi.org/10.1093/gbe/evw190.

Turmel, M., Otis, C., & Lemieux, C. (2017). Divergent copies of the large inverted repeat in the chloroplast genomes of ulvophycean green algae. *Scientific Reports*, 7(1), 994. https://doi.org/10.1038/s41598-017-01144-1.

Turmel, M., Pombert, J. F., Charlebois, P., Otis, C., & Lemieux, C. (2007). The green algal ancestry of land plants as revealed by the chloroplast genome. *International Journal of Plant Sciences*, *168*(5), 679–689. https://doi.org/10.1086/513470.

Umen, J. G. (2014). Green algae and the origins of multicellularity in the plant kingdom. *Cold Spring Harbor Perspectives in Biology*, *6*(11), a016170. https://doi.org/10.1101/cshperspect.a016170.

Wakasugi, T., Nagai, T., Kapoor, M., Sugita, M., Ito, M., Ito, S., et al. (1997). Complete nucleotide sequence of the chloroplast genome from the green alga *Chlorella vulgaris*: The existence of genes possibly involved in chloroplast division. *Proceedings of the National Academy of Sciences of the United States of America*, *94*(11), 5967–5972.

Watanabe, S., Fucikova, K., Lewis, L. A., & Lewis, P. O. (2016). Hiding in plain sight: *Koshicola spirodelophila* gen. et sp. nov. (Chaetopeltidales, Chlorophyceae), a novel green alga associated with the aquatic angiosperm *Spirodela polyrhiza*. *American Journal of Botany*, *103*(5), 865–875. https://doi.org/10.3732/ajb.1500481.

Weng, M. L., Blazier, J. C., Govindu, M., & Jansen, R. K. (2014). Reconstruction of the ancestral plastid genome in Geraniaceae reveals a correlation between genome rearrangements, repeats, and nucleotide substitution rates. *Molecular Biology and Evolution*, *31*(3), 645–659. https://doi.org/10.1093/molbev/mst257.

Wicke, S., Schneeweiss, G. M., dePamphilis, C. W., Muller, K. F., & Quandt, D. (2011). The evolution of the plastid chromosome in land plants: Gene content, gene order, gene function. *Plant Molecular Biology*, *76*(3–5), 273–297. https://doi.org/10.1007/s11103-011-9762-4.

Wodniok, S., Brinkmann, H., Glockner, G., Heidel, A. J., Philippe, H., Melkonian, M., et al. (2011). Origin of land plants: Do conjugating green algae hold the key? *BMC Evolutionary Biology*, *11*, 104. https://doi.org/10.1186/1471-2148-11-104.

Zoschke, R., Nakamura, M., Liere, K., Sugiura, M., Borner, T., & Schmitz-Linneweber, C. (2010). An organellar maturase associates with multiple group II introns. *Proceedings of the National Academy of Sciences of the United States of America*, *107*(7), 3245–3250. https://doi.org/10.1073/pnas.0909400107.

CHAPTER SEVEN

Evolution of Gymnosperm Plastid Genomes

Shu-Miaw Chaw[*,†,1], **Chung-Shien Wu**[*], **Edi Sudianto**[*,†,‡]
[*]Biodiversity Research Center, Academia Sinica, Taipei, Taiwan
[†]Biodiversity Program, Taiwan International Graduate Program, Academia Sinica and National Taiwan Normal University, Taipei, Taiwan
[‡]National Taiwan Normal University, Taipei, Taiwan
[1]Corresponding author: e-mail address: smchaw@sinica.edu.tw

Contents

Abstract

The rapid increase in plastome availability on GenBank has greatly deepened our understanding of plastomic evolution and plastid phylogenomics in gymnosperms. The plastomes of the five extant gymnosperm groups show distinctive evolutionary

Advances in Botanical Research, Volume 85
ISSN 0065-2296
https://doi.org/10.1016/bs.abr.2017.11.018

patterns. For example, those of cycads are conserved in architecture, gene content, and nucleotide substitution rates. Compared to cycads, the plastome of ginkgo has its inverted repeats (IRs) slightly contracted. The IRs of the three gnetophyte genera, represented by *Ephedra*, *Gnetum*, and *Welwitschia*, have undergone multiple expansions, contractions, and inversions. Meanwhile, the highly rearranged plastomes of Pinaceae and cupressophytes lack canonical IRs and contain lineage-specific repeats that trigger the generation of isomeric plastomes. In terms of nucleotide substitution rates, the plastome of ginkgo features an extremely slow rate of nucleotide substitutions, similar to those of cycads. In contrast, the plastomes of gnetophytes have relatively accelerated rates of nucleotide substitutions. Comparatively, nucleotide substitution rates in the plastomes of Pinaceae and cupressophytes are faster than cycads and ginkgo, but slower than those of gnetophytes. In this chapter, we summarize the progression of these findings and discuss potential causes for the variation in gymnosperms. We also review the use of these plastomes for resolving long-standing issues in seed plant and gymnosperm phylogenies. We conclude this chapter with some future directions for plastomic studies in gymnosperms.

1. INTRODUCTION

Gymnosperms, a class of seed-bearing plants, consist of 1079 species in 83 genera and 12 families (Christenhusz & Byng, 2016). In contrast to flowering plants (or angiosperms), the seeds of gymnosperms grow on the surface of scales or leaves, not enclosed within an ovary (which usually develops into fruits), and are therefore called "naked seeds". Molecular studies (Chaw, Zharkikh, Sung, Lau, & Li, 1997; Rai, Reeves, Peakall, Olmstead, & Graham, 2008) divided gymnosperms into five groups—cycads, ginkgo, gnetophytes, pines (conifers I), and cupressophytes (conifers II). Cupressophytes contain about 405 species in 5 families, including Araucariaceae, Cupressaceae sensu *lato* (here including also Taxodiaceae), Podocarpaceae, Sciadopityaceae, and Taxaceae (Gernandt, Willyard, Syring, & Liston, 2011). They dominate huge terrestrial areas in the Northern Hemisphere, but most genera in Araucariaceae and Podocarpaceae are mainly in the Southern Hemisphere (Williams, 2009). Cupressophytes are of economic and ecological importance. Many species of Cupressaceae (cypress family), including arborvitae (*Thuja*), bald cypress (*Taxodium*), China fir (*Cunninghamia*), and false cypress (*Chamaecyparis*), are valuable as timber sources or ornamentals. The Taxaceae (yew family), with about 30 species in 6 genera, are renown for the efficacy of taxane compounds in anticancer therapies.

In 1994, the first complete plastid genome (plastome) of a gymnosperm—*Pinus thunbergii* (black pine)—was sequenced (Wakasugi et al., 1994). However, gymnosperm plastomes were not compared until the first cycad plastome, *Cycas taitungensis*, was reported (Wu, Wang, Liu, & Chaw, 2007). The recent advent of high-throughput next-generation sequencing (NGS) has allowed plastomes to be sequenced and made publicly available at a higher rate. NGS methods particularly facilitated the assembly of plastome sequences from total DNA (Nock et al., 2011; Wu et al., 2012). As of May 25, 2017, 100 gymnosperm plastomes are available on GenBank, representing all 12 recognized families (Fig. 1).

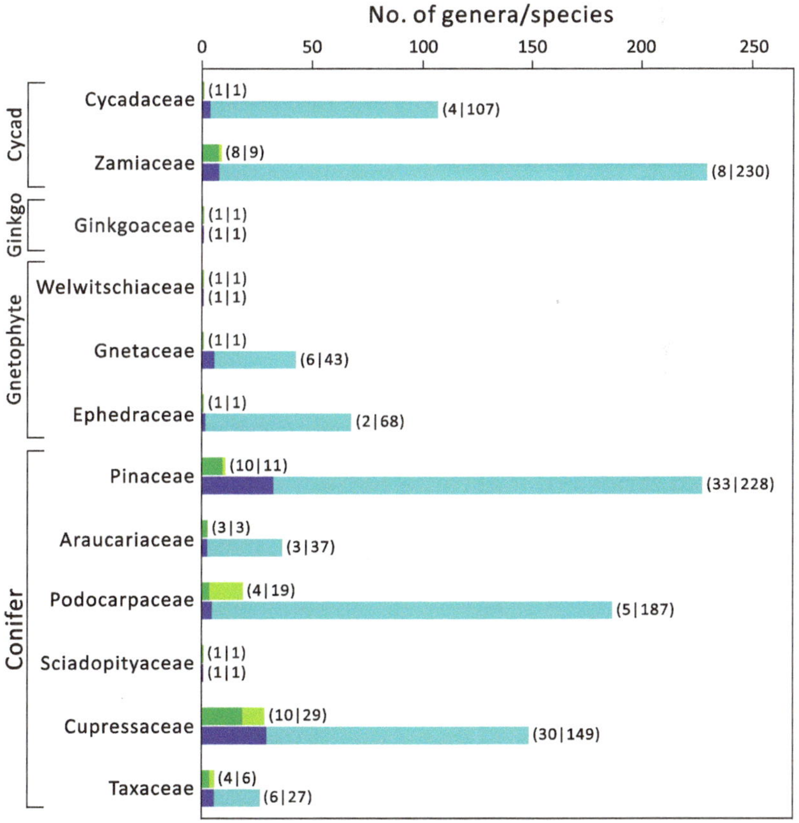

Fig. 1 *Stacked bars* showing the plastomes of gymnosperms publicly available on GenBank. For each family, genera with at least one (*green bar*) or none (*light green bar*) of the representative plastomes are shown. The *blue* and *light blue bars* denote sequenced plastomes and the remaining species, respectively. *Numbers in parentheses* are available genera/species vs total number of genera/species in families.

However, some issues persist. Despite tremendous efforts to determine the gymnosperm plastome, some families remain poorly sampled at the generic/species level. For example, only 10 of the 29 genera in Cupressaceae and 4 of the 19 genera in Podocarpaceae have their plastomes available on GenBank (Fig. 1). At the species level, none of the gymnosperm families have more than 50% of their species sequenced, except for the monotypic families Ginkgoaceae, Welwitschiaceae, and Sciadopityaceae (Fig. 1). In addition, sampling bias was observed within some families. For example, Pinaceae contains 11 genera. However, 17 of the 33 sequenced Pinaceous species were from the genus *Pinus*, while the other 16 species were sampled from 10 other genera.

In the past decade, plastomic characteristics have been reviewed in land plants (Daniell, Lin, Yu, & Chang, 2016; Wicke, Schneeweiss, dePamphilis, Muller, & Quandt, 2011), ferns (Wolf et al., 2011), seed plants (Jansen & Ruhlman, 2012), and flowering plants (Ruhlman & Jansen, 2014). However, we lack an overall review of plastome evolution in gymnosperms. In this chapter, we summarize the advances in sequencing methods, variation in plastome size and architecture, the evolution of nucleotide substitution rates, and plastid phylogenomic approaches in addressing the phylogenies within extant seed plant and gymnosperm lineages.

2. SEQUENCING THE PLASTOMES OF GYMNOSPERMS

2.1 Advances in Plastome Sequencing

Before the advent of NGS technology, sequencing entire plastomes was labour-intensive, with three basic steps: random shearing of plastid DNA (ptDNA), DNA cloning, and sequence determination by use of Maxam–Gilbert (e.g. *P. thunbergii*: Wakasugi et al., 1994) or Sanger sequencing (e.g. *C. taitungensis*: Wu et al., 2007; gnetophytes: McCoy, Kuehl, Boore, & Raubeson, 2008; Wu, Lai, Lin, Wang, & Chaw, 2009; some conifer species: Hirao, Watanabe, Kurita, Kondo, & Takata, 2008; Lin, Huang, Wu, Hsu, & Chaw, 2010; Wu, Lin, Hsu, Wang, & Chaw, 2011; Wu, Wang, Hsu, Lin, & Chaw, 2011). In late 2008 (Cronn et al., 2008), NGS was first used to decipher the plastomes of eight *Pinus* species, and a number of advantages over Sanger sequencing were highlighted, including multiplex sequencing, high sequence depth, and low error rate compared with Sanger sequencing. Here, we outline DNA preparation strategies for NGS, focusing on sequencing plastomes from total genomic DNA (gDNA).

2.2 Enrichment of ptDNA via Plastid Isolation or PCR

Pure plastid DNA (ptDNA) is an ideal resource for sequencing entire plastomes. Extracting pure DNA from plastids relies on three processes: (1) separation of plastids from other cellular organelles and debris, (2) lysis of the plastids, and (3) extraction of DNA. The method for isolating ptDNA was reviewed in Jansen et al. (2005), and more recently, a modified protocol was proposed for isolating conifer ptDNA (Vieira, Faoro, Fraga, et al., 2014). However, obtaining pure ptDNA has proved difficult, even with the modified protocol. For example, although the protocol described in Vieira, Faoro, Fraga, et al. (2014) was adopted for ptDNA isolation of *Callitris sulcate*, only 10% of the Illumina reads came from plastome sequences (Sakaguchi et al., 2017). Some studies used PCR amplicons from gymnosperm ptDNAs to obtain NGS data (e.g. Cronn et al., 2008; Lin, Wu, Huang, & Chaw, 2012; Ruhsam et al., 2015). However, there are limitations to enriching ptDNA via PCR. First, published primers may not apply to all gymnosperm species, so species-specific primers need to be designed. Second, the conifer plastomes are highly rearranged (Wu & Chaw, 2014, 2016; Wu, Wang, et al., 2011), and hence it is often difficult to predict the relative gene order for primer design. Third, the existence of isomeric plastomes (Guo et al., 2014; Qu, Wu, Chaw, & Yi, 2017; Vieira et al., 2016) may be overlooked.

2.3 Recovery of Plastomes From Total gDNA

A number of publicly available gymnosperm plastomes were recovered from gDNA, including some cycads (Jiang, Hinsinger, & Strijk, 2016; Wu & Chaw, 2015), gnetophytes (Zhu, Guo, Gupta, Fan, & Mower, 2016), Pinaceae (Jackman et al., 2015; Sudianto, Wu, Lin, & Chaw, 2016; Sullivan, Schiffthaler, Thompson, Street, & Wang, 2017; Whittall et al., 2010), and cupressophytes (Guo et al., 2014; Hsu, Wu, & Chaw, 2014, 2016; Li, Gao, et al., 2016; Wu & Chaw, 2014, 2016; Yi, Gao, Wang, Su, & Wang, 2013). Nonetheless, recovery of the entire plastomic sequence from gDNA requires intensive sequencing (Du et al., 2015). The minimum number of NGS reads required to recover a high-quality gymnosperm plastome has never been assessed. The nuclear genomes of gymnosperms vary from 2.3 picograms per haploid (pg/C) to 36 pg/C, with a mean of 18.5 pg/C (Gregory et al., 2007). We examined the plastome recovery rate for different numbers of NGS reads sequenced from gDNA; we used *Amentotaxus formosana* as an example because its nuclear genome (30 pg/1C) is one of the largest among gymnosperms.

Fifty nanograms of the gDNA were used to construct a paired-end library with an insertion size of 350 bp. The sequencing was performed on the Illumina HiSeq 2000 with a read length of 2 × 90 bp. After quality trimming, the reads were randomly extracted to create four datasets of 0.36, 0.72, 1.08, and 1.44 gigabases (Gb). De novo assembly and sequence mapping involved the use of CLC Genomics Workbench 4.9 (CLC Bio, Arhus, Denmark). Contigs <1 kb were discarded. The publicly available plastome of *A. formosana* (NC_024945) was used as the reference to identify plastid contigs and the plastid contigs were mapped onto it (Fig. 2). Our results showed that increased data greatly improved the sequence coverage and that all plastid contigs assembled from a dataset of 0.72 Gb had more than 30 × coverage, which is the proposed threshold for plastome assembly

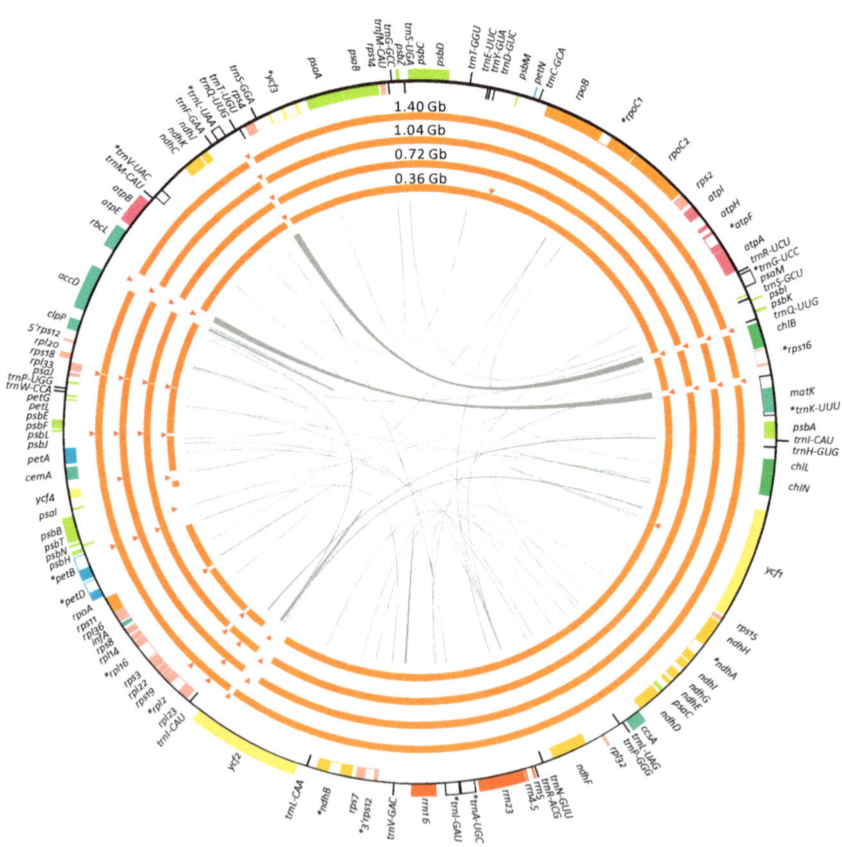

Fig. 2 An assembly of the plastome from *Amentotaxus formosana* using different amounts of paired-end reads. The *outmost circle* is the plastome map. Contigs are denoted by *orange bars*. *Gaps* between contigs are shown with *red triangles*. Dispersed copies of repeats are linked with *grey lines*.

(Straub et al., 2012). With data increase from 0.36 to 1.44 Gb, the yielded plastid contigs accounted for 92.1%–97.56% of the referenced plastome. Gaps were often generated in the regions containing repeats, regardless of how much data there were for assembly (Fig. 2). These gaps are expected to be closed with the use of a longer insertion size or the PacBio sequencing platform.

3. PLASTOME CHARACTERISTICS IN GYMNOSPERMS

3.1 Plastome Architecture

Plastome architecture is variable among the five gymnosperm groups. In cycads, ginkgo, and gnetophytes, it is quadripartite with a pair of large inverted repeats (IRs) separated by a large single-copy (LSC) region and a small single-copy (SSC) region. The IR is characterized by the core unit of four ribosomal RNA (rRNA) genes (i.e. *rrn4.5*, *rrn5*, *rrn16*, and *rrn23*). In addition to the four rRNA genes, the ancestral IR of gymnosperms might also contain *trnN*^{GUU}, *trnR*^{ACG}, *trnA*^{UGC}, *trnI*^{GAU}, *trnV*^{GAC}, *3′rps12*, *rps7*, *ndhB*, *trnL*^{CAA}, *ycf2*, and *trnH*^{GUG} (Zhu et al., 2016). However, all conifer plastomes lack IRs (Raubeson & Jansen, 1992). Comparative analyses of plastomes suggest that Pinaceae and cupressophytes independently lost their IRs (Hao et al., 2016; Wu & Chaw, 2014; Wu, Lin, et al., 2011). In contrast, Yi et al. (2013) suggested that it was difficult to clarify which IR copy was lost from Pinaceae when the presence of plastome isomers were taken into consideration. Therefore, more comprehensive data and methods are required to evaluate the evolutionary process of IR loss in conifers. Some regions were hypothesized to be IR residues in conifer plastomes. For example, in Pinaceae, the region that includes *trnI*^{CAU} and 3′*psbA* was recognized as the highly reduced IR (Lin et al., 2010; Sudianto et al., 2016; Tsudzuki et al., 1992; Wu, Wang, et al., 2011). In cupressophytes such as *Cryptomeria japonica*, the two inverted copies of *trnI*^{CAU} were thought to be the IR residues (Hirao et al., 2008). Because these IR residues are relatively short, ranging from 114 (*C. japonica*) to 495 bp (*P. thunbergii*), the conifer plastomes are not considered to have a quadripartite structure.

Genes encoded in plastomes were classified as protein-coding, rRNA, and transfer RNA (tRNA) genes. Most of these were grouped into several conserved gene clusters to facilitate cotranscription of the genes. Constraints on the gene clusters were proposed to be a stabilizing factor (Wicke et al., 2011). However, it was reported that those conserved gene clusters were disrupted in cupressophytes. In *Taxus mairei*, an 18-kb inversion breaks the S10 gene cluster (i.e. *rpl23*, *rpl2*, *rps19*, *rpl22*, *rps3*, *rpl14*, *rps8*, *infA*, *rpl36*, *rps11*, and *rpoA*) into two dispersed fragments, *rpl23–rps8* and *infA–rpoB* (Hsu et al., 2014). In *Sciadopitys verticillata*, recombination between the gene clusters *rps2* (i.e. *rps2*, *atpI*, *atpH*, *atpF*, and *atpA*) and *psbB* (i.e. *psbB*, *psbT*, *psbH*, *petB*, and *petD*) generated two novel and chimeric gene clusters, in which the relocated genes can be cotranscribed (Hsu et al., 2016). In the highly rearranged *Callitris rhomboidea* plastome, the *rps2* gene cluster is split into two separate fragments, *rps2–atpI* and *atpH–atpA* (Wu & Chaw, 2016). Together, four disruptions of the conserved gene clusters have been documented in the plastomes of cupressophytes.

3.2 Plastome Size and Guanine–Cytosine Content

Gymnosperm plastomes are highly variable in size, ranging from 107,122 bp (*Cathaya argyrophylla*) to 166,341 bp (*Macrozamia mountperriensis*), with a mean of 130,211 bp (Fig. 3). Several major factors contribute to this variation. First, loss of IRs has largely reduced the plastomes of conifers. Second, in gnetophytes, loss of at least 18 genes and shrinkage of introns and intergenic spacers (IGSs) have caused plastome reduction and compaction (McCoy et al., 2008; Wu et al., 2009). Third, *C. argyrophylla* has the smallest gymnosperm plastome because it has lost IRs, 11 plastid *ndh* genes, and a fragment flanked by *ycf2* and *trnV*^{GAC} (Lin et al., 2010). Fourth, ginkgo's IRs do not include *ycf2* (a gene of ~7 kb in length), so its plastome is smaller

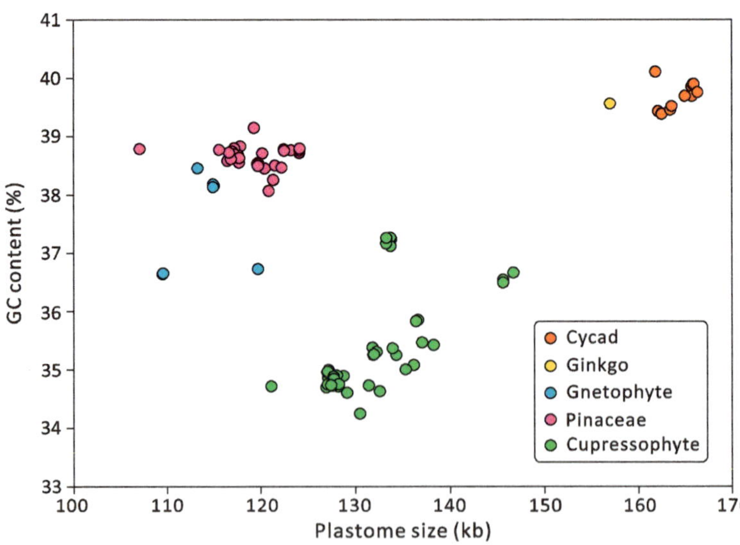

Fig. 3 A summary of plastome size and GC content across gymnosperms.

than those of cycads (Lin et al., 2012). Moreover, the cupressophyte plastome's size variation comes from different degrees of nongenic sequence deletion (Wu & Chaw, 2014, 2016).

Diverse underlying mechanisms have been proposed to explain the evolution of plastome size in gymnosperms. For example, plastome rearrangements and mutation rates together were shown to influence cupressophyte plastome size. It was proposed that the former and the latter are associated with elongating and shortening nongenic loci, respectively (Wu & Chaw, 2014). Comparative analyses across the plastomes of seed plants revealed that only cupressophytes and gnetophytes have compact plastomes (Wu & Chaw, 2016). The smaller and more compact plastomes of gnetophytes were hypothesized to be an outcome of selection for rapid replication (McCoy et al., 2008) or efficient use of crude DNA resources (Wu et al., 2009). In cupressophytes, the degree of plastome compactness is associated with synonymous substitution rates, which suggests that mutation rates play a vital role in shaping the plastome size (Wu & Chaw, 2016).

The guanine–cytosine (GC) content of gymnosperm plastomes varies considerably, from 34.24% to 40.11% (Fig. 3). Thus, the nucleotide composition is overwhelmingly GC-poor in the plastomes. However, GC content is not evenly distributed across the plastomes. Among protein-coding genes in plastids, rRNA and tRNA genes have the highest GC content, followed by protein-coding genes, then introns and IGSs (Chen et al., 2015; Li, Gao, et al., 2016; Wu & Chaw, 2014; Wu et al., 2009; Yap et al., 2015; Yi et al., 2013). In addition, GC content differs among codon positions, with the third codon position usually having a lower GC content than the other two. In terms of codon usage, AT-rich codons are predominant in the plastomes of gymnosperms, gnetophytes being the most intense case (Wu et al., 2009). GC content distribution also depends on the plastomic architecture. In the plastome of cycads, GC content is higher in IR than in LSC and SSC regions (Jiang et al., 2016; Wu & Chaw, 2015). This bias in GC content is associated with two facts. First, each of the IR copies contains four rRNA genes. Second, GC-bias gene conversion acts more frequently on IRs than LSC and SSC regions. As a result, elevated GC content was observed in IRs, but not in other regions (Wu & Chaw, 2015).

3.3 Gene Content

Gymnosperm plastomes contain 66–87 protein-coding genes, 4–8 rRNA genes, and 28–42 tRNA genes (Fig. 4). The total number of plastome genes

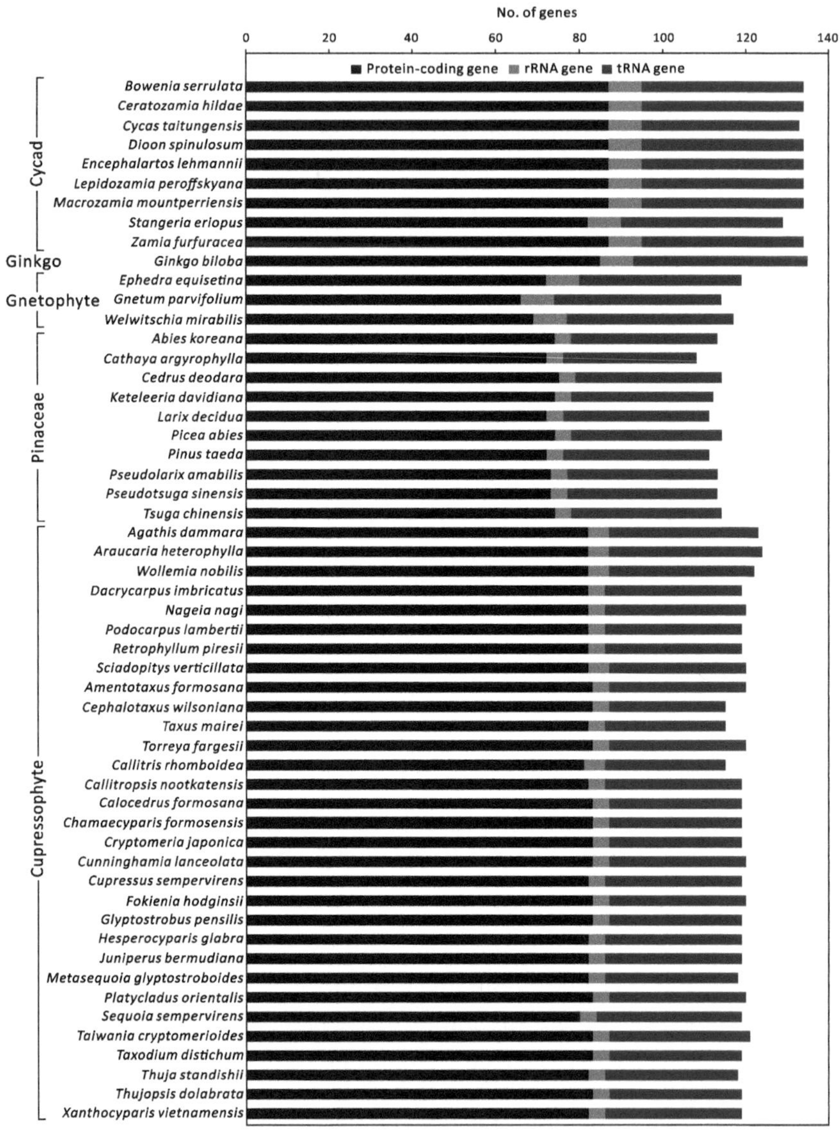

Fig. 4 A comparison of the numbers of protein-coding, rRNA, and tRNA genes among available gymnosperm plastomes.

varies from 108 (*C. argyrophylla*) to 135 (*Ginkgo biloba*). Two major factors contribute to variation in gene content. First, loss of an IR copy resulted in the removal of about 14 genes from the Pinaceae and cupressophytes. Second, gnetophytes and Pinaceae lost all 11 *ndh* genes.

In cycad plastomes, gene content is highly conserved, with two exceptions: *trnT*^{GGU} was lost from *Cycas* (Jiang et al., 2016; Wu et al., 2007) and *chlB*, *chlL*, *chlN*, *psaJ*, and *rpl23* were lost or pseudogenized in *Stangeria eriopus* (Wu & Chaw, 2015). In ginkgo, *rpl23* was pseudogenized and a copy of *ycf2* was lost because of IR contraction (Lin et al., 2012). In addition, ginkgo contains a specific cluster of three novel tRNA genes, possibly derived from the tandem duplication of *trnC*^{GCA} located in the region between *petN* and *rpoB* (Lin et al., 2012). Although plastid *tufA* was lost from the common ancestor of seed plants (Baldauf & Palmer, 1990), residual sequences from this gene were commonly found in the plastomes of cycads and ginkgo (Lin et al., 2012; Wu & Chaw, 2015; Wu et al., 2007). Therefore, pseudo-*tufA* was retained for at least 300 million years (MY) in these two lineages, possibly because the lineages' substitution rates were extremely slow compared to other genomes (Wu & Chaw, 2015).

In gnetophytes, variation in plastid gene content is mostly due to IRs contraction/expansion. The common ancestor of gnetophytes was suggested to have undergone a series of IR expansions to include *chlL*, *chlN*, *rps15*, and *rpl32*, and subsequently the former three genes were lost from the common ancestor of *Gnetum* and *Welwitschia* because of IR contraction (Wu et al., 2009). In addition, genes located in the Pinaceae-specific repeats (the so-called Type 1 repeat) vary in number from one to four (Sudianto et al., 2016; Wu, Wang, et al., 2011), which suggests that expansion/contraction of these repeats altered the gene content.

In cupressophytes, duplication of rRNA and tRNA genes largely contributes to variation in gene content. For example, duplicated *rrn5* was found only in Araucariaceae (Yap et al., 2015) and Sciadopityaceae (Hsu et al., 2016; Li, Gao, et al., 2016). Two or three copies of *trnQ*^{UUG} are present in Cupressaceae, Sciadopityaceae, and Taxaceae (Guo et al., 2014; Hsu et al., 2016; Li, Gao, et al., 2016; Qu et al., 2017), but *trnQ*^{UUG} is present as a single copy in both Araucariaceae and Podocarpaceae. In contrast, there are two or three copies of *trnD*^{GUC} in both Araucariaceae (Wu & Chaw, 2014; Yap et al., 2015) and Podocarpaceae (Vieira et al., 2016), but *trnN*^{GUU} is only duplicated in Podocarpaceae (Vieira, Faoro, Rogalski, et al., 2014; Vieira et al., 2016; Wu & Chaw, 2014, 2016). Loss of protein-coding genes is rare, but it altered gene content within cupressophytes. For example, *rps16* is present in some species of Cupressaceae and Taxaceae but absent from both Araucariaceae and Podocarpaceae (Yap et al., 2015). *S. verticillata* is the only cupressophyte species in which plastid *accD* was lost and might have been functionally complemented by a nuclear counterpart

(Li, Gao, et al., 2016). Moreover, expansion of *accD* with insertions of specific tandem repeats was documented in cupressophytes (Yi et al., 2013) and a Pinaceous species, *Tsuga chinensis* (Sudianto et al., 2016). Nonetheless, the evolutionary significance of these repeat insertions remains unclear.

4. PLASTOME REARRANGEMENTS

4.1 Evolution of IRS

With the accumulation of publicly available plastomes on GenBank, the IRs of the five gymnosperm groups were found to have gone through distinctive evolutionary scenarios. These scenarios include evolutionary stasis, contraction, boundary shift, extreme reduction, and complete loss. Zhu et al. (2016) proposed that the putative ancestral IRs of gymnosperms might comprise 15 genes, which are all retained in the IRs of cycads (Fig. 5). This indicates that the IRs of cycads are evolutionarily static. However, the IR of ginkgo contains only 13 genes (Fig. 5). Lin et al. (2012) discovered that the ginkgo IR was contracted to exclude *ycf2*. Within IRs, duplicated genes facilitate gene conversion to decelerate rates of nucleotide substitutions (Li, Kuo, Pryer, & Rothfels, 2016; Perry & Wolfe, 2002; Wu & Chaw, 2015), and accelerated rates of nucleotide substitutions were observed in genes that moved out of IRs (Zhu et al., 2016). However, in ginkgo, an accelerated rate of nucleotide substitutions was not detected in the retained *ycf2* copy. Therefore, Lin et al. (2012) hypothesized that IR contraction likely occurred recently in ginkgo, which is why it has not accumulated a significant number of nucleotide substitutions.

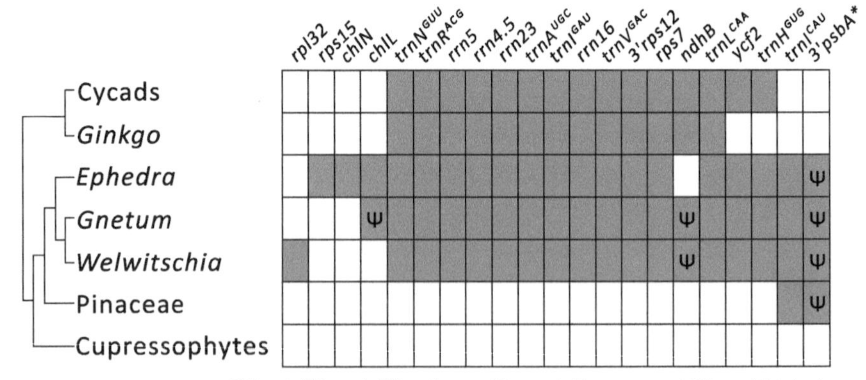

Fig. 5 A comparison of genes located in inverted repeats (IRs).

In gnetophytes, shifts of the IR boundaries are evident (Fig. 5). Gnetophytes were suggested to have experienced multiple steps of expansions, inversions, and gene losses, which resulted in their distinct IR boundaries (Wu et al., 2009). The plastomes of Pinaceous species have an extremely reduced pair of IRs that contains only $trnI^{CAU}$ or $trnI^{CAU}$ and *3′psbA* (Fig. 5). In the plastomes of cupressophytes, it is difficult to identify any IR residue. For example, in *Cryptomeria*, two inverted copies of $trnI^{CAU}$ were hypothesized to be putative residues of IRs (Hirao et al., 2008), but such inverted copies are not present in other cupressophyte genera such as *Nageia* (Wu & Chaw, 2014) and *Podocarpus* (Vieira, Faoro, Rogalski, et al., 2014). Previously, loss of an IR copy was considered a synapomorphic character shared by Pinaceae and cupressophytes (Raubeson & Jansen, 1992). Later, comparative analyses of plastomes suggested that the retained IR copies in Pinaceae and cupressophytes are nonhomologous because their flanking genes are remarkably different (Chen et al., 2015; Hao et al., 2016; Wu, Lin, et al., 2011).

IRs are prevalent in most of the land plants, so they might play an important role in plastomic evolution. IRs may help stabilize plastomes because intramolecular recombination was largely confined to IRs, thus decreasing rearrangements in LSC and SSC regions (Palmer, 1991). Accordingly, loss of IRs would result in accumulations of plastomic rearrangements. This suggestion provides a potential interpretation for the numerous rearrangements found in the IR-lacking plastomes of Pinaceae and cupressophytes.

4.2 Inversions

An inversion is one of the major plastome rearrangements in Pinaceae and cupressophytes. In Pinaceae, plastome inversions are confined to two large fragments, of about 20–21 kb long, which are flanked by Pinaceae-specific repeats (Tsumura, Suyama, & Yoshimura, 2000; Wu, Wang, et al., 2011). These Pinaceae-specific repeats mediate intramolecular homologous recombination to generate four distinct forms of plastomes (Wu, Lin, et al., 2011). Furthermore, interspecific recombinant plastomes were reported in *Picea*, and these chimeric plastomes, generated by recombination of heterogeneous plastomes, might explain why conflicting phylogenetic results were inferred from different plastomic loci (Sullivan et al., 2017).

Inversions have contributed to the highly rearranged plastomes of cupressophytes. For example, *Cryptomeria* was inferred to have had at least 13 plastomic inversions after its split from *Cycas* (Hirao et al., 2008). There are at least 10 plastomic inversions between *Agathis* and *Calocedrus* (Wu & Chaw, 2014). Interspecific plastomic inversions were also reported within the genus *Podocarpus* (Podocarpaceae; Vieira, Faoro, Rogalski, et al., 2014). In Taxaceae, three plastomic inversions separate *Taxus* from *Amentotaxus* (Hsu et al., 2014). In *Sciadopitys*, plastomic inversions have led to recombination of conserved gene clusters (Hsu et al., 2016). The plastomic inversion rate of cupressophytes was estimated at 0.1031 inversions per MY, approximately 3.6 times faster than that in Pinaceae (Hao et al., 2016). Moreover, in cupressophytes, the degree of plastomic inversions is family dependent and positively correlated with mutation rates, which suggests that inversions have evolved in a neutral manner (Wu & Chaw, 2016).

Reconstruction of plastomic inversions was proposed to enable probing the nuclear plastid DNA (*nupt*) by using PCR (Hsu et al., 2014). To prevent amplifying the extant plastomic DNA, primer design should be based on the inferred ancestral gene order of a particular region that has encountered an inversion in the extant plastome. However, this PCR-based approach has some limitations because the examined plastomes have to contain inversions and the obtained sequences likely represent only a small part of the population of *nupt*s.

4.3 Isomeric Plastomes

IRs can trigger homologous recombination, thereby resulting in coexistence of two isomeric plastomes within species (Martin, Baurens, Cardi, Aury, & D'Hont, 2013; Palmer, 1983). Despite lacking IRs, conifers also contain isomeric plastomes generated from their specific repeats of diverse sizes. One of the Pinaceae-specific repeats is likely associated with the coexistence of two isomeric plastomes (i.e. the so-called A and B forms) in *Pseudotsuga* (Wu, Wang, et al., 2011). In cupressophytes, repeats are family-specific. An IR with approximately 250bp that contains *trnQ*^{UUG} (termed *trnQ*-IR) is commonly observed in Cupressaceae and Taxaceae of cupressophytes (Guo et al., 2014; Li, Gao, et al., 2016; Wu & Chaw, 2016). PCR and read mapping analyses have demonstrated that the *trnQ*-IR can induce an inversion that distinguishes the major from the minor isomeric form in both Cupressaceae (Guo et al., 2014; Qu et al., 2017) and Taxaceae (Yi et al., 2013). Notably, the *trnQ*-IR was suggested to be derived from tandem duplicated copies of *trnQ*^{UUG}, as exemplified by the *Sciadopitys* plastome (Li, Gao, et al., 2016). Furthermore, *Sciadopitys* possesses the second specific IRs that contain partial sequences of both *rpoC1* and *rpoC2* (termed *rpoC2*-IR). Isomeric plastomes associated with the *rpoC2*-IR were detected in this species by using PCR (Hsu et al., 2016).

In Podocarpaceae, two specific plastid repeats are recombinationally active. One contains *trnN*^{GUU} (termed *trnN*-IR) and is ubiquitously present in Podocarpaceae (Wu & Chaw, 2016). PCR assays have confirmed the existence of the *trnN*-IR-associated inversion in *Retrophyllum* (Vieira et al., 2016). The other is a *trnD*^{GUC}-containing direct repeat that mediates homologous recombination, thereby resulting in a large fragment deletion (Vieira et al., 2016). Although a pair of *rrn5*-IRs is commonly found in Araucariaceae (Wu & Chaw, 2016; Yap et al., 2015), its recombinant activity has not been assessed.

In summary, the presence of diverse specific repeats has complicated the plastomic evolution in conifers. Wu, Lin, et al. (2011) proposed that the Pinaceae-specific repeats might complement the reduced IR and increase the diversity of plastomic architecture. In cupressophytes, the relative abundance of the isomeric plastomes has shifted among congeneric species, which suggests that the existence of isomeric plastomes and shift in their abundance together contribute to the plastome complexity (Guo et al., 2014; Qu et al., 2017). Nonetheless, the mechanisms that underlie the shift in abundance between isomeric plastomes is still poorly studied.

5. EVOLUTION OF NUCLEOTIDE SUBSTITUTION RATES

5.1 Rates of Nucleotide Substitutions Vary Among Gymnosperm Lineages

Although the first gymnosperm plastome was deciphered more than two decades ago (Wakasugi et al., 1994), plastome-wide comparisons of nucleotide substitution rates among gymnosperms were not conducted until 13 years later (Wu et al., 2007). After analysing 56 concatenated plastid protein-coding genes and conducting relative rate tests, Wu et al. (2007) reported that the nucleotide substitution rates were significantly higher in *Gnetum* than *Cycas*, *Ginkgo*, and *Pinus* for transition and transversion sites in all codon positions. Subsequent reports also documented accelerated nucleotide substitution rates in two other genera of gnetophytes—*Welwitschia* (McCoy et al., 2008) and *Ephedra* (Wu et al., 2009)—which indicates that all gnetophytes have had accelerated rates of nucleotide substitutions. Despite of this, the plastid genes of gnetophytes are under strong functional constraints (Wang, Jiang, Zhou, Su, & Wang, 2015; Wu et al., 2009). The accelerated substitution rates together with functional constraints led to the hypothesis that lineage effects, such as generation time (Wang et al., 2015; Wu et al., 2009) and tree height (Wang et al., 2015), drive nucleotide evolution in the plastomes of gnetophytes.

In contrast, plastid nucleotide substitution rates are relatively slower in cycads and ginkgo than other gymnosperms (Wu & Chaw, 2015; Wu, Lin, et al., 2011; Zhu et al., 2016). For example, the nonsynonymous (*dN*) and synonymous (*dS*) substitution rates of cycad plastomes are about 1.7 and 2.3 times slower, respectively, than those of gnetophytes (Wu & Chaw, 2015). In ginkgo and cycads, stasis in nucleotide substitution rates might explain why their plastomes still retain a residual sequence of the elongation factor *tufA* (Lin et al., 2012; Wu & Chaw, 2015; Wu et al., 2007), although this gene has been transferred to the nucleus (Baldauf & Palmer, 1990).

Nucleotide substitution rates vary greatly within the cupressophytes. After diverging from their common ancestor, cupressophytes have evolved a wide range of nucleotide substitution rates among different genera: from 0.122 to 0.348 substitutions per site (Wu & Chaw, 2016). Taller tree species may have longer generation times and slower rates of mitosis in their apical meristems (Lanfear et al., 2013), so tree heights were proposed to account for the rate heterogeneity in cupressophytes (Wu & Chaw, 2016). Moreover, a positive association between the nucleotide substitution rates and the plastome compactness was demonstrated in both gnetophyte and cupressophyte plastomes (Wu & Chaw, 2016; Wu et al., 2009), which implies that the two lineages have convergent plastomic evolutionary trends.

5.2 Different Mutational Trends Between IR and SC Regions

The studies by Wolfe, Li, and Sharp (1987) and Gaut (1998) first documented higher nucleotide substitution rates in IR than SC regions. The lower substitution rates detected in IR were suggested to result from copy-correction activity because two copies of the same genes reside in IRs (Perry & Wolfe, 2002; Wolfe et al., 1987). Experimental assays have verified that nucleotide mutations in plastomes could be corrected via gene conversion (Khakhlova & Bock, 2006). A recent large-scale comparison across the plastomes of land plants showed that *dS* rates of the IR genes are on average 3.7 times slower than those of the SC genes (Zhu et al., 2016).

In cycad plastomes, the nucleotide substitution rates in the nongenic regions of the IR are about half of those in the SC regions (Wu & Chaw, 2015). The IR and SC regions also have contrasting patterns of nucleotide substitutions, the former being GC-biased and the latter AT-biased, which explains the disequilibrium of GC content at silent sites between IR and SC regions (Wu & Chaw, 2015).

The IR-lacking plastomes of conifers provide opportunities to measure nucleotide substitution rate changes in genes that were relocated from IR to SC regions. Zhu et al. (2016) found that, in conifers, genes that moved out of the IR show accelerated rates of nucleotide substitutions. However, decelerated rates of nucleotide substitutions were observed in the genes that moved into the IR in diverse land plant species (Li, Kuo, et al., 2016; Zhu et al., 2016) with the exception of one flowering plant genus, *Pelargonium* (Weng, Ruhlman, & Jansen, 2017). In conclusion, mounting evidence has shown that nucleotide substitution rates are generally decelerated in IRs, which suggests that a copy number-dependent effect has shaped the rate and tempo of plastid genes.

5.3 Plastid Mutational Hotspots in Gymnosperms

A mutational hotspot is a locus that is more prone to mutate than other loci (Rogozin & Pavlov, 2003). Previously, studies of mutational hotspots were rarely conducted in the gymnosperm plastome. A locus of the *Pseudotsuga* (Pinaceae) plastome that contains tandem repeats was recognized as a mutational hotspot, which separates Asian from North American species (Hipkins, Marshall, Neale, Rottmann, & Strauss, 1995). Mutational hotspots involving tandem repeats were also reported in other gymnosperm plastomes. For example, in cupressophytes, insertions of different types of tandem repeats have caused much length variation in *accD* (Yi et al., 2013). The *accD*-coding frame in *Tsuga* (Pinaceae) was expanded with a unique insertion of tandem repeats, suggesting that this gene is a good marker for distinguishing *Tsuga* from other genera (Sudianto et al., 2016).

Plastomic loci that contain simple sequence repeats (SSRs) are also potential markers for resolving different populations or ecotypes in the same species. In gymnosperm plastomes, the SSR-containing loci are mainly found in the IGS regions (Chen et al., 2015; Hao et al., 2016; Jiang et al., 2016; Vieira, Faoro, Rogalski, et al., 2014; Yap et al., 2015; Zhang et al., 2014). Comparative analyses have identified three plastid loci, i.e. *5′clpP*, *5′ycf1*, and the IGS between *rrn16* and *rrn23*, for the population genetics study of *T. mairei* because those loci contain the most abundant SSRs, indels, and single-nucleotide polymorphisms (Hsu et al., 2014). In cupressophytes, *ycf1*, *accD*, *ycf2*, *clpP*, and *rpl32* were reported to vary greatly in length and nucleotide composition, which suggests that they are valuable as phylogenetic markers as well (Chen et al., 2015). In *Cycas*, *rpoB*, *psbC*, *ycf1*, *ycf2*, introns of *clpP*, *psbA–trnH*, and *trnL–trnF* showed a great level of interspecific variations and were proposed to be useful for DNA-barcoding and phylogenetic reconstruction (Jiang et al., 2016).

To date, plastid mutational hotspots have not been evaluated in gnetophytes, although all of their three genera have representative plastomes available on GenBank. Similarly, many genera of cupressophyte families also have representative plastomes elucidated (Wu & Chaw, 2016). However, a systematic analysis of their mutational hotspots at the familial, generic, and specific ranks is wanting. More plastome data from the Podocaparceae and Cupressacae are most desirable to comprehend such an analytic study. As a consequence, it is unclear whether all gymnosperms shared common mutational hotspots, or alternatively, the mutational hotspots have evolved independently among the five gymnosperm groups.

6. PLASTID PHYLOGENOMICS OF GYMNOSPERMS

Phylogenetic relationships among the five major groups of living gymnosperms—cycads, ginkgo, Pinaceae, cupressophytes, and gnetophytes—have been hotly debated since the early 20th century. This section presents three examples of plastid phylogenomics used to reexamine gymnosperm phylogenies.

6.1 Are Extant Gymnosperms Monophyletic?

Previously, whether angiosperms diverged from one of the five gymnosperm groups or are sister to all gymnosperms was one of the oldest unresolved issues in seed plant systematics. Early morphological and fossil-based studies suggest that angiosperms and gymnosperms formed a monophyletic clade with the former nested within the latter (Crane, 1985; Hill & Crane, 1982; Loconte & Stevenson, 1990; Parenti, 1980). This point of view was later challenged by the emergence of the anthophyte hypothesis—whereby angiosperms and gnetophytes formed a monophyletic clade sister to the remaining gymnosperm lineages. The anthophyte hypothesis held by Doyle and Donoghue (1986) has been discarded because it was countered by most molecular analyses based on various loci, such as 5S *rRNA* (Hori, Lim, & Osawa, 1985), *rbcL* (Hasebe, Ito, Kofuji, Iwatsuki, & Ueda, 1992), chloroplast ITS (Goremykin et al., 1996), 18S *rRNA* (Chaw et al., 1997), and multiple genes from three genomes (Bowe, Coat, & de Pamphilis, 2000; Chaw, Parkinson, Cheng, Vincent, & Palmer, 2000; Soltis, Soltis, & Zanis, 2002). However, many were still sceptical about the monophyly of extant gymnosperms (see review by Donoghue & Doyle, 2000), despite the multiple lines of evidence mentioned earlier.

Phylogenetic trees inferred from the common plastid genes of 37 (Wu et al., 2007), 13 (Zhong, Yonezawa, Zhong, & Hasegawa, 2010), and 23 (Xi, Rest, & Davis, 2013) land plant plastomes consistently reaffirm the monophyly of extant gymnosperms and its sisterhood to angiosperms. Plastid phylogenomic analyses also resolved two well-supported clades, conifers–gnetophytes and ginkgo–cycads (Wu, Lin, et al., 2011; Zhong et al., 2010), in gymnosperms. A large-scale phylogenomic analysis of 360 green-plant plastomes also held that all extant gymnosperms constitute a monophyletic group that is sister to angiosperms (Ruhfel, Gitzendanner, Soltis, Soltis, & Burleigh, 2014). The monophyly of extant gymnosperms is currently indisputable in plastid phylogenomics, but the phylogenetic placements of gnetophytes and ginkgo are still not totally resolved.

6.2 Are Conifers Monophyletic or Paraphyletic?

Conifers, the largest and most diverse group of living gymnosperms, were traditionally treated as a monophyletic clade (Chaw et al., 1997; Rydin, Kallersjo, & Friis, 2002; Stefanovic, Jager, Deutsch, Broutin, & Masselot, 1998). There are about 670 species and 71 genera in 6 conifer families: Pinaceae, Cupressaceae, Taxaceae, Sciadopityaceae, Podocarpaceae, and Araucariaceae (Farjon & Filer, 2013; Gernandt et al., 2011). Early studies based on cladistics and molecular phylogeny maintained that all conifers formed a monophyletic group, with Pinaceae as the earliest divergent family (Chaw et al., 1997; Stefanovic et al., 1998). Thus, Pinaceae and the remaining five families are designated as the conifers I and conifers II (or cupressophytes) clades, respectively. To date, there have been three major competing molecular phylogenetic hypotheses for the placement of gnetophytes relative to Pinaceae and cupressophytes (Fig. 6). The "gnetifers" hypothesis held that Pinaceae and cupressophytes are monophyletic and sister to gnetophytes (Fig. 6A; Chaw et al., 1997). In contrast, gnetophytes were also considered sister to Pinaceae (i.e. the "gnepines" hypothesis; Fig. 6B; Bowe et al., 2000; Chaw et al., 2000) or to cupressophytes (the "gnecup" hypothesis; Fig. 6C; Nickrent, Parkinson, Palmer, & Duff, 2000). Notably, despite the incongruent placements of

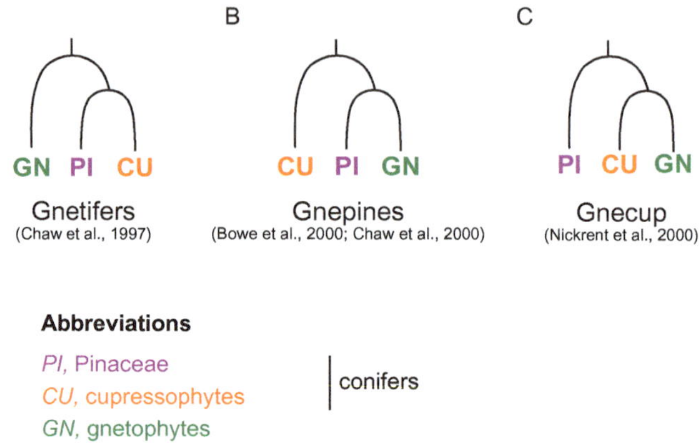

Fig. 6 Three competing molecular-based trees regarding the placement of gnetophytes in the gymnosperm phylogeny: (A) sister to conifers (gnetifer hypothesis), (B) sister to Pinaceae (gnepines hypothesis), or (C) sister to cupressophytes (gnecup hypothesis). References where these relationships were first reported are listed below each tree.

gnetophytes, both gnepines and gnecup hypotheses suggested that conifers are paraphyletic.

Plastid phylogenomic studies have agreed with the gnepines (Wu, Lin, et al., 2011; Zhong et al., 2011, 2010) or gnecup (Ruhfel et al., 2014) hypothesis and rejected the gnetifer hypothesis. Previously, restriction mapping analyses suggested the common loss of IRs from Pinaceae and cupressophytes, which led to the view that conifers were monophyletic (Raubeson & Jansen, 1992). Later, comparative plastome analyses suggested that conifers lost IRs on two separate occasions because the IR copies retained in Pinaceae and cupressophytes are different (Wu & Chaw, 2014; Wu, Lin, et al., 2011). Therefore, plastomic structural changes allow for conifer paraphyly (as suggested in the gnepines or gnecup hypothesis).

We argue that the overall data support the gnepines hypothesis. Gnetophytes have considerably accelerated rates of nucleotide substitutions that may cause long-branch attraction (LBA) artefacts (Wu, Lin, et al., 2011; Wu, Wang, et al., 2011; Wu et al., 2007; Zhong et al., 2011, 2010). Multiple measures were previously used to alleviate the LBA effect, including removal of fast-evolving genes or sites (Wu et al., 2007; Zhong et al., 2011, 2010), the addition of more taxa (Zhong et al., 2010), and exclusion of high heterotachous genes from datasets (Wu, Lin, et al., 2011). All of these efforts have consistently recovered the gnepines clade (Wu, Lin, et al., 2011; Wu et al., 2007; Zhong et al., 2011, 2010). In addition to evidence from sequence analyses, the gnepines clade was supported by some unique plastomic characteristics shared by Pinaceae and gnetophytes, such as loss of *rps16* (Wu et al., 2007) and all *ndh* genes (Braukmann, Kuzmina, & Stefanovic, 2009) and expansion of IRs to include the 3′*psbA* gene (Wu et al., 2009, 2007). More recently, utilizing 106 nuclear single-copy genes, Li et al. (2017) claimed that substitutions of the third codon positions were saturated, and removal of the third codon positions from the datasets resulted in the recovery of the gnepines clade. Altogether, there are overwhelming data to support the gnepines clade.

6.3 Which Taxon Is Sister to Ginkgo?

G. biloba (common name: ginkgo), the only surviving species of Ginkgoales, has been widely considered a living fossil because its leaves and ovules resembled those of the extinct *Ginkgo* species that lived more than 100 MY ago (Zhou & Zheng, 2003). Based on morphological evidence, ginkgo was previously proposed to be closely related to conifers (Norstog, Gifford, & Stevenson, 2004) or an intermediate between cycads and conifers (Wang et al., 2011). With a few molecular loci used for phylogenetic analyses, ginkgo was also placed as sister to the clade consisting of conifers and gnetophytes (Bowe et al., 2000; Chaw et al., 2000; Lu, Ran, Guo, Yang, & Wang, 2014; Soltis et al., 2002). Other placements were also proposed, such as being sister to (1) other gymnosperms as a whole, (2) the clade comprising cycads and conifers, (3) cycads, (4) conifers, or (5) the clade including cycads and angiosperms (see review in Wu, Chaw, & Huang, 2013).

Wu et al. (2013) found that when DNA sequences were used for tree construction, the differences in codon positions, breadth of taxon sampling, tree-building methods, or exclusion of gnetophytes from datasets contributed to the conflicting placements of ginkgo. In contrast, trees inferred from amino acids congruently supported the sisterhood of ginkgo and cycads, regardless of which datasets or methods were used in phylogenetic analyses. Moreover, the sisterhood of ginkgo and cycads was also recovered in nuclear phylotranscriptomic (Finet, Timme, Delwiche, & Marlétaz, 2010; Lee et al., 2011; Wickett et al., 2014) and nuclear phylogenomic analyses (Li et al., 2017). Ginkgo and cycads commonly contain haustorial pollen tubes (Friedman, 1993), multiflagellated sperm cells (Brenner, Stevenson, & Twigg, 2003), simple strobili (Rudall & Bateman, 2010), and some particular patterns during embryogenesis (Wang et al., 2011). These traits add a morphological line of evidence to reinforce the sister relationship between ginkgo and cycads.

7. CONCLUSIONS AND FUTURE DIRECTIONS

Over the last decade, tremendous efforts have been made to decipher gymnosperm plastomes, which has substantially expanded the available plastomic data and also given us a better picture of the gymnosperm plastome evolution. The elucidated plastomes of the five gymnosperm groups vary in architecture, IR evolution, and nucleotide substitution rates. The IR-lacking plastomes of cupressophytes are particularly interesting because they exhibit several unusual features, such as varied size, numerous rearrangements, diverse repeats, disruptions of several conserved gene clusters, and the existence of major and minor isomeric plastomes.

However, we are still missing some representative genera from two families in cupressophytes: Cupressaceae (30 genera and about 133 species) and Podocarpaceae (more than 17 genera and 125–165 species). Sequencing them and including their plastomes in comparative analyses will provide more comprehensive insights into the evolution of gymnosperm plastomes. In addition, the evolutionary impact on disrupted gene clusters and relocated genes has not been investigated. Comparative transcriptomic analysis is also needed, as is investigating whether gene expression mechanisms are altered in relocated genes—particularly those in the disrupted gene clusters.

ACKNOWLEDGMENTS

S.-M.C. is a Distinguished Research Fellow of the Biodiversity Research Center, Academia Sinica (AS), and appreciates the Ministry of Science and Technology of Taiwan and AS for grant supports related to the chapter topic. C.-S.W. greatly acknowledges the Postdoctoral Fellowship of AS and S.M.C.'s grants through AS. E.S. is a graduate student of the Taiwan International Graduate Program (TIGP), AS, and appreciates the TIGP fellowship. We thank Dr J.P. Mower for his critical comments, which helped us to improve this chapter.

REFERENCES

Baldauf, S. L., & Palmer, J. D. (1990). Evolutionary transfer of the chloroplast *tufA* gene to the nucleus. *Nature, 15*, 262–265.

Bowe, L. M., Coat, G., & de Pamphilis, C. W. (2000). Phylogeny of seed plants based on all three genomic compartments: Extant gymnosperms are monophyletic and Gnetales' closest relatives are conifers. *Proceedings of the National Academy of Sciences of the United States of America, 97*, 4092–4097.

Braukmann, T. W., Kuzmina, M., & Stefanovic, S. (2009). Loss of all plastid *ndh* genes in Gnetales and conifers: Extent and evolutionary significance for the seed plant phylogeny. *Current Genetics, 55*, 323–337.

Brenner, E. D., Stevenson, D. W., & Twigg, R. W. (2003). Cycads: Evolutionary innovations and the role of plant-derived neurotoxins. *Trends in Plant Science, 8*, 446–452.

Chaw, S. M., Zharkikh, A., Sung, H. M., Lau, T. C., & Li, W. H. (1997). Molecular phylogeny of extant gymnosperms and seed plant evolution: Analysis of nuclear 18S rRNA sequences. *Molecular Biology and Evolution, 14*, 56–68.

Chaw, S. M., Parkinson, C. L., Cheng, Y., Vincent, T. M., & Palmer, J. D. (2000). Seed plant phylogeny inferred from all three plant genomes: Monophyly of extant gymnosperms and origin of Gnetales from conifers. *Proceedings of the National Academy of Sciences of the United States of America, 97*, 4086–4091.

Chen, J., Hao, Z., Xu, H., Yang, L., Liu, G., Sheng, Y., et al. (2015). The complete chloroplast genome sequence of the relict woody plant *Metasequoia glyptostroboides* Hu et Cheng. *Frontiers in Plant Science, 6*, 447.

Christenhusz, M. J. M., & Byng, J. W. (2016). The number of known plants species in the world and its annual increase. *Phytotaxa, 261*, 201–217.

Crane, P. R. (1985). Phylogenetic analysis of seed plants and the origin of angiosperms. *Annals of the Missouri Botanical Garden, 72*, 716–793.

Cronn, R., Liston, A., Parks, M., Gernandt, D. S., Shen, R., & Mockler, T. (2008). Multiplex sequencing of plant chloroplast genomes using Solexa sequencing-by-synthesis technology. *Nucleic Acids Research, 36*, e122.

Daniell, H., Lin, C. S., Yu, M., & Chang, W. J. (2016). Chloroplast genomes: Diversity, evolution, and applications in genetic engineering. *Genome Biology, 17*, 134.

Donoghue, M. J., & Doyle, J. A. (2000). Seed plant phylogeny: Demise of the anthophyte hypothesis? *Current Biology, 10*, R106–R109.

Doyle, J. A., & Donoghue, M. J. (1986). Seed plant phylogeny and the origin of angiosperms: An experimental cladistics approach. *The Botanical Review, 52*, 321–431.

Du, F. K., Lang, T. G., Lu, S. H., Wang, Y. Y., Li, J. Q., & Yin, K. Q. (2015). An improved method for chloroplast genome sequencing in non-model forest tree species. *Tree Genetics and Genomes, 11*, 114.

Farjon, A., & Filer, D. (2013). *Creating the atlas of the world's conifers: An analysis of their distribution, biogeography, diversity and conservation status*. Leiden: Brill Publishers.

Finet, C., Timme, R. E., Delwiche, C. F., & Marlétaz, F. (2010). Multigene phylogeny of the green lineage reveals the origin and diversification of land plants. *Current Biology, 20*, 2217–2222.

Friedman, W. E. (1993). The evolutionary history of the seed plant male gametophyte. *Trends in Ecology and Evolution, 8*, 15–21.

Gaut, B. S. (1998). Molecular clocks and nucleotide substitution rates in higher plants. In M. K. Hecht et al. (Eds.) *Evolutionary biology: Vol. 30*. (pp. 93–120). New York: Plenum Press.

Gernandt, D. S., Willyard, A., Syring, J. V., & Liston, A. (2011). The conifers (Pinophyta). In C. Plomion, J. Bousquet, & C. Kole (Eds.), *Genetics, genomics and breeding of conifers* (pp. 1–39). New Hampshire: Science Publishers.

Goremykin, V., Bobrova, V., Pahnke, J., Troitsky, A., Antonov, A., & Martin, W. (1996). Noncoding sequences from the slowly evolving chloroplast inverted repeat in addition *rbcL* data do not support gnetalean affinities of angiosperms. *Molecular Biology and Evolution, 13*, 383–396.

Gregory, T. R., Nicol, J. A., Tamm, H., Kullman, B., Kullman, K., Leitch, I. J., et al. (2007). Eukaryotic genome size databases. *Nucleic Acids Research, 35*, D332–D338.

Guo, W., Grewe, F., Cobo-Clark, A., Fan, W., Duan, Z., Adams, R. P., et al. (2014). Predominant and substoichiometric isomers of the plastid genome coexist within *Juniperus* plants and have shifted multiple times during cupressophyte evolution. *Genome Biology and Evolution, 6*, 580–590.

Hao, Z., Cheng, T., Zheng, R., Xu, H., Zhou, Y., Li, M., et al. (2016). The complete chloroplast genome sequence of a relict conifer *Glyptostrobus pensilis*: Comparative analysis and insights into dynamics of chloroplast genome rearrangement in cupressophytes and Pinaceae. *PLoS One*, *11*, e0161809.

Hasebe, M., Ito, M., Kofuji, R., Iwatsuki, K., & Ueda, K. (1992). Phylogeny of gymnosperms inferred from *rbcL* gene sequences. *The Botanical Magazine (Tokyo)*, *105*, 673–679.

Hill, C. R., & Crane, P. R. (1982). Evolutionary cladistics and the origin of angiosperms. In K. A. Joysey & A. E. Friday (Eds.), *Problems of phylogenetic reconstruction* (pp. 269–361). London: Academic Press.

Hipkins, V. D., Marshall, K. A., Neale, D. B., Rottmann, W. H., & Strauss, S. H. (1995). A mutation hotspot in the chloroplast genome of a conifer (Douglas-fir: *Pseudotsuga*) is caused by variability in the number of direct repeats derived from a partially duplicated tRNA gene. *Current Genetics*, *27*, 572–579.

Hirao, T., Watanabe, A., Kurita, M., Kondo, T., & Takata, K. (2008). Complete nucleotide sequence of the *Cryptomeria japonica* D. Don. chloroplast genome and comparative chloroplast genomics: Diversified genomic structure of coniferous species. *BMC Plant Biology*, *8*, 70.

Hori, H., Lim, B. L., & Osawa, S. (1985). Evolution of green plants as deduced from 5S rRNA sequences. *Proceedings of the National Academy of Sciences of the United States of America*, *82*, 820–823.

Hsu, C. Y., Wu, C. S., & Chaw, S. M. (2014). Ancient nuclear plastid DNA in the yew family (Taxaceae). *Genome Biology and Evolution*, *6*, 2111–2121.

Hsu, C. Y., Wu, C. S., & Chaw, S. M. (2016). Birth of four chimeric plastid gene clusters in Japanese umbrella pine. *Genome Biology and Evolution*, *8*, 1776–1784.

Jackman, S. D., Warren, R. L., Gibb, E. A., Vandervalk, B. P., Mohamadi, H., Chu, J., et al. (2015). Organellar genomes of white spruce (*Picea glauca*): Assembly and annotation. *Genome Biology and Evolution*, *8*, 29–41.

Jansen, R. K., Raubeson, L. A., Boore, J. L., dePamphilis, C. W., Chumley, T. W., Haberle, R. C., et al. (2005). Methods for obtaining and analyzing whole chloroplast genome sequences. *Methods in Enzymology*, *395*, 348–384.

Jansen, R. K., & Ruhlman, T. A. (2012). Plastid genomes of seed plants. In R. Bock & V. Knoop (Eds.), *Genomics of chloroplasts and mitochondria* (pp. 103–126). Dordrecht: Springer.

Jiang, G. F., Hinsinger, D. D., & Strijk, J. S. (2016). Comparison of intraspecific, interspecific and intergeneric chloroplast diversity in cycads. *Scientific Reports*, *6*, 31473.

Khakhlova, O., & Bock, R. (2006). Elimination of deleterious mutations in plastid genomes by gene conversion. *The Plant Journal*, *46*, 85–94.

Lanfear, R., Ho, S. Y., Davies, T. J., Moles, A. T., Aarssen, L., Swenson, N. G., et al. (2013). Taller plants have lower rates of molecular evolution. *Nature Communications*, *4*, 1879.

Lee, E. K., Cibrian-Jaramillo, A., Kolokotronis, S. O., Katari, M. S., Stamatakis, A., Ott, M., et al. (2011). A functional phylogenomic view of the seed plants. *PLoS Genetics*, *7*, e1002411.

Li, F. W., Kuo, L. Y., Pryer, K. M., & Rothfels, C. J. (2016). Genes translocated into the plastid inverted repeat show decelerated substitution rates and elevated GC content. *Genome Biology and Evolution*, *8*, 2452–2458.

Li, J., Gao, L., Chen, S., Tao, K., Su, Y., & Wang, T. (2016). Evolution of short inverted repeat in cupressophytes, transfer of accD to nucleus in *Sciadopitys verticillata* and phylogenetic position of Sciadopityaceae. *Scientific Reports*, *6*, 20934.

Li, Z., De La Torre, A. R., Sterck, L., Canovas, F. M., Avila, C., Merino, I., et al. (2017). Single-copy genes as molecular markers for phylogenomic studies in seed plants. *Genome Biology and Evolution*, *9*, 1130–1147.

Lin, C. P., Huang, J. P., Wu, C. S., Hsu, C. Y., & Chaw, S. M. (2010). Comparative chloroplast genomics reveals the evolution of Pinaceae genera and subfamilies. *Genome Biology and Evolution, 2*, 504–517.

Lin, C. P., Wu, C. S., Huang, Y. Y., & Chaw, S. M. (2012). The complete chloroplast genome of *Ginkgo biloba* reveals the mechanism of inverted repeat contraction. *Genome Biology and Evolution, 4*, 374–381.

Loconte, H., & Stevenson, D. W. (1990). Cladistics of the Spermatophyta. *Brittonia, 42*, 197–211.

Lu, Y., Ran, J. H., Guo, D. M., Yang, Z. Y., & Wang, X. Q. (2014). Phylogeny and divergence times of gymnosperms inferred from single-copy nuclear genes. *PLoS One, 9*, e107679.

Martin, G., Baurens, F. C., Cardi, C., Aury, J. M., & D'Hont, A. (2013). The complete chloroplast genome of banana (*Musa acuminata*, Zingiberales): Insight into plastid monocotyledon evolution. *PLoS One, 8*, e67350.

McCoy, S. R., Kuehl, J. V., Boore, J. L., & Raubeson, L. A. (2008). The complete plastid genome sequence of *Welwitschia mirabilis*: An unusually compact plastome with accelerated divergence rates. *BMC Evolutionary Biology, 8*, 130.

Nickrent, D. L., Parkinson, D. L., Palmer, J. D., & Duff, R. J. (2000). Multigene phylogeny of land plants with special reference to bryophytes and the earliest land plants. *Molecular Biology and Evolution, 17*, 1885–1895.

Nock, C. J., Waters, D. L., Edwards, M. A., Bowen, S. G., Rice, N., Cordeiro, G. M., et al. (2011). Chloroplast genome sequences from total DNA for plant identification. *Plant Biotechnology Journal, 9*, 328–333.

Norstog, K. J., Gifford, E. M., & Stevenson, D. W. M. (2004). Comparative development of the spermatozoids of cycads and *Ginkgo biloba*. *The Botanical Review, 70*, 5–15.

Palmer, J. D. (1983). Chloroplast DNA exists in two orientations. *Nature, 301*, 92–93.

Palmer, J. D. (1991). Plastid chromosomes: Structure and evolution. In L. Bogorad & I. K. Vasil (Eds.), *Molecular biology of plastids: Vol. 7A. Cell culture and somatic cell genetics of plants* (pp. 5–53). San Diego: Academic Press.

Parenti, L. R. (1980). A phylogenetic analysis of the land plants. *Biological Journal of the Linnean Society, 13*, 225–242.

Perry, A. S., & Wolfe, K. H. (2002). Nucleotide substitution rates in legume chloroplast DNA depend on the presence of the inverted repeat. *Journal of Molecular Evolution, 55*, 501–508.

Qu, X. J., Wu, C. S., Chaw, S. M., & Yi, T. S. (2017). Insights into the existence of isomeric plastomes in Cupressoideae (Cupressaceae). *Genome Biology and Evolution, 9*, 1110–1119.

Rai, H. S., Reeves, P. A., Peakall, R., Olmstead, R. G., & Graham, S. W. (2008). Inference of higher-order conifer relationships from multi-locus plastid dataset. *Botany, 86*, 658–669.

Raubeson, L. A., & Jansen, R. K. (1992). A rare chloroplast DNA structure mutation is shared by all conifers. *Biochemical Systematics and Ecology, 20*, 17–24.

Rogozin, I. B., & Pavlov, Y. I. (2003). Theoretical analysis of mutation hotspots and their DNA sequence context specificity. *Mutation Research, 544*, 65–85.

Rudall, P. J., & Bateman, R. M. (2010). Defining the limits of flowers: The challenge of distinguishing between the evolutionary products of simple versus compound strobili. *Philosophical Transactions of the Royal Society B: Biological Sciences, 365*, 397–409.

Ruhfel, B. R., Gitzendanner, M. A., Soltis, P. S., Soltis, D. E., & Burleigh, J. G. (2014). From algae to angiosperms—Inferring the phylogeny of green plants (Viridiplantae) from 360 plastid genomes. *BMC Evolutionary Biology, 14*, 23.

Ruhlman, T., & Jansen, R. (2014). The plastid genomes of flowering plants. In P. Maliga (Ed.), *Chloroplast biotechnology* (pp. 3–38). New York: Humana Press.

Ruhsam, M., Rai, H. S., Mathews, S., Ross, T. G., Graham, S. W., Raubeson, L. A., et al. (2015). Does complete plastid genome sequencing improve species discrimination and phylogenetic resolution in *Araucaria*? *Molecular Ecology Resources*, *15*, 1067–1078.

Rydin, C., Kallersjo, M., & Friis, E. M. (2002). Seed plant relationships and the systematic position of Gnetales based on nuclear and chloroplast DNA: Conflicting data, rooting problems, and the monophyly of conifers. *International Journal of Plant Sciences*, *163*, 197–214.

Sakaguchi, S., Ueno, S., Tsumura, Y., Setoguchi, H., Ito, M., Hattori, C., et al. (2017). Application of a simplified method of chloroplast enrichment to small amounts of tissue for chloroplast genome sequencing. *Applications in Plant Sciences*, *5*, apps.1700002. https://doi.org/10.3732/apps.1700002.

Soltis, D. E., Soltis, P. S., & Zanis, M. J. (2002). Phylogeny of seed plants based on evidence from eight genes. *American Journal of Botany*, *89*, 1670–1681.

Stefanovic, S., Jager, M., Deutsch, J., Broutin, J., & Masselot, M. (1998). Phylogenetic relationships of conifers inferred from partial 28S rRNA gene sequences. *American Journal of Botany*, *85*, 688–697.

Straub, S. C., Parks, M., Weitemier, K., Fishbein, M., Cronn, R. C., & Liston, A. (2012). Navigating the tip of the genomic iceberg: Next-generation sequencing for plant systematics. *American Journal of Botany*, *99*, 349–364.

Sudianto, E., Wu, C. S., Lin, C. P., & Chaw, S. M. (2016). Revisiting the plastid phylogenomics of Pinaceae with two complete plastomes of *Pseudolarix* and *Tsuga*. *Genome Biology and Evolution*, *8*, 1804–1811.

Sullivan, A. R., Schiffthaler, B., Thompson, S. L., Street, N. R., & Wang, X. R. (2017). Interspecific plastome recombination reflects ancient reticulate evolution in *Picea* (Pinaceae). *Molecular Biology and Evolution*, *34*, 1689–1701.

Tsudzuki, J., Nakashima, K., Tsudzuki, T., Hiratsuka, J., Shibata, M., Wakasugi, T., et al. (1992). Chloroplast DNA of black pine retains a residual inverted repeat lacking rRNA genes: Nucleotide sequences of *trnQ, trnK, psbA, trnI* and *trnH* and the absence of *rps16*. *Molecular Genetics and Genomics*, *232*, 206–214.

Tsumura, Y., Suyama, Y., & Yoshimura, K. (2000). Chloroplast DNA inversion polymorphism in populations of *Abies* and *Tsuga*. *Molecular Biology and Evolution*, *17*, 1302–1312.

Vieira, L. N., Faoro, H., Fraga, H. P. F., Rogalski, M., de Souza, E. M., de Oliveira Pedrosa, F., et al. (2014). An improved protocol for intact chloroplasts and cpDNA isolation in conifers. *PLoS One*, *9*, e84792.

Vieira, L. N., Faoro, H., Rogalski, M., Fraga, H. P. F., Cardoso, R. L., de Souza, E. M., et al. (2014). The complete chloroplast genome sequence of *Podocarpus lambertii*: Genome structure, evolutionary aspects, gene content and SSR detection. *PLoS One*, *9*, e90618.

Vieira, L. N., Rogalski, M., Faoro, H., Fraga, H. P. F., Anjos, K. G., Picchi, G. F. A., et al. (2016). The plastome sequence of the endemic Amazonian conifer, *Retrophyllum piresii* (Silba) C.N. Page, reveals different recombination events and plastome isoforms. *Tree Genetics and Genomes*, *12*, 10.

Wakasugi, T., Tsudzuki, J., Ito, S., Nakashima, K., Tsudzuki, T., & Sugiura, M. (1994). Loss of all *ndh* genes as determined by sequencing the entire chloroplast genome of the black pine *Pinus thunbergii*. *Proceedings of the National Academy of Sciences of the United States of America*, *91*, 9794–9798.

Wang, B., Jiang, B., Zhou, Y., Su, Y. J., & Wang, T. (2015). Higher substitution rates and lower *dN/dS* for the plastid genes in Gnetales than other gymnosperms. *Biochemical Systematics and Ecology*, *59*, 278–287.

Wang, L., Wang, D., Lin, M. M., Lu, Y., Jiang, X. X., & Jin, B. (2011). An embryological study and systematic significance of the primitive gymnosperm *Ginkgo biloba*. *Journal of Systematics and Evolution*, *49*, 353–361.

Weng, M. L., Ruhlman, T. A., & Jansen, R. K. (2017). Expansion of inverted repeat does not decrease substitution rates in *Pelargonium* plastid genomes. *New Phytologist*, *214*, 842–851.

Whittall, J. B., Syring, J., Parks, M., Buenrostro, J., Dick, C., Liston, A., et al. (2010). Finding a (pine) needle in a haystack: Chloroplast genome sequence divergence in rare and widespread pines. *Molecular Ecology*, *19*, 100–114.

Wicke, S., Schneeweiss, G. M., dePamphilis, C. W., Muller, K. F., & Quandt, D. (2011). The evolution of the plastid chromosome in land plants: Gene content, gene order, gene function. *Plant Molecular Biology*, *76*, 273–297.

Wickett, N. J., Mirarab, S., Nguyen, N., Warnow, T., Carpenter, E., Matasci, N., et al. (2014). Phylotranscriptomic analysis of the origin and early diversification of land plants. *Proceedings of the National Academy of Sciences of the United States of America*, *111*, E4859–4868.

Williams, C. G. (2009). *Conifer reproductive biology*. Dordrecht: Springer Netherlands.

Wolf, P. G., Der, J. P., Duffy, A. M., Davidson, J. B., Grusz, A. L., & Pryer, K. M. (2011). The evolution of chloroplast genes and genomes in ferns. *Plant Molecular Biology*, *76*, 251–261.

Wolfe, K. H., Li, W. H., & Sharp, P. M. (1987). Rates of nucleotide substitution vary greatly among plant mitochondrial, chloroplast and nuclear DNAs. *Proceedings of the National Academy of Sciences of the United States of America*, *84*, 9054–9058.

Wu, C. S., Chaw, S. M., & Huang, Y. Y. (2013). Chloroplast phylogenomics indicates that *Ginkgo biloba* is sister to cycads. *Genome Biology and Evolution*, *5*, 243–254.

Wu, C. S., & Chaw, S. M. (2014). Highly rearranged and size-variable chloroplast genomes in conifers II clade (cupressophytes): Evolution towards shorter intergenic spacers. *Plant Biotechnology Journal*, *12*, 344–353.

Wu, C. S., & Chaw, S. M. (2015). Evolutionary stasis in cycad plastomes and the first case of plastome GC-biased gene conversion. *Genome Biology and Evolution*, *7*, 2000–2009.

Wu, C. S., & Chaw, S. M. (2016). Large-scale comparative analysis reveals the mechanisms driving plastomic compaction, reduction, and inversions in conifers II (cupressophytes). *Genome Biology and Evolution*, *8*, 3740–3750.

Wu, C. S., Lai, Y. T., Lin, C. P., Wang, Y. N., & Chaw, S. M. (2009). Evolution of reduced and compact chloroplast genomes (cpDNAs) in gnetophytes: Selection toward a lower-cost strategy. *Molecular Phylogenetics and Evolution*, *52*, 115–124.

Wu, C. S., Lin, C. P., Hsu, C. Y., Wang, R. J., & Chaw, S. M. (2011). Comparative chloroplast genomes of Pinaceae: Insights into the mechanism of diversified genomic organizations. *Genome Biology and Evolution*, *3*, 309–319.

Wu, C. S., Wang, Y. N., Hsu, C. Y., Lin, C. P., & Chaw, S. M. (2011). Loss of different inverted repeat copies from the chloroplast genomes of Pinaceae and cupressophytes and influence of heterotachy on the evaluation of gymnosperm phylogeny. *Genome Biology and Evolution*, *3*, 1284–1295.

Wu, C. S., Wang, Y. N., Liu, S. M., & Chaw, S. M. (2007). Chloroplast genome (cpDNA) of *Cycas taitungensis* and 56 cp protein-coding genes of *Gnetum parvifolium*: Insights into cpDNA evolution and phylogeny of extant seed plants. *Molecular Biology and Evolution*, *24*, 1366–1379.

Wu, J., Liu, B., Cheng, F., Ramchiary, N., Choi, S. R., Lim, Y. P., et al. (2012). Sequencing of chloroplast genome using whole cellular DNA and Solexa sequencing technology. *Frontiers in Plant Science*, *3*, 243.

Xi, Z., Rest, J. S., & Davis, C. C. (2013). Phylogenomics and coalescent analyses resolve extant seed plant relationships. *PLoS One*, *8*, e80870.

Yap, J. Y., Rohner, T., Greenfield, A., Van Der Merwe, M., McPherson, H., Glenn, W., et al. (2015). Complete chloroplast genome of the wollemi pine (*Wollemia nobilis*): Structure and evolution. *PLoS One*, *10*, e0128126.

Yi, X., Gao, L., Wang, B., Su, Y. J., & Wang, T. (2013). The complete chloroplast genome sequence of *Cephalotaxus oliveri* (Cephalotaxaceae): Evolutionary comparison of *Cephalotaxus* chloroplast DNAs and insights into the loss of inverted repeat copies in gymnosperms. *Genome Biology and Evolution, 5*, 688–698.

Zhang, Y., Ma, J., Yang, B., Li, R., Zhu, W., Sun, L., et al. (2014). The complete chloroplast genome sequence of *Taxus chinensis var. mairei* (Taxaceae): Loss of an inverted repeat region and comparative analysis with related species. *Gene, 540*, 201–209.

Zhong, B., Deusch, O., Goremykin, V. V., Penny, D., Biggs, P. J., Atherton, R. A., et al. (2011). Systematic error in seed plant phylogenomics. *Genome Biology and Evolution, 3*, 1340–1348.

Zhong, B., Yonezawa, T., Zhong, Y., & Hasegawa, M. (2010). The position of gnetales among seed plants: Overcoming pitfalls of chloroplast phylogenomics. *Molecular Biology and Evolution, 27*, 2855–2863.

Zhou, Z., & Zheng, S. (2003). The missing link in *Ginkgo* evolution. *Nature, 423*, 821–822.

Zhu, A., Guo, W., Gupta, S., Fan, W., & Mower, J. P. (2016). Evolutionary dynamics of the plastid inverted repeat: The effects of expansion, contraction, and loss on substitution rates. *New Phytologist, 209*, 1747–1756.

CHAPTER EIGHT

Aberration or Analogy? The Atypical Plastomes of Geraniaceae

Tracey A. Ruhlman[*,1], **Robert K. Jansen**[*,†]

[*]Department of Integrative Biology, University of Texas at Austin, Austin, TX, United States
[†]Genomics and Biotechnology Research Group, Faculty of Science, King Abdulaziz University, Jeddah, Saudi Arabia
[1]Corresponding author: e-mail address: truhlman@austin.utexas.edu

Contents

Abstract

A number of plant groups have been proposed as ideal systems to explore plastid inheritance, plastome evolution and plastome-nuclear genome coevolution. Quick generation times and a compact nuclear genome in *Arabidopsis thaliana*, the relative ease of plastid isolation from *Spinacia oleracea* and the tractability of plastid transformation in *Nicotiana tabacum* are all desirable attributes in a model system; however, these and most other groups all lack novelty in terms of plastome structure and nucleotide sequence evolution. Contemporary sequencing and assembly technologies have facilitated analyses of atypical plastomes and, as predicted by early investigations, Geraniaceae plastomes have experienced unprecedented rearrangements relative to the canonical structure and exhibit remarkably high rates of synonymous and nonsynonymous nucleotide substitutions. While not the only lineage with unusual plastome features, likely no other group represents the array of aberrant phenomena recorded for the family. In this chapter, Geraniaceae plastomes will be discussed and, where possible, compared with other taxa.

Advances in Botanical Research, Volume 85
ISSN 0065-2296
https://doi.org/10.1016/bs.abr.2017.11.017

1. INTRODUCTION

Plastid genomes (plastomes) have been the subject of study since the recognition of their existence in plant and algal cells. Today nearly 2000 seed plant plastomes have been sequenced and analysed revealing genome-sized units (unit-genome) with highly conserved structure and gene content and limited variation in evolutionary rates. Typical angiosperm plastomes are maternally inherited and comprise many copies of the unit-genome, each containing a large inverted repeat (IR), a large and small single copy region (LSC and SSC, respectively) and approximately 120–130 genes mostly encoding ribosomal RNAs, transfer RNAs and proteins integral to plastid gene expression and photosynthesis. The genes are densely arrayed on both strands of plastome DNA, which typically has very low repetitive content. Variation in the order of genes is uncommon and more than half of the coding sequences are transcribed as polycistronic pre-mRNAs (Ruhlman & Jansen, 2014). Although the vast majority of seed plant plastomes conform to this description, there are several lineages that have experienced acceleration in nucleotide substitution rates and/or structural changes, including inversion, gene and intron loss, IR loss and accumulation of repetitive DNA (Jansen & Ruhlman, 2012).

Long before the genomic age and the advent of next-generation sequencing gave unprecedented access to plastome sequences, Geraniaceae were garnering attention. At the turn of the last century, Baur (1909) was exploring non-Mendelian inheritance patterns that he observed in the progeny of crosses between different *Pelargonium zonale* cultivars. Plastid inheritance is biparental in *Pelargonium* therefore hybrid zygotes can contain either maternal or paternal plastids or a mixture of both parental plastid types (Birky, 1995). Hybrid variegation can arise from disharmony between the hybrid nucleus and the plastome of one parent and is observed where both parents contribute plastids to the progeny, as in *Pelargonium* (Metzlaf, Pohlheim, Börner, & Hagemann, 1982; Metzlaff, Borner, & Hagemann, 1981; Weihe, Apitz, Pohlheim, Salinas-Hartwig, & Börner, 2009). Because plastid development and function is dependent on the nuclear genome, plastids bearing an incompatible plastome fail to develop in the hybrid, giving rise to white or yellow sectors on green leaves (Kirk & Tilney-Bassett, 1967).

Variegated congeneric hybrids, or so-called chimeras, of *Pelargonium* and *Geranium* were studied through the 1920s and 1930s (Hagemann, 2010), and among a very few other taxa were the workhorses of the evolving theories of organelle inheritance and extranuclear genetics. That trend continued through the 20th century and new techniques to examine plastid DNA and its inheritance were employed with *Pelargonium*. Southern blotting of digested plastid DNA revealed variation in the *Eco*RI fragments among *P. zonale* hybrids and found that the parental plastome genotypes (plasmotype) could be identified in the progeny using this technique (Metzlaff et al., 1981), providing a molecular link to the variegated phenotypes examined by Baur. Employing similar approaches, plastomes from four Geraniaceae genera were examined including *P.* × *hortorum* (Palmer, Nugent, & Herbon, 1987), *Monsonia* (formerly *Sarcocolon*), *Geranium* and *Erodium* (Palmer, 1991). The results suggested that an entire suite of plastome anomalies were present within the family.

Technology has permitted the sequencing and assembly of genomes and Geraniaceae plastomes are no exception. While the tantalizing results of early Southern analyses hinted at the unusual, contemporary methods have uncovered some of the most bizarre plastomes among seed plants. Here, the unusual features of Geraniaceae plastomes will be discussed and, where possible, compared with other taxa. Many of the changes in the family may be found in other lineages (Table 1); however, it appears likely that no other group of plants represents the range of plastome variation seen in Geraniaceae.

2. THE GREAT AND THE SMALL

Among photosynthetic angiosperms Geraniaceae plastomes occupy extremes with regard to size, with a collection of phenomena that have inflated and diminutized them. As is the case in many groups, substantial changes in overall nucleotide content involve expansion and contraction of the IR. Plastomes have also been expanded through seemingly IR-independent repeat accumulation in the family and elsewhere. Although rare, incorporation of extraplastomic DNA (native mitochondrial) has influenced plastome size, as has deletion of canonical sequences from single copy and IR regions.

2.1 The Dynamic Plastome IR

Unsurprisingly, the smallest plastomes lack the IR; however, not all IR-less plastomes are small. Among Geraniaceae three genera contain species that lack the IR, *Monsonia*, *Geranium* and *Erodium* (Blazier, Jansen, et al., 2016; Guisinger, Kuehl, Boore, & Jansen, 2011; Ruhlman, Zhang, Blazier, Sabir, & Jansen, 2017; Zhang et al., 2016). These losses appear independent

Table 1 Plastome Anomalies Across Geraniaceae and Selected Angiosperms

Features / Genera	*Hypseocharis*	*Pelargonium*	*Monsonia*	*Geranium*	*Erodium*	*California*	*Oenothera*	*Vaccinium*	*Trifolium*	*Passiflora*	*Campanulastrum*	*Silene*	*Carnegiea*
IR expansion[a]	X	X	X	X	X	X		X		X	U	X	
IR contraction[a]	X	X	X	X						X	U		
IR loss			X	X	X				X		U		X
Repeats													
>100 bp	X	X	X	X	X	X		X	X	X	U		X
>1000 bp		X	X	X	X				X		U		
General rate acceleration		X	X	X	X				X	X	X	X	X
≥3 Inversions	X	X	X	X	X			X	X	X	U		X
PEP genes[b]													
rpoA		X		X						X			
rpoB		X		X									
rpoC1		X								X			
rpoC2		X											
accD[b]	X	X	X	X	X	X	X	X	X	X	X	X	X
clpP[b]		X	X	X	X	X	X	X	X	X	X	X	X
ycf1[b]		X	X	X	X	X	X	X	X	X	X	X	X
ycf2[b]		X	X	X	X	X	X	X		X	X	X	
NDH genes[b]		~			X		~	~					X
RPL		X	X	X	X				X	X	X		X
RPS		X	X	X	X			X	X	X			
Disrupted TU[c]		X	X	X	X				X		U		
Biparental inheritance[d]	*	D	*	D	*	*	D	*	P	D	D	P	*
PGI	*	X	*	X	*	*	X		X	X	X	X	

[a]IR expansion/contraction >5 kb.
[b]Divergent, pseudogenized or absent from plastome. For NDH, X indicates loss of all plastid-encoded NDH sequences, ~ indicates that variable individual NDH sequences are affected.
[c]TU = transcriptional units; broken or fused.
[d]Potential (P) or demonstrated (D) biparental inheritance. *Asterisk* indicates another family member was assayed.
Geraniaceae genera are indicated in *light green*. Each *colour* indicates that the representative genus belongs to a different family (key below).
■, Onagraceae; ■, Ericaceae; ■, Fabaceae; ■, Passifloraceae; ■, Campanulaceae; ■, Caryopyllaceae; ■, Cactaceae.
PEP, plastid-encoded polymerase; PGI, plastome-genome incompatibility; demonstrated or inferred from variegated hybrid progeny; RPL, large subunit ribosomal protein; RPS, small subunit ribosomal protein; U, complete plastome sequence is lacking; unknown; X, feature is present in the genus.

and no example suggests a mechanism. *Monsonia speciosa* has the smallest plastome sequenced in the genus at 128 kb; however, it retains a modest IR (7313 bp), encoding four protein genes and the ribosomal operon, but excluding *rrn16* (Guisinger et al., 2011). The finding that the highly rearranged sequences of *M. speciosa* included a drastically abbreviated IR seemed to support a hypothesis based on early studies of legume plastomes that lack the large repeat (IRLC; Wojciechowski, Lavin, & Sanderson, 2004): that the IR plays a role in conservation of plastome structure (Palmer, Osorio, Aldrich, & Thompson, 1987). The sequencing of two additional species in the genus, *M. marlothii* (~134 kb) and *M. emarginata* (~157 kb; Zhang et al., 2016), delivered more scrambled plastomes both of which completely lacked the canonical IR and bore little resemblance to their sister *M. speciosa* in terms of gene order. Indeed these two, which are sister species, bore little resemblance to each other. Like the IR-less *Trifolium subterraneum* (~144 kb), massive accumulation of repeats has inflated the larger *Monsonia* plastomes (Ruhlman et al., 2017; Zhang et al., 2016).

The first *Geranium* plastome sequenced contained both a reduced IR (~11 kb) and rampant accumulation of non-IR repeats including those that exceed 1 kb. The SSC of *G. palmatum* (~48 kb) comprises some genes typical of the region and many that are more commonly found in the LSC. While the IR does contain the ribosomal operon, the pseudogenized *ycf2* and the sequences that typically surround it are situated in the LSC (~85 kb; Guisinger et al., 2011). The plastome of *G. palmatum* (~156 kb) is typically sized but the similarities between the plastomes of this genus and those of most plants certainly end there. In *Geranium*, the presence or absence of the IR has little influence on plastome stability as both *G. palmatum* and the IR-less *G. incanum* are wildly rearranged. Lacking the IR has not reduced overall size in *G. incanum;* again arrays of tandem and dispersed repeats have bulked this plastome to more than 166 kb (Zhang et al., 2016).

Currently the smallest plastome sequences from photosynthetic angiosperms include IR-less *Carnegiea gigantea* (Cactaceae) at ~113 kb (Sanderson et al., 2015), members of two IR-containing Poaceae genera (*Triticum* and *Aegilops*; NCBI accessed 9/25/2017) ranging from ~113 to ~115 kb and *Erodium foetidum* at ~115 kb. Plastome sizes in *Erodium* clade II (Fig. 1) range from this low up to ~124 kb with 7 of 10 sequenced members below 120 kb. Loss of the IR has reduced the size of clade II plastomes in *Erodium*, nonetheless with regard to gene order and non-IR repeats accumulation they are rather uniform and display none of the anomalies predicted by hypotheses of IR stabilization (Blazier, Jansen, et al., 2016). The long branch clade

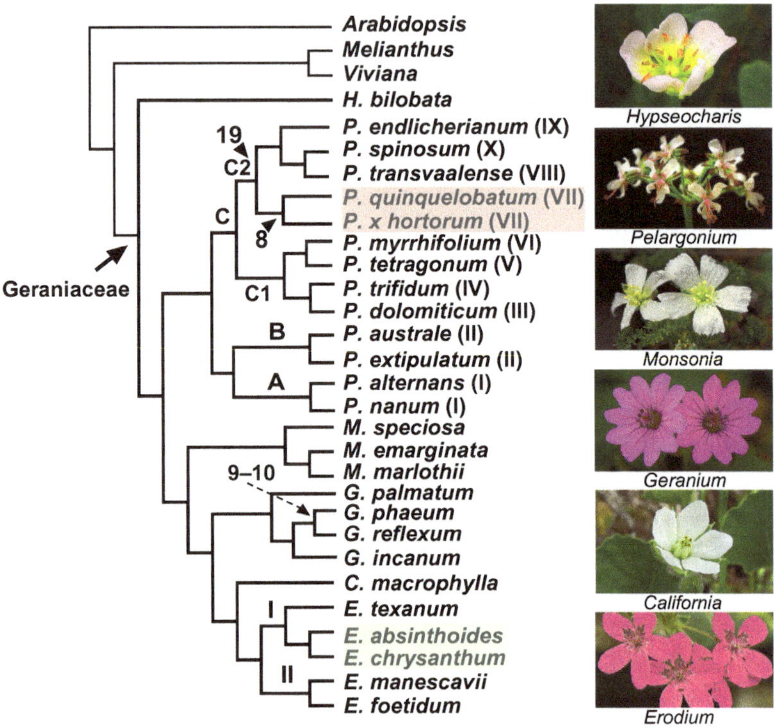

Fig. 1 Relationships among major clades of Geraniaceae. The cladogram is based on published molecular phylogenies (Blazier, Jansen, et al., 2016; Park et al., 2017; Ruhlman et al., 2017; Weng, Ruhlman, Gibby, & Jansen, 2012) and depicts the well-supported, established relationships in Geraniaceae. Clades within *Pelargonium* (A, B, C, C1, C2) and *Erodium* (I, II) are indicated at the nodes. The *Erodium* LBC is highlighted with a *green box* and *Pelargonium* section *Ciconium* taxa are highlighted in *pink*. *Roman numerals* following *Pelargonium* species names indicate plastome types according to Weng, Ruhlman, and Jansen (2017). Approximate divergence time estimates for clades mentioned in the text are indicated with an *arrowhead* and the value in MYA (million year ago; Bakker, Culham, de Marais, & Gibby, 2005; Park et al., 2015). All photos except *California* were contributed by Mao-Lun Weng. *The* California *image is freely available for academic purposes from ©2009 Andrew Borcher.*

(LBC; Blazier, Guisinger, & Jansen, 2011), named for its long branches in phylogenetic trees from plastid data, occurs within *Erodium* clade I. Sequenced representatives of the LBC contain pseudogene copies of all the plastid genes that encode subunits of the NAD(P)H dehydrogenase complex (NDH) and have a novel, large IR (~25.5–47.5 kb) that contains the entire ribosomal operon plus a range of protein- and tRNA-coding genes (Blazier, Jansen, et al., 2016). While the large IR identified in the LBC inflated their plastomes up to

~169 kb, its presence has not detectably influenced stability in these plastomes as they are the most highly rearranged among sequenced *Erodium*. Loss of the canonical IR as well as the NDH genes was reported for *C. gigantea* (Sanderson et al., 2015), a loss that contributed to shrinking its plastome. The relationship between IR loss and concomitant NDH gene loss is tenuous; however, as both autotrophic and heterotrophic orchids contain the IR and lack plastid NDH genes (Luo et al., 2014; Ruhlman et al., 2015).

The twofold influence of IR size variation on unit-genome size ranges in both directions in Geraniaceae. The genus *Pelargonium* contains species with the largest plastomes by a good measure and IR expansion has played a major role. Sequencing of *P.* × *hortorum* revealed a very large plastome (~218 kb) that included an IR that had expanded to encompass nearly 76 kb. Expansion predominately duplicated LSC sequence but also fully duplicated the *ycf1* pseudogene and included several other genes typically found in SSC (Chumley et al., 2006). The smallest *Pelargonium* IRs, and consequently plastomes, are nonetheless large relative to the average IR and plastome size for more than 1800 angiosperms, at ~25 and ~151 kb (NCBI accessed 9/25/17), respectively. Smaller *Pelargonium* plastomes range from ~165 to ~173 kb, while the larger representatives reach up to ~242 kb. In addition IR sizes range from ~36 kb up to nearly 88 kb. Plastome size scales with IR size fairly consistently in *Pelargonium*; the proportion of the plastome represented by IR sequence ranges from ~42% up to ~76% (Weng et al., 2017).

Both the earliest and the latest diverging genera of Geraniaceae (Fig. 1) resemble the inferred ancestral plastomes for the family. *Hypseocharis bilobata* shares the highly reduced ancestral SSC (~6.7 kb), while the exclusion of *ycf2* shortened the IR. The *H. bilobata* IR (~29 kb) contains the ribosomal operon and *ycf1* along with a few genes encoding NDH subunits, ribosomal proteins (RPs) and tRNAs (Weng, Blazier, Govindu, & Jansen, 2014). The large IR contributes ~35% of the *H. bilobata* plastome (~165 kb), unlike *Pelargonium* where plastomes of this size typically carry about 42% of their sequence as IR.

Perhaps the most surprising plastome in Geraniaceae comes from the monotypic genus *California*. Sister to *Erodium*, one would predict a highly rearranged plastome riddled with large repeats and pseudogenization events. On the contrary, the *C. macrophylla* (formerly *E. macrophyllum*) plastome (~149 kb) has just one inversion relative to the inferred ancestral arrangement. A relatively normal IR (~22 kb) is somewhat reduced by the loss of *ycf1* and a shortened *ycf2* pseudogene (Weng et al., 2014).

2.2 Accumulating Non-IR Repeats

Any discussion of plastome size must consider the contribution of IR expansion and contraction. However, as discussed, the presence and extent of the IR does not always account for size expansion of plastomes in Geraniaceae. Overall angiosperm plastomes are characterized as repeat poor. The gene dense unit-genome has evolved small intergenic sequences (IGSs) in which simple sequence repeats (commonly ranging from mono- to trinucleotide, rarely from tetra- to hexanucleotide) are found in typical plastomes. It is unlikely that small repeats only arise in IGS regions, rather it is in these regions of reduced functional constraint that they persist and perpetuate. In Geraniaceae the accumulation of relatively large, non-IR repeats has been unprecedented.

Tandem and dispersed repeats ranging in size from 16 to 3095 bp represent nearly one-third of the plastome in *M. emarginata* (Ruhlman et al., 2017). Repeats ranging in size from 15 to >2000 bp represent ~17% of *P.* × *hortorum* and *E. texanum* plastomes and ~27% of the *G. palmatum* plastome (Guisinger et al., 2011). Previous estimates of repeat content in *P.* × *hortorum* were more conservative, at 9% and only considered repeats ≥30 bp (Cai et al., 2008). It was suggested that the approach used previously to calculate the *T. subterraneum* repeat content (19.5%; Cai et al., 2008) was an underestimate (Guisinger et al., 2011). However, in that study repeat content of several groups of angiosperms was estimated by the same method for comparison and showed that other IRLC legumes had approximately threefold fewer repeats than *Trifolium* while *Vitis* contained 10-fold fewer (Cai et al., 2008). Both studies report ~2% repeat content for *Vitis* and ~4% for *Arabidopsis* plastomes. The different parameters used to evaluate repeat content, especially large repeat content, could have produced the incongruence in the studies. The most recent and comprehensive study of *Erodium* plastomes suggested that ~17% was likely an underestimate of the repeat content in *E. texanum*. Across the genus repeat content ranged from ≤1% in the compact *E. manescavii* plastome (~117 kb) to more than 23% in *E. texanum* (~131 kb). As expected, the less rearranged plastomes of clade II (Fig. 1) maintain repeat content in normal ranges; however, *E. guttatum* (clade I) reached ~18% and the LBC plastomes varied from relatively low (4.6%) to moderate (8.6%) and high (16.2%). As several studies have demonstrated, repeat content correlates positively with plastome rearrangement (Guisinger et al., 2011; Schwarz et al., 2015; Weng et al., 2014).

In plastomes where repeat content is low and repeats are generally small, genes are duplicated by the virtue of their inclusion in the IR and gene order appears static. When longer repeats arise, coding regions may be duplicated in SC regions or in the IR. In the repeat rich plastomes of Geraniaceae and others genes are often duplicated, likely by stochastic mechanisms indifferent to the genic nature of the sequence.

Repeat-mediated recombination was predicted to occur between the IR copies (Kolodner & Tewari, 1979) and was suggested to explain the presence of SSC inversion isomers observed in lettuce and spinach. Although the mechanism of inversion was misinterpreted and incorrectly dubbed 'flip-flop' recombination (Palmer, 1983), variation in plastome structure and conservation of sequence identity in plastomes are intimately connected to recombination. The recombination-dependent replication (RDR) DNA repair pathway could account for inversions and drive repeat accumulation in plastomes with large non-IR repeats (Maréchal & Brisson, 2010; Oldenburg & Bendich, 2015). *Monsonia emarginata* contains a number of large repeats, in fact ~22% of its plastome sequence comprises repeats ranging from ~1 to >3 kb making it a candidate in which to detect alternative plastome arrangements derived from RDR. The application of PacBio SMRT long read sequencing identified alternative arrangements as assembly of plastomes like that of *M. emarginata* with ~150 bp Illumina reads from ~750 bp insert libraries does not have the power to detect low level arrangement heteroplasmy as the vast majority of reads will be of the predominant type (Ruhlman et al., 2017). The core of the R20 repeat system of *M. emarginata* comprises five copies that share ~97% nucleotide identity over 2022 bp. PacBio reads that included the repeat and adjacent sequences revealed variation of adjacencies indicative of alternative sequence arrangement around the repeats. Replication initiation via recombination between the repeats within the same or different unit-genome copies could cause further expansion of repeat content, dispersal of sequence blocks and inversion of adjacent sequences (Ruhlman et al., 2017). In the absence of large repeats, most recombination is focused in the IR limiting rearrangements and repeat extension and accumulation.

3. CHANGE OR STAY THE SAME

Recombination between IR copies within a unit-genome, or between any part of the unit-genome and another copy in the highly iterative plastome is thought to maintain uniformity in typical angiosperm plastomes. When a mutation arises, it does so at a single locus. Persistence of the mutation or a return to the original state depends on recombination between individual copies (or alleles) of the locus. Given the large number of unit-genome copies that are available to template copy correction, it is curious that plastid genes diverge at all. However, so-called illegitimate recombination between homeologous sequences that accumulate in atypical plastomes may promote divergence in structure while homogenizing nucleotide sequences among repeat copies.

3.1 The Homogenizing Effect of Gene Conversion

One hallmark of RDR is GC-biased gene conversion, the copy correction mechanism that limits divergence between IR copies (Maréchal & Brisson, 2010; Oldenburg & Bendich, 2015). Gene conversion was identified among the R20 repeats of *M. emarginata* and likely explains the high identity they share across coding and noncoding portions (Ruhlman et al., 2017). Although situated in the very large IR of *Pelargonium*, the *rpoA* genes have not only strongly diverged from other Geraniaceae but are highly divergent both within and between clades in the genus (Blazier, Ruhlman, et al., 2016). Species in clades A, B and C1 (Fig. 1) each contain a single *rpoA* gene while clade C2 representatives, excluding section *Ciconum*, contain two or six copies of this IR sequence so that *P. transvaalense* has accumulated a total of 12 *rpoA* sequences per unit-genome. Maximum likelihood (ML) phylogenies of the *rpoA* open reading frame (ORF) copies suggested that gene conversion had occurred among the clade C2 ORFs. For *P. spinosum*, *P. endlicherianum* and *P. transvaalense* the *rpoA* ORFs grouped by species rather than with their ortholog(s) indicating that these sequences have not evolved independently since their duplication in the ancestor of C2 taxa. Conversely the three *rpoA* ORFs of other section *Ciconium* taxa grouped with their paralogs. Both manual inspection and OrgConv (Hao, 2010) analysis predicted recombination and gene conversion among the *Ciconium* *rpoA* ORFs and their adjacent sequences. It may be that the relatively recent divergence of these taxa influenced the phylogenetic placement of the ORFs in ML trees (Blazier, Ruhlman, et al., 2016).

While the situation with *rpoA* is highly unusual in *Pelargonium*, there are other groups where this gene is divergent including *Annona*, *Passiflora* and *Berberis*. Common to all four groups are particularly fluid IR boundaries, including expansions that duplicated *rpoA*. Despite their overall lack of conservation, the *rpoA* genes in these cases were predicted to encode all functional domains required to serve as the alpha subunit of the plastid-encoded RNA polymerase (PEP; Blazier, Ruhlman, et al., 2016). Sequences encoding the alpha subunit of PEP have also been duplicated in common ancestor of *Geranium phaeum* (~182 kb) and *G. reflexum* (incomplete draft; Park et al., 2017). These highly rearranged plastomes lack the typical IR and the direct repeat that duplicated *rpoA* is situated between sequences that flank this gene across typical plastomes (*psbB* to *petD*, *rpl16* to *rpoA*) near the ancestral LSC-IR_B boundary (J_{LB}). The *rpoA* paralogs in both species share ~88% nucleotide identity. This is considerably lower than the between paralog nucleotide identities observed among the *Pelargonium* ORFs, which approached 100% in *P.* × *hortorum* (Blazier, Ruhlman, et al., 2016). Homogenization of repeated sequences is another hallmark of gene conversion and likely maintains the high identity among the *rpoA* ORFs in clade C2 of *Pelargonium* (Fig. 1). However, gene conversion analysis did not predict recombination between the repeats that duplicated *rpoA* in *G. phaeum* and *G. reflexum* (Park et al., 2017). As mentioned, recent divergence of the species in *Pelargonium* clade C2 (Blazier, Ruhlman, et al., 2016) could also account for their high identity relative to the *Geranium rpoA* duplicates. Divergence time estimates for the lineage that includes *G. phaeum* suggest that these species diverged from *Geranium* species with a single *rpoA* sequence in their plastomes ~9–10 MYA (Park et al., 2015). *Pelargonium* clade C2, where repeats have produced up to six copies of the *rpoA* sequence, diverged from those that do not contain this feature (i.e. clade C1) more than 18.9 MYA and section *Ciconium* from other C2 species around 8.4 MYA (Bakker et al., 2005).

The duplication of sequences including *rpoA* likely occurred in the common ancestor of clade C2 species indicating that the event in *Pelargonium* is nearly twice as old as the lineage that includes *G. phaeum*. Therefore, it is reasonable to infer that gene conversion is responsible for the remarkably high sequence identity among *rpoA* repeats in *Pelargonium* while the *Geranium* examples are diverging in the absence of recombination between the repeats. The examples uncovered thus far that demonstrate very high nucleotide identities between *rpoA* paralogs are cases where the repeated sequences are inside the expanded IR. It could be that their IR location contributes to their propensity to undergo gene conversion as the *Geranium* plastomes with *rpoA* duplicates lack the IR.

3.2 Divergent or Missing Genes

In the previous section, the focus on repeats highlighted the strong sequence identity among paralogs undergoing gene conversion. In the case of *rpoA*, however, it is the extremely low identity to typical *rpoA* sequences that at one time was cited to suggest its loss as a functional gene. In silico identification of conserved domains within divergent sequences, rather than pairwise nucleotide identity, may turn out to be a better predictor of functionality in plastid genes. Like the PEP alpha subunit gene, several other plastid genes that had previously been assessed as putatively lost may indeed prove to be competent. Also like *rpoA*, there are examples that seem to turn up across disparate lineages.

3.2.1 ycf1 *and* ycf2

The plastid genes *ycf1* and *ycf2* are yet to be conclusively assigned a function in plant cells; however, both genes appear to be indispensable (de Vries, Sousa, Bölter, Soll, & Gould, 2015; Drescher, Ruf, Calsa, Carrer, & Bock, 2000). In typical plastomes these are the two largest coding sequences. In *Arabidopsis thaliana* plastomes the complete *ycf1* and *ycf2* span 5360 and 6884 bp, respectively. Given that *ycf2* is situated in the IR, this sequence alone represents ~9% of the unit-genome. As such, the loss of the bases that encode *ycf2* can significantly impact IR size as seen in Poaceae (Guisinger, Chumley, Kuehl, Boore, & Jansen, 2010) where the unit-genome size (~113–140 kb) is reduced relative to IR-containing nongrass species. Among Geraniaceae, *H. bilobata* contains full-length copies of both *ycf1* and *ycf2* although IR boundary movement has resulted in complete incorporation of *ycf1* in the IR and complete exclusion of *ycf2* (Weng et al., 2014) as in Campanulaceae (Cheon, Kim, & Yoo, 2017; Haberle, Fourcade, Boore, & Jansen, 2008). Both ORFs are located in the IR and highly divergent in *P.* × *hortorum*. The *ycf1* ORF is expanded to 7659 bp and nearly impossible to align outside of its terminal sequences and both sequences contain numerous indels. Overlapping and nested repeats of ≥31 bp were identified in both genes (Chumley et al., 2006; Downie, Katz-Downie, Wolfe, Calie, & Palmer, 1994). Despite their highly variable sequences, *ycf1* and *ycf2* are annotated as genes in the plastomes of *Pelargonium* clades A, B and C2 (Fig. 1). However, the two largest plastome types belonging to clade C1, *P. dolomiticum* (~192 kb, type III) and *P. trifidum* (~200 kb, type IV), each contain pseudogenized copies of these two sequences in expanded IRs that represent 76% and 75% of each plastome, respectively (Weng et al., 2017).

Given that *ycf1* and *ycf2* are listed among either pseudogenized or missing sequences for all other investigated taxa in *Monsonia*, *Geranium*, *Erodium* and *California*, their persistence in all but one small *Pelargonium* lineage is curious. *Passiflora* subgenera also show variation in the presence of full-length *ycf1* and *ycf2*, and concomitant variation in IR size despite IR expansion. Degradation of these sequences accounts for most IR size variation in subgenera *Passiflora* and *Decoloba*, but *P. pittieri* (subgenus *Astrophea*) contains putatively functional genes (Rabah et al., 2017). Several other lineages harbour highly divergent, degraded and pseudogenized copies of *ycf1* and *ycf2* accompanied by IR expansion (Table 1), often into the SSC including Campanulaceae (Cheon et al., 2017; Haberle et al., 2008), Poaceae (Guisinger et al., 2010) and Ericaceae (Fajardo et al., 2013; Martínez-Alberola et al., 2013).

The situation is reversed for IR loss in *C. gigantea*. This small plastome contains *ycf2* intact; however, the state of *ycf1* is less clear. The two are nearly adjacent on the same strand and while the *ycf1* reading frame is preserved, it is 'littered' with small tandem repeats with no homology to other plastome *ycf1* sequences (Sanderson et al., 2015). Although no allusion to the function of the *ycf2* gene product has been reported, one proposition for the product of *ycf1* was a role in plastid protein import as Tic subunit (Kikuchi et al., 2013). Others have speculated that its tendency to be lost or highly divergent along with the plastid *accD* gene, as in grasses, Geraniaceae and *Passiflora*, could suggest a role in assembly of the plastid ACCase holoenzyme (de Vries et al., 2015). Other than coincident divergence/loss, however, there is no indication that the products of these two genes have any interaction at all.

3.2.2 accD

While questions remain regarding the function of the two large *ycf* ORFs the gene product of *accD* and its role in fatty acid metabolism is well characterized. Although few reports describe efforts to uncover a nuclear transfer or substitution serving plastomes that apparently lack *ycf1* and/or *ycf2*, both scenarios have been identified for *accD*, sometimes in the same species (Jansen & Ruhlman, 2012).

All Geraniaceae in which this gene can be detected contain divergent *accD* sequences with a disrupted C-terminal functional domain (Zhang, Yang, Shen, & Tong, 2003) indicating that the insertion occurred in the common ancestor of the family (Fig. 2). The divergent sequences have retained the reading frame and while the conserved domain was truncated at the N-end, it retains the putative catalytic site sequence downstream of the disruption (Park et al., 2017). The size of the insertion is fairly conserved in the family and would represent ~550 amino acids in a translated protein. The sequence that disrupts the *Geranium accD* comprises tandem repeats and it is unclear whether the repeats expanded in individual lineages or in the ancestor.

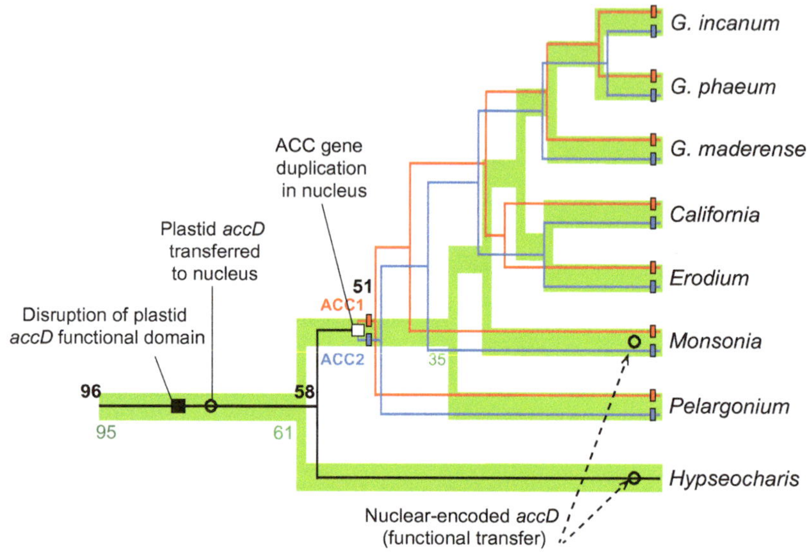

Fig. 2 Summary of nuclear *ACC* duplication and *accD* nuclear transfer events in Geraniaceae. Events of interest are indicated on the tree. *Numbers* at nodes indicate divergence time estimates in MYA (million years ago). The *green numbers* correspond to the background phylogeny (*green*) of the family. *Black branches* and divergence estimate trace time prior to the duplication of nuclear ACC. *Red* and *blue branches* correspond to the *ACC1* and *ACC2* orthologs, respectively. More detailed divergence time estimates for Geraniaceae are available in Park et al. (2017). *G, Geranium. Adapted from Park, S., Grewe, F., Zhu, A., Ruhlman, T.A., Sabir, J., & Mower, J.P., (2015) Dynamic evolution of* Geranium *mitochondrial genomes through multiple horizontal and intracellular gene transfers.* New Phytologist, 208, *570–583. Figure was redrawn and adapted from Park, S., Ruhlman, T.A., Weng, M-L., Hajrah, N.H., Sabir, J.S.M., & Jansen, R.K. (2017). Contrasting patterns of nucleotide substitution rates provide insight into dynamic evolution of plastid and mitochondrial genomes of* Geranium. Genome Biology and Evolution, 9, *1766–1780.*

Probing Geraniaceae transcriptome data identified a putatively functional *accD* in the nuclear genome of *Hypseocharis* and *Monsonia* exclusively, which lack a detectable plastome copy of the gene. The nuclear copies contain the truncated functional domain of the divergent plastid gene of other Geraniaceae, each with its own transit peptide. However, the bulk of the sequence upstream of the functional domain is missing in these nuclear copies, which differ in length by just two amino acids (Park et al., 2017). The presence of the truncated domain supports the disruption of plastid *accD* in a common ancestor of Geraniaceae but the lack of upstream sequences obscures the timing of repeat expansion.

Transcriptome data also provided convincing evidence for a duplication of the nuclear gene encoding the cytosol targeted eukaryotic ACCase across Geraniaceae, excluding *Hypseocharis* (Fig. 2). In all cases one copy carries a predicted N-terminal extension for plastid targeting allowing for substitution of the multisubunit prokaryotic type holoenzyme by the single polypeptide, nuclear-encoded protein in the family as suggested previously for *Poaceae* (Konishi & Sasaki, 1994) and some papilionoid legumes (Magee et al., 2010; Sabir et al., 2014). *Arabidopsis* plastomes encode *accD* and its nuclear genome houses both the duplicated, plastid-targeted ACCase along with the three other subunits of the prokaryotic holoenzyme (Babiychuk et al., 2011).

The detection of the *accD* nuclear transfer suggests that both ACCase types encoded in the nucleus may function in *Monsonia* (Fig. 2). If indeed the divergent *accD* gene encodes a functional subunit in other Geraniaceae genera, it is plausible that both forms are active in plastids across the species that contain them. Transcriptome data suggest this is the case as all three nuclear-encoded, plastid-targeted subunit genes (*accA*, *accB* and *accC*) were identified in all examined Geraniaceae (Park et al., 2017). For any plastome sequence encoding a required function there must logically be a period of time where both the gene's product and its nuclear-encoded replacement must be active in the plastid. Otherwise there would be a lapse in the required function and the evolution of a protein substitution would be halted. The small subunit ribosomal protein Rps16 provides an illustration of this phenomenon. The gene *rps16* is typically encoded in the plastome, but was missing from the plastome of *Medicago truncatula* and *Populus alba*, where its function has been substituted by the dual targeted mitochondrial *rps16* encoded in the nucleus (Ueda et al., 2008). As it turned out, several other species were examined and showed that there were at least two additional species that imported the nuclear-encoded protein, yet retained expression of the plastid-encoded gene. Additional species that contain a plastid-encoded *rps16* were also predicted to import the dual-targeted protein (Ueda et al., 2008).

The unusual evolution of the plastid *accD* gene is intriguing in that it is highly divergent or lost across disparate lineages of angiosperms (Table 1). Where the divergent plastome genes are characterized, there are often internal tandem repeat-mediated length variation that disrupts the functional domain but conserves the reading frame (Cai et al., 2008; Gurdon & Maliga, 2014; Magee et al., 2010; Nagano, Matsuno, & Sasaki, 1991;

Park et al., 2017; Rabah et al., 2017). In *M. truncatula* recombinationally active repeats in *accD* produced unique genes in each of 24 ecotypes examined. Ten lines were sequenced and showed the reading frame was preserved in each and, as in other cases, the C-terminal catalytic site remained intact. A similar pattern of repeat variation was observed for *ycf1*, although to a lesser extent (Gurdon & Maliga, 2014). An extended and highly divergent copy of *accD* was identified in 15 species of *Passiflora* representing three subgenera (Rabah et al., 2017). Also present in all examined species is an inversion of ~10 kb that includes *accD* at one end. The number of tandem repeat sequences upstream of functional domain is variable between species but in all cases the reading frame is conserved and there was high sequence identity across the catalytic site (Rabah et al., 2017). A presumed *accD* pseudogene was annotated for the *C. gigantea* plastome in GenBank. Although the ORF is of a typical size for angiosperms at 1493 bp there appear to be two *accD* fragments depicted on the unit-genome map. It is unclear whether the tandem repeats that intervene the two fragments on the map are included in the sequence reported for *C. gigantea accD* (Sanderson et al., 2015). The sequence is situated on the opposite strand some distance from *rbcL*, a result produced through inversion of the DNA including *rbcL* through *atpB*. Fragments of the *accD* sequence were detected in the regions between *rbcL* and *psaI* in the *Jasminum* and *Menodora* (Oleaceae) plastomes. This region usually comprises ~3 kb in diverse angiosperms but ranges from nearly 3 kb to more than 5 kb among the five Jasmineae examined (Lee, Jansen, Chumley, & Kim, 2007).

A nuclear-encoded AccD, the product of a gene of plastid origin, is imported by plastids of Campanulaceae where expression of the plastome copy is lost. Although there is no evidence to support the involvement of repeats, the nuclear copy is truncated at the N-terminus relative to typical *accD* genes resulting in a 311 amino acid polypeptide in *Trachelium caeruleum* (Hong et al., 2017; Rousseau-Gueutin et al., 2013).

3.2.3 clpP

Divergence or loss of the plastid gene *clpP* encoding a subunit of ATP-dependent caseinolytic protease has been documented in several unrelated lineages and for the most part these are the same groups discussed earlier. In Geraniaceae, there is variation in each genus with respect to *clpP*. Although the sequence looks typical in *Hypseocharis*, across the family one or both introns have been lost and substitution rates are accelerated; duplications, insertions and deletions have all played a role in *clpP* variation and/or pseudogenization. A genome-wide analysis of substitution rates in the family excluded *clpP* sequences as they were unalignable either as nucleotide or amino acid sequences (Guisinger, Kuehl, Boore, & Jansen, 2008).

In *Pelargonium* the *clpP* introns are present in clades A, B and C2; however, some clade C1 species lack both. Within clade C1 type III and IV plastomes (Fig. 1) have *clpP* duplicated in their expanded IR, as does clade C2, while clades A (type 1) and B (type II) and one lineage in clade C1 (types V and VI) carry a single copy of *clpP* (Weng et al., 2017). Regardless of its position relative to the unit-genome map, the proximal sequences up- and downstream of *clpP* are invariant in *Pelargonium* despite structural changes within the gene. The loss of *clpP* intron I was homoplasious in *Erodium* where it is absent some clade I taxa including LBC plastomes and lost in three distinct lineages of clade II (Fig. 1). In *E. texanum* (clade I) a fragment containing *clpP* and *rps12_5′* was duplicated and lies ~35 kb upstream in reverse orientation relative to the sequence found in the more common locus (Blazier, Jansen, et al., 2016). Both introns are missing from the highly divergent *clpP* gene in the three *Monsonia* plastomes available. Although lacking both introns in *M. emarginata*, the *clpP* coding region is expanded to 1044 bp and duplicated with ~84% and ~91% nucleotide identity over the entire sequence and the caseinolytic protease catalytic domain (516 bp). However, four identical smaller fragments of *clpP* 3′ sequence (63 bp) were identified that had >93% identity to the 3′ end of the duplicated gene. The 63 bp repeats are each contained within larger repeats with three copies arrayed between *trnA-UGC* and *rrn23* and interspersed with repeats of *trnI-CAU*, all of which are IR genes in IR-containing plants (Ruhlman et al., 2017). One of the large *clpP* sequences precedes *rps12_5′* and *rpl20*, preserving the predicted transcriptional unit (Hattori, Miyake, & Sugita, 2007; Kuroda & Maliga, 2003); however, the other lies upstream of the *rps12_3′-rps7-ndhB* transcriptional unit present in the IR of unrearranged plastomes and downstream from *ycf3*. In this example, as throughout the Geraniaceae, transcriptional units are broken and, ultimately, new ones potentially created. Whether either plastid-encoded copy of the *clpP* is functional in *M. emarginata* would have to be assessed at the protein level as the presence of transcripts and the production and accumulation of encoded proteins are largely uncoupled in plastids (Deng & Gruissem, 1987; Gruissem, Barkan, Deng, & Stern, 1988; Quesada-Vargas, Ruiz, & Daniell, 2005; Ruhlman, Verma, Samson, & Daniell, 2010). Positive signals in reverse transcription-PCR experiments could reflect read-through products and lack the specificity that defines operon transcription (Blazier, Ruhlman, et al., 2016; Lima & Smith, 2017; Shi et al., 2016; Stern & Gruissem, 1987).

Although all *Passiflora* examined shared the inversion that included *accD*, loss of the *clpP* intron was only detected for representatives of subgenera *Passiflora* and *Decoloba*. Subgenus *Astrophea* retains both introns; however, both subgenera *Astrophea* and *Passiflora* share an inversion that included the *clpP* and *rps12_5′* sequence at each end. In *P. auriculata* (subgenus *Decoloba*) the IR has expanded in both directions to include the *clpP* sequence (Rabah et al., 2017). Its IR location in this species would facilitate gene conversion between the two sequences, unlike in the *Geranium* cases; however, given the identical sequence of IR copies it would be challenging to detect specific sites. The intronless *clpP* ORF along with *rps12_5′* sequences also separated from *rpl20* in *C. gigantea*, while another copy of the *rps12_5′* gene lies between *rpl20* and a fragment of *clpP* (Sanderson et al., 2015). This fragment is described as a partial duplication in GenBank yet it is this region that has the canonical up- and downstream genes.

While the first intron of *clpP* was lost in the branch leading to the IRLC Fabaceae (Jansen et al., 2007), intron 2 has also been lost in *Glycyrrhiza glabra* (Sabir et al., 2014). The missing intron 2 sequence coincides precisely with the established exon borders of the *clpP* coding region, and the conserved C–U editing site in the proximal 3′ region, characterized in *Arabidopsis* (Chateigner-Boutin et al., 2008). The *clpP* sequences encoded by IRLC legumes are not exceptionally divergent except with respect to structure and are thought to be functional proteins. Like other IRLC plastomes, *Trifolium* lacks intron 1; however, the *clpP* sequence has been isolated from its transcription unit; *rps12_5′* and *rpl20* situated ~70 kb away. The region surrounding *clpP* in *Trifolium* was described as novel DNA in that it lacked homology to any sequence in the NCBI databank at that time, plastome or otherwise (Cai et al., 2008). Subsequent investigation took advantage of a much improved database and found that the vast majority of the more than 20 kb of novel DNA in fact showed identity to other plastome sequences (Sabir et al., 2014).

Two branches were indicated with accelerations of nonsynonymous substitution rates in *clpP*, one leading to the IRLC and again in *Lathyrus* (Schwarz et al., 2017). The typically more conservative plastomes of mimosoid legumes have also experienced nonsynonymous rate acceleration in *clpP* (Dugas et al., 2015; Schwarz et al., 2017; Williams, Boykin, Howell, Nevill, & Small, 2015), particularly on the branch leading to *Acacia* and to *Inga* (Dugas et al., 2015). The gene in *A. ligulata* retains two introns and reverse transcription PCR suggested that these were accurately spliced following transcription, but it lacks an invariant aspartate required for catalytic activity. Furthermore tests of selection concluded that unlike other mimiosoid *clpP* sequences under purifying selection, the signal detected for the *A. ligulata* branch in the *clpP* phylogeny suggested that this sequence is not under selection ($dN/dS = 1.07$; Williams et al., 2015). The *Inga* gene has lost intron 1 but still maintains a *clpP* ORF of more than 600 bp. Analyses of evolutionary rates indicated greater nonsynonymous change in the *Inga* gene relative to *Acacia* and revealed that the branch subtending the two taxa may have experienced positive selection (Dugas et al., 2015).

Geranium plastomes contain an ORF that retains the *clpP* functional domains but lacks introns. Like the *rpoA* sequence in *G. phaeum* and *G. reflexum*, the fragment containing *clpP* and *rps12_5′* was tandemly duplicated in those plastomes. Although the duplicated *Erodium* fragments encoding the same genes were not examined for recombinant activity (Guisinger et al., 2011), in *Geranium* GENCONV analysis did not detect gene conversion among the paralogs (Park et al., 2017).

In *Sileneae* (Caryophyllaceae), duplication, repeat insertion, intron loss and rate acceleration in *clpP* genes were associated with positive selection (Erixon & Oxelman, 2008). Positive selection was also proposed for *clpP* in *Campanulastrum americanum* and *T. caeruleum* (Campanulaceae) based on the elevated substitution rate ratios and comparison between intron and exon ML phylogenies (Barnard-Kubow, Sloan, & Galloway, 2014). Likewise, *Oenothera flava* (Onagraceae), which contains the *clpP* exons, displayed more variability in those sequences than in introns (Erixon & Oxelman, 2008). The most extreme rate accelerations in *Silene* occurred in lineages that lacked both introns precluding the comparison. Additional copies of *clpP* sequences were detected in *Lychnis chalcedonica* and *S. fruticosa*. Of the four copies examined in *L. chalcedonica*, one (Lc1) was identifiable as the functional copy, although one (Lc3) appeared less divergent than the others. The authors propose an ancient duplication of *clpP* gave rise to at least Lc3 and the event preceded the rate increase and subsequent positive selection. A similar observation was made for *clpP* repeats in the *S. shafta* plastome. Positive selection on *clpP* in *Silene* and *Lychnis* plastomes, where repeat expansion preserved the ORFs, could suggest that repetitive insertions have some benefit, possibly providing a new source of variation (Erixon & Oxelman, 2008). Where *clpP*, or indeed any gene, has been duplicated, relaxation of purifying selection due to the removal of functional constraint could permit at least one copy to diverge (Hahn, 2009). Studies supporting a role for positive selection on plastome sequences (Hu et al., 2015), particularly in lineages with divergent or repeated sequences including *clpP* (Erixon &

Oxelman, 2008; Guisinger et al., 2008; Rockenbach et al., 2016; Weng, Ruhlman, & Jansen, 2016), are dispelling the notion that plastome sequences are immune to nonneutral evolution.

The many examples of *clpP* loss or divergence among distantly related groups typically occur in plastomes that have experienced upheaval in their structure and are often accompanied by rate accelerations in *clpP* and other plastome sequences. Plastomes that are less rearranged nonetheless experience occasional gene loss or divergence. Although some species of *Sileneae* are rearranged relative to typical plastomes (Sloan, Triant, Forrester, et al., 2014) they do not show the same degree structural divergence as Geraniaceae, *Trifolium* or some *Passiflora*. *Jasminum* presents another intermediate case as the plastid *clpP* gene lacks both introns and contains insertions, and its *accD* gene was reported as lost in one lineage (Lee et al., 2007), but overall the Jasmineae plastomes lack highly variable rearrangements and substitution rate acceleration.

4. KEEPING UP WITH THE RATE RACE: ACCELERATION AND COEVOLUTION

Geraniaceae plastomes have experienced structural changes that have enlarged and diminished them including IR boundary changes and IR loss, accumulation of repeated sequence and sequence loss (Weng et al., 2014). Because the plastid unit-genome is iterative, gene conversion, one of the mechanisms responsible for maintaining plastome uniformity, can also participate in elevating evolutionary rates or driving mutations to fixation. Repeat content has been linked to rate acceleration in Geraniaceae and suggests a role for recombination between repeated sequences in the acceleration process (Guisinger et al., 2011; Weng et al., 2014). While the plastome wide mean for the rate of nonsynonymous substitutions (*dN*) is significantly higher in Geraniaceae than in other angiosperms (Guisinger et al., 2008), within the family two classes of genes were identified that are significantly accelerated relative to genes involved in photosynthesis.

4.1 Ribosomal Proteins

Early in the evolution of plant cells the vast majority of endosymbiont genes were transferred to the host nucleus in a process that remains ongoing (Kleine, Maier, & Leister, 2009; Matsuo, Ito, Yamauchi, & Obokata, 2005; Noutsos, Richly, & Leister, 2005; Stegemann & Bock, 2006; Stegemann, Hartmann, Ruf, & Bock, 2003; Timmis, Ayliffe, Huang, & Martin, 2004). More recent transfer or loss-and-substitution events may have included divergent genes like *accD*, with both scenarios proposed for different species in Geraniaceae. Far and away the most common group of genes lost from angiosperm plastomes are those that encode ribosomal proteins (RPs). This class of genes has been studied both in rearranged plastomes and those depauperate in plastome structural changes. A number of RP losses are found at deeper nodes and represent synapomorphies, while others are scattered across the angiosperm phylogeny (Ruhlman & Jansen, 2014). With respect to the structure or loss of this specific class of proteins, Geraniaceae and other groups with atypical plastomes are not outstanding. With regard to evolutionary rates, however, RPs in Geraniaceae do stand out demonstrating both lineage-specific and locus-specific accelerations (Blazier, Jansen, et al., 2016; Guisinger et al., 2008; Park et al., 2017; Weng et al., 2016).

Both large and small subunit RP genes were affected by accelerations in synonymous substitution rate (*dS*) and *dN*. The lack of acceleration in photosynthetic genes suggested a locus-specific effect. ML trees for the fastest evolving RP genes indicated a lineage-specific effect where the branches leading to the most recent common ancestor of the family and to *E. chrysanthum* (LBC; Fig. 1) had rapidly accumulated both synonymous and nonsynonymous mutations (Guisinger et al., 2008). A model-based ML analysis of evolutionary rates in *Erodium* that employed sequences for two large and two small subunit RPs confirmed significant acceleration on the branch leading to the LBC (Fig. 1). Acceleration of *dS* was confirmed for all four RPs and all but one gene showed significant acceleration of *dN* (Blazier, Jansen, et al., 2016).

Nonsynonymous substitutions in two interacting proteins should occur sequentially under a model of coevolution. ML reconstruction of ancestral sequences was used to evaluate 49 protein pairs that had nonsynonymous substitutions on the same branches of the Geraniaceae phylogeny (Weng et al., 2016). The three plastid-encoded RPs (Rps4, Rps12 and Rpl2) that conformed to the model for coevolution with nuclear-encoded RPs are involved in ribosome assembly in the plastid suggesting that their required function is constraining their divergence. Branch-site tests, which allow detection of codon-specific positive selection ($dN/dS > 1$) in prespecified lineages, detected positive selection in 5 of 20 plastid-encoded RPs but not in photosynthetic genes in Geraniaceae. Sites under selection in three small subunit proteins, Rps2, Rps4 and Rps7, lie within 10 Å of a residue in another subunit in the ribosome (Weng et al., 2016). Branch-site tests did not detect positive selection among the nuclear-encoded genes evaluated in the study, including subunits of the plastid ribosome, cytoplasmic RP and non-RP genes whose products are targeted to the plastid nor in the plastid-encoded photosynthetic genes. With regard to *dN/dS* Geraniaceae was significantly higher than the outgroups for plastid-encoded and nuclear-encoded plastid-targeted RP suggesting that there may have been a relaxation of purifying selection in the family (Weng et al., 2016).

When amino acid substitutions, indicated by *dN*, occur in a protein that assembles into a multisubunit complex the interacting proteins could be driven to make accommodating, or compensatory, changes to preserve the function of the complex. Given the very different rates of nucleotide substitution in plastids and the nucleus, and that plastid ribosomes comprise both nuclear- and plastid-encoded subunits, there is likely to be coevolution between the sequences of interacting proteins. Physical interaction was predicted to be a driver of coevolutionary change; however, the proximal residues in plastid- and nuclear-encoded ribosomal subunits that showed positive selection did not show corresponding signatures of coevolution. Compensatory coevolution may be occurring between nonproximal residues or could be driven by the ribosomal RNA constituents as they provide the foundation for ribosome assembly and displayed a high degree of indel variation that could account for changes in both plastid- and nuclear-encoded subunits (Weng et al., 2016).

Comparison between nuclear-encoded cytosolic and plastid-localized ribosomal subunits in *A. thaliana* showed that genes encoding organelle-targeted subunits had significantly higher *dN/dS* and the difference was driven by nonsynonymous changes (Sloan, Triant, Wu, & Taylor, 2014). The absence of significant variation in *dS* among the nuclear-encoded RPs indicates that all had similar underlying mutation rates and suggests that the variation in *dN/dS* arose through differential selection on amino acid substitutions. Interspecific divergence between *A. thaliana* and *A. lyrata* relative to intraspecific polymorphism was examined to clarify whether positive selection or relaxed purifying selection was differentiating *dN/dS* between cytoplasmic and organelle RPs (Sloan, Triant, Wu, et al., 2014). If nonsynonymous substitutions are fixed rapidly, as under positive selection, the ratio of nonsynonymous to synonymous substitutions should be elevated relative to intraspecific polymorphism (MK test; McDonald & Kreitman, 1991). Both cytosolic and plastid RPs exhibited similar ratios in polymorphism and divergence indicating that relaxed purifying selection, rather than positive selection has shaped the evolution of RPs in *Arabidopsis*. Similar comparisons carried out using pairs of species in *Silene*, where plastome substitution rates are accelerated in *S. conica* and *S. noctiflora*, suggested rapid divergence in plastid-targeted RPs but not in cytosolic RPs (Sloan, Triant, Wu, et al., 2014). Although the findings support the hypothesis that changes in plastid-encoded RPs have driven compensatory mutation in nuclear-encoded RPs the evolutionary forces driving coordinated acceleration in were not investigated within a phylogenetic framework to account for the effects of shared ancestry.

4.2 Plastid-Encoded RNA Polymerase Subunits

The plastid ribosome assembles with more nuclear-encoded proteins than any other complex comprising constituents encoded in both compartments; 9 small and 22 large subunit RPs are imported (Yamaguchi & Subramanian, 2000, 2003). Smaller assemblages, such as the plastid-encoded RNA polymerase (PEP) holoenzyme, provide another platform to examine plastid-nuclear coevolution, particularly where PEP subunit genes are highly divergent as in Geraniaceae. Rate comparisons for the family demonstrated significant rate accelerations in nonsynonymous substitutions in all four plastid-encoded PEP subunits (*rpoA*, *rpoB*, *rpoC1* and *rpoC2*; Guisinger et al., 2008). In the *Pelargonium* C2 (Fig. 1) clade the *rpoA* genes have undergone duplication and rapid divergence in both structure (Section 2.1) and nucleotide sequence (Blazier, Ruhlman, et al., 2016).

Early studies predicted that the *rpoA* sequence was missing or highly divergent in *P.* × *hortorum* (Palmer, Nugent, et al., 1987). Indeed, subsequent sequencing of the *P.* × *hortorum* and other *Pelargonium* plastomes depicted a range of structural divergence (Chumley et al., 2006), accumulation of repeated copies and ongoing gene conversion for *rpoA* sequences from *Pelargonium* clade C2 (Fig. 1; Blazier, Ruhlman, et al., 2016). Duplication and divergence of *rpoA* was also demonstrated for one clade in *Geranium*. The *rpoA* copies have diverged from each other giving rise to *dN* branches up to ~15 times longer for one copy and *dS* up to ~22 times longer. Although six branches in the *Geranium* phylogeny for *rpoA* showed signals of positive selection ($dN/dS > 1$), only one branch was significant after likelihood ratio test (LRT), the *phaeum*/*reflexum* branch ($dN/dS = 4.7$). Lineage-specific accelerations were detected in the three other PEP subunit genes but only *rpoB* remained significant after LRT (Park et al., 2017).

LRTs confirmed that dN/dS were significantly different in Geraniaceae taxa relative to other angiosperms when *rpoB*, *rpoC1* and *rpoC2* are considered, but not *rpoA*. Elevated dN/dS indicated that positive selection or relaxed purifying selection had acted on the Geraniaceae genes (Guisinger et al., 2008). Using analyses based on sequence alignment is problematic when considering genes like *Pelargonium* clade C2 *rpoA* as the extended and highly divergent ORFs permit only a portion of the clade C2 sequences to be aligned and analysed. A comprehensive analysis that included all four plastid-encoded PEP genes from *Pelargonium* but excluding clade C2 (Fig. 1) returned uniformly low dN/dS for *rpoA* regardless of the alignment method or outgroups employed and suggests purifying selection had acted along the branches of interest. Several branches had $dN/dS > 1$ for the other subunit genes. Sequences of *rpoB*, *rpoC1* and *rpoC2* all had elevated dN/dS on the branches leading to clades A and B. The branches leading to each clade were very similar except for *rpoC2* on the branch leading to clade A, where $dN/dS \geq 1$. On the branch leading to clade C1 dN/dS was elevated for *rpoC1* and *rpoC2* but not *rpoB* (Blazier, Ruhlman, et al., 2016).

Comparison of nucleotide substitution rates in the introns and exons of *rpoC1* for *Pelargonium*, revealed that the rate in exons was 1.3–10.6 times higher than for intron sequences across the genus. In clade C1, 30/33 branches had higher rates in exons relative to the intron. Differences between plastid genes were predominantly recorded for dN, and the *rpoC1* gene had the highest dN. LRTs confirmed significant acceleration of dN on the branches leading to Geraniaceae, *Pelargonium*, *Pelargonium* clades A and B, and branches within clade C (Fig. 1; Weng et al., 2012).

The curious state of the PEP genes encoded in Geraniaceae and especially *Pelargonium* plastomes led to speculation over the fate of the *rpoA* gene, which was at one time considered to be a pseudogene in *P.* × *hortorum*. Perhaps plastid DNA, containing a copy of the *rpoA* coding region was transferred to the nucleus where it gained the required constituents for expression and plastid transport. This would leave the plastome copy free to diverge as its encoded function would then be redundant. The nuclear-encoded alpha subunit could have evolved new ways of associating with the other subunits, thereby driving rate acceleration in those genes as part of a compensatory mechanism. Several studies have shifted the prevailing view of *rpoA* and PEP evolution in *Pelargonium* plastids. Transcriptome data representing nuclear gene space were generated through deep sequencing and analysed for *P.* × *hortorum* (Zhang, Ruhlman, Mower, & Jansen, 2013). Despite the identification of all six sigma factors that direct PEP-mediated transcription via promoter recognition and subunit recruitment, no sequences with homology to *rpoA* were detected suggesting that a protein encoded by one of the plastome ORFs may indeed assemble with the other PEP subunits, as previously suggested (Chumley et al., 2006; Zhang et al., 2013).

Assembly of the PEP holoenzyme upstream of the transcription start site is preceded by sigma factor binding to a PEP promoter element. The specificity of sigma factor recognition varies; some sigma factors recognize and bind to a range of sequences while others recognize the promoter sequences for a single gene (Chi, He, Mao, Jiang, & Zhang, 2015; Lysenko, 2007). Transcriptomes from two *Erodium* species lacked the *sig4* transcript (Ruhlman et al., 2015). However, these LBC *Erodium* no longer encode functional NDH genes including *ndhF* (Blazier et al., 2011), for which Sig4 activates transcription (Favory et al., 2005), nor were the nuclear-encoded NDH subunit sequences detected (Ruhlman et al., 2015). The loss of nuclear and plastid-encoded genes for the NDH complex is an example of coevolution among proteins encoded in each compartment.

Several other strong correlations of *dN*, but not *dS*, were detected between the plastid- and nuclear-encoded PEP subunits, but not between the PEP subunit sequences and nuclear or plastid control genes unrelated to the polymerase holoenzyme (Zhang, Ruhlman, Sabir, Blazier, & Jansen, 2015). In this study, the phylogenetic context of analyses allowed the effect of shared phylogeny to be considered in predictions and inferences of coevolution. Despite overall rate differences between the two compartments, correlations of *dN* were identified between *rpoB* and *sig1*, and between *rpoC2* and *sig2*. Significant correlations of *dN*/*dS* were detected between subsets of PEP genes encoded in the different compartments, but the rate and ratio for *rpoA* were uncorrelated. Structurally mediated coevolution would act at interaction residues in the assembled complex; however, none of the amino acid pairs predicted to be coevolved were involved in direct interaction between the plastid- and nuclear-encoded subunits (Zhang et al., 2015).

Unlike the study using interacting RPs encoded in different cellular compartments (Weng et al., 2016), the PEP study did not include analyses to predict the direction of compensatory coevolution. The exact nature of interaction and coevolution between the PEP sequences will have to be more fully explored. It could be that the divergence in *rpoA* drives compensatory changes in the other plastid-encoded subunits, with subsequent changes arising in nuclear-encoded sigma factors, although the role of contact site evolution between interacting residues may not be a major factor (Zhang et al., 2015).

Background mutation rate, as inferred from synonymous substitutions, suggests that angiosperm nuclear genes should drive coevolutionary processes as their protein-coding sequences have the faster rates by approximately fivefold (Drouin, Daoud, & Xia, 2008). Both *dN* and *dS* were compared between 59 plastome and 102 nuclear gene sequences common to 27 species of Geraniales (Zhang et al., 2016). The average *dN* of nuclear genes was approximately three times that of plastid genes and *dS* was about four times higher in nuclear genes. Using the same gene sets to represent nuclear and plastid genes from Brassicales for comparison demonstrated that the Geraniales genes were accelerated in *dN* for both compartments, while acceleration of *dS* was only identified for plastid-encoded genes in Geraniales relative to Brassicales (Zhang et al., 2016). So while rate acceleration is seen for *dN* and *dS* in Geraniales plastomes, it nonetheless appears to be fairly consistent with other angiosperms with respect to ratio for mutation rates between plastome and nuclear sequences. The Geraniaceae substitution rates were calculated on a dataset collected with another evolutionary question in mind, which could be influencing the outcome of the comparison. Considered were 59 plastome genes, 33 nuclear genes that are targeted to the plastome, 19 to the mitochondrion and 20 to other cellular locations. Approximately half of the sequences included in the nuclear gene set are targeted to the plastome, where they will interact with accelerated and divergent sequences in Geraniaceae. In fact *dN* values that were significantly accelerated included only plastid-encoded and plastid-targeted sequences (Zhang et al., 2016).

4.3 More Perplexing Complexes

Nucleotide substitution rate variation was used to investigate the constituent sequences of other plastid-localized complexes that have already been introduced in previous sections on structural divergence. Studies have demonstrated correlation between rearrangements and acceleration of substitution rates in Geraniaceae plastomes. A relationship between structural divergence and substitution rates in specific gene sequences is intuitive, and in Geraniaceae this seems to be the case.

4.3.1 ATP-Dependent Caseinolytic Protease

Like *rpoA*, *clpP* sequences were duplicated in one lineage of *Geranium*, in the common ancestor of *G. phaeum* and *G. reflexum*, and episodes of repeat expansion have disrupted *accD* across the family. Substitution rates are elevated in the intronless *clpP* gene across *Geranium*; *dN* estimates were

~threefold higher than other plastid-encoded genes. The duplicated *clpP* and *rpoA* copies in the *phaeum*/*reflexum* lineage are diverging from their paralogs with respect to rates. Branch length for one *clpP* copy was double that of the other in *dN* and more than seven times longer for *dS* (Park et al., 2017).

The relationship between structural evolution and rate acceleration in *clpP* was examined using intronless *clpP* ORFs from *Geranium* and *Monsonia*. Conserved domain sequences from representative angiosperms were selected and aligned for estimation of substitution rates. Multiple lineage-specific accelerations were recorded, all in groups where *clpP* genes have experienced intron loss, and *dN* and *dS* were more strongly correlated in groups that contain structural alterations. LRTs supported significant difference on several branches within *Geranium* and leading to *California*. Lineage-specific rate accelerations were detected for *clpP* sequences from *Vaccinium*, *Viviana*, IRLC legumes and *Oryza* (Park et al., 2017), all of which have experienced intron loss in this gene. The relationship between structural change and rate acceleration seems to extend beyond Geraniaceae.

Pursuing earlier findings in *Silene*, the nuclear-encoded subunits of the Clp protease were retrieved from the assembled transcriptomes of six species (Rockenbach et al., 2016). Three of the species had structural changes and highly accelerated *dN* in the plastid *clpP* while the other three species lacked structural changes and substitutions rates were low (Sloan, Alverson, Wu, Palmer, & Taylor, 2012; Sloan, Triant, Forrester, et al., 2014). The species with rapidly evolving plastid *clpP* sequences also contain nuclear-encoded subunit genes that have experienced accelerated rates of nucleotide substitution leading to elevated values in *dN*/*dS* (Rockenbach et al., 2016). In both *S. conica* and *S. noctiflora* *dN*/*dS* of the concatenated nuclear gene set was significantly >1 and in *S. paradoxa* the value was ~1. The acceleration of *dN* drove the increase in *dN*/*dS*; *dS* was nearly constant across all species while those with typical *clpP* genes had values from 0.05 to 0.16 for *dN*/*dS* in nuclear-encoded subunit genes. Striking observations included rate differences in genes encoding the ClpR subunit, which assembles in the same structural ring as the *clpP* gene product, and amino acid substitutions that were predominantly situated within domains that interact most closely with the plastid-encoded subunit. Population level data for *S. conica* were implemented in MK tests to investigate the prediction of positive selection in nuclear sequences encoding Clp protease subunits. Unlike the instance described earlier for *Arabidopsis*, significantly more nonsynonymous divergence from *S. latifolia* was detected relative to levels of nonsynonymous and synonymous polymorphism within *S. conica* (Rockenbach et al., 2016).

Similar to *Geranium* (see Section 3.2.2) the relationship of *accD* sequence divergence and changes in structure were noted for *Viviana marifolia* (Vivianaceae, Geraniales) and one lineage in *Pelargonium*, which each had disruptions in the *accD* conserved domain (Park et al., 2017). Higher *dN* and *dS* values were estimated for branches leading to Geraniaceae and to *Viviana*, suggesting the coupling of rate acceleration and structural divergence extends beyond the family and may not be confined to a specific gene or functional group.

4.3.2 Acetyl-Coenzyme A Carboxylase

When confronted with plastome genes that have experienced a great deal of change so as to make their functionality suspect the hypothesis of nuclear transfer or substitution is often proposed. During the transition from dependence on a plastid-encoded to a nuclear-encoded function both gene products must for a time be acting in the plastid. During this time and under a paradigm that is not well understood the redundant plastid gene may functionally diverge or degrade into a pseudogene. Although no gene that could represent a functional replacement for *rpoA* was uncovered in deep sequenced transcriptomes from *Pelargonium* (Zhang et al., 2013), searches for an *accD* transfer or replacement turned up evidence to support both phenomena are likely present in Geraniaceae (Fig. 2). In all species examined from the family the three genes encoding the other subunits of the heteromeric ACCase holoenzyme were discovered (Park et al., 2017). The substitution by the monomeric ACCase in *S. noctiflora* was also predicted when a duplication of the nuclear gene was identified and the nuclear-encoded subunits that assemble with AccD in typical angiosperms were riddled with nonsynonymous substitutions (Rockenbach et al., 2016).

Multisubunit proteins with constituents encoded in different compartments must coevolve to enable continued interaction and functionality. In species with structurally divergent or accelerated gene sequences in their plastomes, the dependence on coevolved complexes can lead to incompatibility in hybrids with different genetic constitutions (Greiner & Bock, 2013; Greiner, Rauwolf, Meurer, & Herrmann, 2011). When new parental combinations in the nucleus yield offspring that lack coevolved components for plastid complexes, the result can range from embryonic lethality to much more subtle effects. Sterility, hybrid variegation and other phenotypes inevident under nonstressful conditions can contribute to and reinforce reproductive barriers and ultimately participate in speciation processes.

5. STAYING IN SYNC: HYBRID HARMONY OR DISSONANCE

Hypotheses have suggested that a major impetus for the ongoing transfer of genes formerly encoded in the endosymbiont to the host nucleus could be related to the different mutation rates between compartments (Brandvain & Wade, 2009), although in typical angiosperms plastome rates are approximately one-fifth that of the nucleus (Drouin et al., 2008). Perhaps the ameliorating effect of sexual recombination on deleterious mutations has driven transfer of organelle genes to the nuclear genome. In groups like Geraniaceae, where substitution rates are elevated overall and especially for some protein coding genes involved in the formation multisubunit complexes, it may be that the nuclear environment limits divergence from nuclear-encoded subunits. Duplication of gene sequences occurs in nuclear genomes as well, by segmental or whole genome duplication (Bennetzen, 2000; Jiao et al., 2012). Plastid-targeted duplicates are often eventually reduced to single copy sequences (De Smet et al., 2013) and retention of paralogs in different genomic contexts or those that have diverged from their progenitor may be retained in different species or cultivars/ecotypes (Scannell, Byrne, Gordon, Wong, & Wolfe, 2006). Over time, coevolution of nuclear and plastid-encoded subunits is established and maintained.

The hybrid variegation documented so long ago by Baur is symptomatic of plastome-genome incompatibility (PGI) in *Pelargonium*, which inherits its plastids from both parents. Offspring of conspecific matings within local communities should not give rise to variegated leaf sectors so long as the plasmotype of each parent is compatible with the hybrid nucleus (Tilney-Bassett, 1973; Tilney-Bassett & Almouslem, 1989) and thereby able to reconstitute plastome complexes comprising subunits from both compartments. Hybrid variegation and other abnormalities have been attributed to PGI in several lineages, including representatives of *Oenothera* (Chiu, Stubbe, & Sears, 1988; Kirk & Tilney-Bassett, 1978), *Zantedeschia* (Yao, Cohen, & Rowland, 1994), *Medicago* (Lesins, 1961; Lilienfeld, 1962, 1965; Masoud, Johnson, & Sorensen, 1990; Schumann & Hancock, 1989), *Passiflora* (Hansen, Escobar, Gilbert, & Jansen, 2007; Rabah et al., 2017), *Pisum* (Bogdanova, 2007; Bogdanova & Kosterin, 2006; Bogdanova et al., 2015), *Campanulastrum* (Barnard-Kubow, McCoy, & Galloway, 2017; Barnard-Kubow, So, & Galloway, 2016) and *Campanula* (see Greiner et al., 2011).

Among these examples some were thought to inherit their plastids in a predominantly uniparental, maternal fashion and others biparentally (Corriveau & Coleman, 1988; Zhang, Liu, & Sodmergen, 2003). Whether the biparental inheritance of plastids is occasional or typical in a given lineage, the frequency with which the presence of underrepresented plasmotypes is detected in hybrid plants depends on the type of plastome markers and experimental protocols employed. Although many examples of hybrid variegation and PGI come from taxa that are known to inherit both parental types, intraspecific crosses from populations isolated in different geographic locations can induce paternal contribution of plastids to overcome incompatibilities in the cells of plants that were thought to inherit their plastid predominantly from the maternal parent (e.g. *Campanulastrum*; Barnard-Kubow et al., 2017). Inheritance studies in *Oenothera* (Chiu et al., 1988) explored the frequency of biparental transmission using a constant nuclear background as host to the four of the five major plasmotypes recognized in the genus (Greiner et al., 2008), which displays a consistent and strong maternal inheritance bias (Kirk & Tilney-Bassett, 1978). Depending on the plasmotype introduced through reciprocal crosses the frequency of biparental transmission of plastomes ranged from 0% to 56% (Chiu et al., 1988). A similar mechanism may be at work in *Passiflora* where all interspecific crosses primarily transmitted plastomes to progeny paternally while all intraspecific crosses had primarily maternal inheritance (Hansen et al., 2007).

These observations may illuminate why some lineages that have divergent plasmotypes have reverted to biparental transmission of organelles, which is thought to be the angiosperm ancestral state (Zhang & Sodmergen, 2010). Coevolution of interacting sequences supports speciation by providing a reproductive barrier to limit successful hybridization. Oppositely, evolution of biparental inheritance of plastids provides a mechanism to stabilize the hybrid nucleus by alleviating incompatible interactions between the hybrid nucleus and the maternal plasmotype.

Along with the PEP genes (Zhang et al., 2015), other genes that are atypical in Geraniaceae plastomes have been implicated in hybrid incompatibility including *clpP*, *ycf1* and *ycf2* in *Campanulastrum* (Barnard-Kubow et al., 2014), an intergenic region upstream of *clpP* in *Oenothera* (Greiner et al., 2008) and *accD* in *Pisum* (Bogdanova, 2007; Bogdanova et al., 2015). Early work with *Pelargonium* focused on nuclear loci involved in plastome replication suggested that plastid inheritance patterns were predominantly controlled by maternal alleles (reviewed in Tilney-Bassett & Abdel-Wahab, 1979). Investigation of plastid inheritance among different genotypes of

Medicago sativa where paternal transmission predominates also suggested dependence on parental nuclear genotypes (Smith, 1989). However, plastids themselves are likely to have some influence on transmission patterns as incompatible interactions involve constituents from both compartments, either at the protein–protein or protein–nucleotide level.

Studies in *Pisum sativum* have illustrated how PGI receives input from both plastids and the nucleus with an example that comes as no surprise. In initial experiments producing F1 hybrids and reciprocal cross progeny, Bogdanova and Berdnikov (2001) observed variegated leaf sectors suggestive of PGI that were shown to contain paternal plasmotypes in the green sectors vs maternal in the chlorophyll-deficient sectors (Bogdanova & Kosterin, 2006). Detection of the paternal plasmotype in cotyledons, roots and leaves indicated that photosynthesis per se was not likely a major player in the incompatibility despite the chlorophyll deficiency (Bogdanova, 2007). Selfing of an 'almost entirely green' F1 plant yielded five progeny. Three carried the paternal type and were phenotypically normal. Of the two F2 progeny that carried the maternal plasmotype, one was variegated while the other was fully green but completely sterile suggesting that segregating nuclear alleles are also involved (Bogdanova, 2007).

Two unlinked nuclear alleles were identified using a mapping population of recombinant inbred lines (Yadrikhinskiy & Bogdanova, 2011) and subsequent plastome sequencing of one cultivated tester line and four wild *P. sativum* accessions differing in cross-compatibility. Four plastid loci, *accD*, *rpoB*, *ycf1* and *ycf2*, contained nonsynonymous substitutions and a high degree of variability was noted in *accD* sequences from the different accessions. Considering three of the four plastid genes are predicted to encode subunits of plastid complexes that include nuclear-encoded constituents, regions of the *M. truncatula* nuclear genome corresponding to the previously indicated nuclear alleles were searched. One locus, corresponding to the gene encoding Biotin carboxyl carrier protein of acetyl-CoA carboxylase (*Bccp3*; *accB*) was identified. Evaluation of variable residues in the *Bccp3* and considering the pattern of incompatibilities among the different accessions led the authors to propose that interaction sites in nuclear-encoded products are coadapted to sites in the plastid-encoded products and that differences in the nuclear alleles are concomitant with differences in the plastid genes (Bogdanova et al., 2015). Although there are likely other nuclear-encoded constituents that contribute to PGI in *P. sativum*, like the other allele identified in mapping populations, the case for the involvement of multisubunit ACCase is indeed convincing.

6. ABERRATION OR ANALOGY?

Apart from the incorporation of foreign DNA by intracellular or horizontal transfer virtually every type of plastome abnormality has been detected in Geraniaceae. Although not necessarily analogical, there are other groups that display one or several of the phenomena exhibited in Geraniaceae plastomes (Table 1). Where there are similar outcomes in terms of the nucleotide substitution rate acceleration or structural divergence, it is nonetheless difficult to postulate an overarching mechanism or common evolutionary force that unifies the phenomena in the family let alone among unrelated angiosperm lineages. The data collected thus far from across Geraniaceae have shown that the group is indeed unique, but unravelling the processes that underlie this nonpareil system will certainly require much further study. Early hypotheses speculated that the array of alterations in Geraniaceae plastomes reflect deficiencies in plastome recombination, replication and repair (RRR) systems (Guisinger et al., 2008). Between species comparisons among 25 Geraniaceae and 2 outgroups in the Geraniales investigated correlation between measures of plastome complexity and *dN* of plastid-targeted DNA–RRR genes. While a number of DNA–RRR proteins were not included because they were not uniformly identified in all taxa, among those that were correlated was Whirly1, the plastid-targeted DNA-binding protein that suppresses illegitimate recombination between small repeats. Other likely candidates such as RecA1 and OSB1 were identified in all species but uncorrelated to plastome anomalies (Zhang et al., 2016). Several studies in the family have used the well-resolved Geraniaceae phylogeny to structure analyses and address the effects of shared ancestry on substitution rate variation (Blazier, Ruhlman, et al., 2016; Weng et al., 2016; Zhang et al., 2015, 2016). This approach allows consideration of events that likely occurred over millions of years. Going forward it will be valuable to also sample within species of Geraniaceae and other groups, gathering variation data at the population level. Long generation times, complexity of nuclear genomes and limited ability to employ reverse genetics impedes experimental evolution approaches, like those used in bacterial systems, in Geraniaceae species. Population level data to compare within and between species could be a proxy for experimental evolution in Geraniaceae and others, providing insight into the selective forces shaping their genomes in real time.

ACKNOWLEDGEMENTS

The authors gratefully acknowledge support from the National Science Foundation (IOS-1027259 to R.K.J. and T.A.R.) and from Vice President for Educational Affairs Abdulrahman O. Alyoubi at King Abdulaziz University, Jeddah, Saudi Arabia (to R.K.J.). We also thank the many students, graduates and undergraduates, over more than 10 years whose efforts and contributions made the Geraniaceae Genomes Project possible including Tim Chumley, Mary Guisinger, J. Chris Blazier, Mao-Lun Weng, Jin Zhang, Seongjun Park, Bikash Shrestha, Sumaiyah Rehman, Madhumita Govindu, Bernice Yau and Chia-Chi Angela Fu. Geraniaceae photographs for Fig. 1 were generously contributed by Mao-Lun Weng except *California*, which is freely available for academic purposes from ©2009 Andrew Borcher (https://calphotos.berkeley.edu/cgi/img_query?seq_num=279854&one=T).

REFERENCES

Babiychuk, E., Vandepoele, K., Wissing, J., Garcia-Diaz, M., De Rycke, R., Akbari, H., et al. (2011). Plastid gene expression and plant development require a plastidic protein of the mitochondrial transcription termination factor family. *Proceedings of the National Academy of Sciences of the United States of America, 108*, 6674–6679.

Bakker, F. T., Culham, A., de Marais, A. B., & Gibby, M. (2005). Nested radiation in Cape Pelargonium. In F. T. Bakker & L. W. Chatrou (Eds.), *Plant species-level systematics: New perspectives on pattern & process* (pp. 75–100). Germany: Koeltz Scientific Books.

Barnard-Kubow, K. B., McCoy, M. A., & Galloway, L. F. (2017). Biparental chloroplast inheritance leads to rescue from cytonuclear incompatibility. *New Phytologist, 213*, 1466–1476.

Barnard-Kubow, K. B., Sloan, D. B., & Galloway, L. F. (2014). Correlation between sequence divergence and polymorphism reveals similar evolutionary mechanisms acting across multiple timescales in a rapidly evolving plastid genome. *BMC Evolutionary Biology, 14*, 268.

Barnard-Kubow, K. B., So, N., & Galloway, L. F. (2016). Cytonuclear incompatibility contributes to the early stages of speciation. *Evolution, 70*, 2752–2766.

Baur, E. (1909). Das Wesen und die Erblichkeitsverhältnisse der "Varietates albomarginatae hort" von *Pelargonium zonale. Zeitschrift für Induktive Abstammungs- und Vererbungslehre, 1*, 330–351.

Bennetzen, J. L. (2000). Comparative sequence analysis of plant nuclear genomes. *The Plant Cell, 12*, 1021–1030.

Birky, C. W. (1995). Uniparental inheritance of mitochondrial and chloroplast genes: Mechanisms and evolution. *Proceedings of the National Academy of Sciences of the United States of America, 92*, 11331–11338.

Blazier, J. C., Guisinger, M. M., & Jansen, R. K. (2011). Recent loss of plastid-encoded *ndh* genes within *Erodium* (Geraniaceae). *Plant Molecular Biology, 76*, 263–272.

Blazier, J. C., Jansen, R. K., Mower, J. P., Govindu, M., Zhang, J., Weng, M. L., et al. (2016). Variable presence of the inverted repeat and plastome stability in *Erodium*. *Annals of Botany, 117*, 1209–1220.

Blazier, J. C., Ruhlman, T. A., Weng, M.-L., Rehman, S. K., Sabir, J. S. M., & Jansen, R. K. (2016). Divergence of RNA polymerase α subunits in angiosperm plastid genomes is mediated by genomic rearrangement. *Scientific Reports, 6*, 24595.

Bogdanova, V. S. (2007). Inheritance of organelle DNA markers in a pea cross associated with nuclear-cytoplasmic incompatibility. *Theoretical and Applied Genetics, 114*, 333–339.

Bogdanova, V. S., & Berdnikov, V. A. (2001). Observation of a phenomenon resembling hybrid dysgenesis, in a wild pea subspecies *Pisum sativum* ssp. *elatius*. *Pisum Genetics*, *33*, 5–8.
Bogdanova, V. S., & Kosterin, O. E. (2006). A case of anomalous chloroplast inheritance in crosses of garden pea involving an accession of wild subspecies. *Doklady Biological Sciences*, *406*, 44–46.
Bogdanova, V. S., Zaytseva, O. O., Mglinets, A. V., Shatskaya, N. V., Kosterin, O. E., & Vasiliev, G. V. (2015). Nuclear-cytoplasmic conflict in pea (*Pisum sativum* L.) is associated with nuclear and plastidic candidate genes encoding Acetyl-CoA carboxylase subunits. *PLoS One*, *10*, e0119835.
Brandvain, Y., & Wade, M. J. (2009). The functional transfer of genes from the mitochondria to the nucleus: The effects of selection, mutation, population size and rate of self-fertilization. *Genetics*, *182*, 1129–1139.
Cai, Z., Guisinger, M., Kim, H.-G., Ruck, E., Blazier, J. C., McMurtry, V., et al. (2008). Extensive reorganization of the plastid genome of *Trifolium subterraneum* (Fabaceae) is associated with numerous repeated sequences and novel DNA insertions. *Journal of Molecular Evolution*, *67*, 696–704.
Chateigner-Boutin, A.-L., Ramos-Vega, M., Guevara-García, A., Andrés, C., de la Luz Gutiérrez-Nava, M., Cantero, A., et al. (2008). CLB19, a pentatricopeptide repeat protein required for editing of *rpoA* and *clpP* chloroplast transcripts. *The Plant Journal*, *56*, 590–602.
Cheon, K.-S., Kim, K.-A., & Yoo, K.-O. (2017). The complete chloroplast genome sequences of three *Adenophora* species and comparative analysis with Campanuloid species (Campanulaceae). *PLoS One*, *12*, e0183652.
Chi, W., He, B., Mao, J., Jiang, J., & Zhang, L. (2015). Plastid sigma factors: Their individual functions and regulation in transcription. *Biochimica et Biophysica Acta—Bioenergetics*, *1847*, 770–778.
Chiu, W.-L., Stubbe, W., & Sears, B. B. (1988). Plastid inheritance in *Oenothera*: Organelle genome modifies the extent of biparental plastid transmission. *Current Genetics*, *13*, 181–189.
Chumley, T. W., Palmer, J. D., Mower, J. P., Fourcade, H. M., Calie, P. J., Boore, J. L., et al. (2006). The complete chloroplast genome sequence of *Pelargonium* × *hortorum*: Organization and evolution of the largest and most highly rearranged chloroplast genome of land plants. *Molecular Biology and Evolution*, *23*, 2175–2190.
Corriveau, J. L., & Coleman, A. W. (1988). Rapid screening method to detect potential biparental inheritance of plastid DNA and results for over 200 angiosperm species. *American Journal of Botany*, *75*, 1443–1458.
De Smet, R. D., Adams, K. L., Vandepoele, K., Montagu, M. C. E. V., Maere, S., & de Peer, Y. V. (2013). Convergent gene loss following gene and genome duplications creates single-copy families in flowering plants. *Proceedings of the National Academy of Sciences of the United States of America*, *110*, 2898–2903.
de Vries, J., Sousa, F. L., Bölter, B., Soll, J., & Gould, S. B. (2015). YCF1: A green TIC? *The Plant Cell*, *27*, 1827–1833.
Deng, X.-W., & Gruissem, W. (1987). Control of plastid gene expression during development: The limited role of transcriptional regulation. *Cell*, *49*, 379–387.
Downie, S. R., Katz-Downie, D. S., Wolfe, K. H., Calie, P. J., & Palmer, J. D. (1994). Structure and evolution of the largest chloroplast gene (ORF2280): Internal plasticity and multiple gene loss during angiosperm evolution. *Current Genetics*, *25*, 367–378.
Drescher, A., Ruf, S., Calsa, T., Carrer, H., & Bock, R. (2000). The two largest chloroplast genome-encoded open reading frames of higher plants are essential genes. *The Plant Journal*, *22*, 97–104.
Drouin, G., Daoud, H., & Xia, J. (2008). Relative rates of synonymous substitutions in the mitochondrial, chloroplast and nuclear genomes of seed plants. *Molecular Phylogenetics and Evolution*, *49*, 827–831.

Dugas, D. V., Hernandez, D., Koenen, E. J. M., Schwarz, E., Straub, S., Hughes, C. E., et al. (2015). Mimosoid legume plastome evolution: IR expansion, tandem repeat expansions, and accelerated rate of evolution in *clpP*. *Scientific Reports*, *5*, 16958.
Erixon, P., & Oxelman, B. (2008). Whole-gene positive selection, elevated synonymous substitution rates, duplication, and indel evolution of the chloroplast *clpP1* gene. *PLoS One*, *3*, e1386.
Fajardo, D., Senalik, D., Ames, M., Zhu, H., Steffan, S. A., Harbut, R., et al. (2013). Complete plastid genome sequence of *Vaccinium macrocarpon*: Structure, gene content, and rearrangements revealed by next generation sequencing. *Tree Genetics & Genomes*, *9*, 489–498.
Favory, J.-J., Kobayshi, M., Tanaka, K., Peltier, G., Kreis, M., Valay, J.-G., et al. (2005). Specific function of a plastid sigma factor for *ndhF* gene transcription. *Nucleic Acids Research*, *33*, 5991–5999.
Greiner, S., & Bock, R. (2013). Tuning a ménage à trois: Co-evolution and co-adaptation of nuclear and organellar genomes in plants. *BioEssays*, *35*, 354–365.
Greiner, S., Rauwolf, U., Meurer, J., & Herrmann, R. G. (2011). The role of plastids in plant speciation. *Molecular Ecology*, *20*, 671–691.
Greiner, S., Wang, X., Rauwolf, U., Silber, M. V., Mayer, K., Meurer, J., et al. (2008). The complete nucleotide sequences of the five genetically distinct plastid genomes of *Oenothera*, subsection *Oenothera*: I. Sequence evaluation and plastome evolution. *Nucleic Acids Research*, *36*, 2366–2378.
Gruissem, W., Barkan, A., Deng, X., & Stern, D. (1988). Transcriptional and post-transcriptional control of plastid mRNA levels in higher plants. *Trends in Genetics*, *4*, 258–263.
Guisinger, M. M., Chumley, T. W., Kuehl, J. V., Boore, J. L., & Jansen, R. K. (2010). Implications of the plastid genome sequence of *Typha* (Typhaceae, Poales) for understanding genome evolution in Poaceae. *Journal of Molecular Evolution*, *70*, 149–166.
Guisinger, M. M., Kuehl, J. V., Boore, J. L., & Jansen, R. K. (2008). Genome-wide analyses of Geraniaceae plastid DNA reveal unprecedented patterns of increased nucleotide substitutions. *Proceedings of the National Academy of Sciences of the United States of America*, *105*, 18424–18429.
Guisinger, M. M., Kuehl, J. V., Boore, J. L., & Jansen, R. K. (2011). Extreme reconfiguration of plastid genomes in the angiosperm family Geraniaceae: Rearrangements, repeats, and codon usage. *Molecular Biology and Evolution*, *28*, 583–600.
Gurdon, C., & Maliga, P. (2014). Two distinct plastid genome configurations and unprecedented intraspecies length variation in the *accD* coding region in *Medicago truncatula*. *DNA Research*, *21*, 417–427.
Haberle, R. C., Fourcade, H. M., Boore, J. L., & Jansen, R. K. (2008). Extensive rearrangements in the chloroplast genome of *Trachelium caeruleum* are associated with repeats and tRNA genes. *Journal of Molecular Evolution*, *66*, 350–361.
Hagemann, R. (2010). The foundation of extranuclear inheritance: Plastid and mitochondrial genetics. *Molecular Genetics and Genomics*, *283*, 199–209.
Hahn, M. W. (2009). Distinguishing among evolutionary models for the maintenance of gene duplicates. *Journal of Heredity*, *100*, 605–617.
Hansen, A. K., Escobar, L. K., Gilbert, L. E., & Jansen, R. K. (2007). Paternal, maternal, and biparental inheritance of the chloroplast genome in *Passiflora* (Passifloraceae): Implications for phylogenetic studies. *American Journal of Botany*, *94*, 42–46.
Hao, W. (2010). OrgConv: Detection of gene conversion using consensus sequences and its application in plant mitochondrial and chloroplast homologs. *BMC Bioinformatics*, *11*, 114.
Hattori, M., Miyake, H., & Sugita, M. (2007). A pentatricopeptide repeat protein is required for RNA processing of *clpP* pre-mRNA in moss chloroplasts. *Journal of Biological Chemistry*, *282*, 10773–10782.

Hong, C. P., Park, J., Lee, Y., Lee, M., Parkm, S. G., Uhm, Y., et al. (2017). *accD* nuclear transfer of *Platycodon grandiflorum* and the plastid of early Campanulaceae. *BMC Genomics, 18*, 607.

Hu, S., Sablok, G., Wang, B., Qu, D., Barbaro, E., Viola, R., et al. (2015). Plastome organization and evolution of chloroplast genes in *Cardamine* species adapted to contrasting habitats. *BMC Genomics, 16*, 306.

Jansen, R. K., Cai, Z., Raubeson, L. A., Daniell, H., Depamphilis, C. W., Leebens-Mack, J., et al. (2007). Analysis of 81 genes from 64 plastid genomes resolves relationships in angiosperms and identifies genome-scale evolutionary patterns. *Proceedings of the National Academy of Sciences of the United States of America, 104*, 19369–19374.

Jansen, R. K., & Ruhlman, T. A. (2012). Plastid genomes of seed plants. In R. Bock & V. Knoop (Eds.), *Genomics of chloroplasts and mitochondria: Vol. 35. Advances in photosynthesis and respiration* (pp. 103–126). The Netherlands: Springer.

Jiao, Y., Leebens-Mack, J., Ayyampalayam, S., Bowers, J. E., McKain, M. R., McNeal, J., et al. (2012). A genome triplication associated with early diversification of the core eudicots. *Genome Biology, 13*, R3.

Kikuchi, S., Bédard, J., Hirano, M., Hirabayashi, Y., Oishi, M., Imai, M., et al. (2013). Uncovering the protein translocon at the chloroplast inner envelope membrane. *Science, 339*, 571–574.

Kirk, J. T., & Tilney-Bassett, R. A. (1967). *The plastids: Their chemistry, structure, growth and inheritance* (1st ed., p. 608). New York: W.H. Freeman and Company.

Kirk, J. T., & Tilney-Bassett, R. A. (1978). *The plastids: Their chemistry, structure, growth and inheritance* (2nd ed., p. 980). Amsterdam: Elsevier/North-Holland Press.

Kleine, T., Maier, U. G., & Leister, D. (2009). DNA transfer from organelles to the nucleus: The idiosyncratic genetics of endosymbiosis. *Annual Review of Plant Biology, 60*, 115–138.

Kolodner, R., & Tewari, K. K. (1979). Inverted repeats in chloroplast DNA from higher plants. *Proceedings of the National Academy of Sciences of the United States of America, 76*, 41–45.

Konishi, T., & Sasaki, Y. (1994). Compartmentalization of two forms of Acetyl-CoA carboxylase in plants and the origin of their tolerance toward herbicides. *Proceedings of the National Academy of Sciences of the United States of America, 91*, 3598–3601.

Kuroda, H., & Maliga, P. (2003). The plastid *clpP1* protease gene is essential for plant development. *Nature, 425*, 86–89.

Lee, H.-L., Jansen, R. K., Chumley, T. W., & Kim, K.-J. (2007). Gene relocations within chloroplast genomes of *Jasminum* and *Menodora* (Oleaceae) are due to multiple, overlapping inversions. *Molecular Biology and Evolution, 24*, 1161–1180.

Lesins, K. (1961). Interspecific crosses involving alfalfa: I. *Medicago dzhawakhetica* (Bordz.) Vass. x *M. sativa* L. and its peculiarities. *Canadian Journal of Genetics and Cytology, 3*, 135–152.

Lilienfeld, F. A. (1962). Plastid behavior in reciprocally different crosses between two races of *Medicago truncatula* Gaertn. *Seiken Ziho, 13*, 3–38.

Lilienfeld, F. A. (1965). A case of malfunctioning plastids in *Medicago truncatula* Gaertn. *Japan Journal of Genetics, 40*, 261–274.

Lima, M. S., & Smith, D. R. (2017). Pervasive transcription of mitochondrial, plastid, and nucleomorph genomes across diverse plastid-bearing species. *Genome Biology and Evolution, 9*, 2560–2657.

Luo, J., Hou, B.-W., Niu, Z.-T., Liu, W., Xue, Q.-Y., & Ding, X.-Y. (2014). Comparative chloroplast genomes of photosynthetic orchids: Insights into evolution of the Orchidaceae and development of molecular markers for phylogenetic applications. *PLoS One, 9*, e99016.

Lysenko, E. A. (2007). Plant sigma factors and their role in plastid transcription. *Plant Cell Reports, 26*, 845–859.

Magee, A. M., Aspinall, S., Rice, D. W., Cusack, B. P., Sémon, M., Perry, A. S., et al. (2010). Localized hypermutation and associated gene losses in legume chloroplast genomes. *Genome Research, 20*, 1700–1710.

Maréchal, A., & Brisson, N. (2010). Recombination and the maintenance of plant organelle genome stability. *The New Phytologist, 186*, 299–317.

Martínez-Alberola, F., Del Campo, E. M., Lázaro-Gimeno, D., Mezquita-Claramonte, S., Molins, A., Mateu-Andrés, I., et al. (2013). Balanced gene losses, duplications and intensive rearrangements led to an unusual regularly sized genome in *Arbutus unedo* chloroplasts. *PLoS One, 8*, e79685.

Masoud, S. A., Johnson, L. B., & Sorensen, E. L. (1990). High transmission of paternal plastid DNA in alfalfa plants demonstrated by restriction fragment polymorphic analysis. *Theoretical and Applied Genetics, 79*, 49–55.

Matsuo, M., Ito, Y., Yamauchi, R., & Obokata, J. (2005). The rice nuclear genome continuously integrates, shuffles, and eliminates the chloroplast genome to cause chloroplast-nuclear DNA flux. *Plant Cell, 17*, 665–675.

McDonald, J. H., & Kreitman, M. (1991). Adaptive protein evolution at the Adh locus in *Drosophila*. *Nature, 351*, 652–654.

Metzlaf, M., Pohlheim, F., Börner, T., & Hagemann, R. (1982). Hybrid variegation in the genus *Pelargonium*. *Current Genetics, 5*, 245–249.

Metzlaff, M., Borner, T., & Hagemann, R. (1981). Variations of chloroplast DNAs in the genus *Pelargonium* and their biparental inheritance. *Theoretical and Applied Genetics, 60*, 37–41.

Nagano, Y., Matsuno, R., & Sasaki, Y. (1991). Sequence and transcriptional analysis of the gene cluster *trnQ-zfpA-psaI-ORF231-petA* in pea chloroplasts. *Current Genetics, 20*, 431–436.

Noutsos, C., Richly, E., & Leister, D. (2005). Generation and evolutionary fate of insertions of organelle DNA in the nuclear genomes of flowering plants. *Genome Research, 15*, 616–628.

Oldenburg, D. J., & Bendich, A. J. (2015). DNA maintenance in plastids and mitochondria of plants. *Frontiers in Plant Science, 6*, 883.

Palmer, J. D. (1983). Chloroplast DNA exists in two orientations. *Nature, 301*, 92–93.

Palmer, J. D. (1991). Plastid chromosomes: Structure and evolution. In L. Bogorad & I. K. Vasil (Eds.), *The molecular biology of plastids: Cell culture and somatic cell genetics of plants* (pp. 5–53). Vienna: Springer.

Palmer, J. D., Nugent, J. M., & Herbon, L. A. (1987). Unusual structure of geranium chloroplast DNA: A triple-sized inverted repeat, extensive gene duplications, multiple inversions, and two repeat families. *Proceedings of the National Academy of Sciences of the United States of America, 84*, 769–773.

Palmer, J. D., Osorio, B., Aldrich, J., & Thompson, W. F. (1987). Chloroplast DNA evolution among legumes: Loss of a large inverted repeat occurred prior to other sequence rearrangements. *Current Genetics, 11*, 275–286.

Park, S., Grewe, F., Zhu, A., Ruhlman, T. A., Sabir, J., Mower, J. P., et al. (2015). Dynamic evolution of *Geranium* mitochondrial genomes through multiple horizontal and intracellular gene transfers. *New Phytologist, 208*, 570–583.

Park, S., Ruhlman, T. A., Weng, M.-L., Hajrah, N. H., Sabir, J. S. M., & Jansen, R. K. (2017). Contrasting patterns of nucleotide substitution rates provide insight into dynamic evolution of plastid and mitochondrial genomes of *Geranium*. *Genome Biology and Evolution, 9*, 1766–1780.

Quesada-Vargas, T., Ruiz, O. N., & Daniell, H. (2005). Characterization of heterologous multigene operons in transgenic chloroplasts: Transcription, processing, and translation. *Plant Physiology, 138*, 1746–1762.

Rabah, S. O., Shrestha, B., Hajrah, N. H., Sabir Mumdoh, J., Alharby, H. F., Sabir Mernan, J., et al. (2017). Passiflora plastome sequencing reveals widespread homoplasy in genomic rearrangements. *American Journal of Botany*. in revision.

Rockenbach, K., Havird, J. C., Monroe, J. G., Triant, D. A., Taylor, D. R., & Sloan, D. B. (2016). Phylogenetic and population genetic evidence for positive selection in rapidly evolving plastid-nuclear enzyme complexes. *Genetics*, *204*, 1507–1522.

Rousseau-Gueutin, M., Huang, X., Higginson, E., Ayliffe, M., Day, A., & Timmis, J. N. (2013). Potential functional replacement of the plastidic Acetyl-CoA carboxylase subunit (*accD*) gene by recent transfers to the nucleus in some angiosperm lineages. *Plant Physiology*, *161*, 1918–1929.

Ruhlman, T. A., Chang, W.-J., Chen, J. J., Huang, Y. T., Chan, M. T., Zhang, J., et al. (2015). NDH expression marks major transitions in plant evolution and reveals coordinate intracellular gene loss. *BMC Plant Biology*, *15*, 100.

Ruhlman, T. A., & Jansen, R. K. (2014). The plastid genomes of flowering plants. In P. Maliga (Ed.), *Chloroplast biotechnology: Methods and protocols* (pp. 3–38). New York: Springer Science and Business Media, LLC.

Ruhlman, T., Verma, D., Samson, N., & Daniell, H. (2010). The role of heterologous chloroplast sequence elements in transgene integration and expression. *Plant Physiology*, *152*, 2088–2104.

Ruhlman, T. A., Zhang, J., Blazier, J. C., Sabir, J. S. M., & Jansen, R. K. (2017). Recombination-dependent replication and gene conversion homogenize repeat sequences and diversify plastid genome structure. *American Journal of Botany*, *104*, 559–572.

Sabir, J., Schwarz, E., Ellison, N., Zhang, J., Baeshen, N. A., Mutwakil, M., et al. (2014). Evolutionary and biotechnology implications of plastid genome variation in the inverted-repeat-lacking clade of legumes. *Plant Biotechnology Journal*, *12*, 743–754.

Sanderson, M. J., Copetti, D., Búrquez, A., Bustamante, E., Charboneau, J. L., Eguiarte, L. E., et al. (2015). Exceptional reduction of the plastid genome of saguaro cactus (*Carnegiea gigantea*): Loss of the *ndh* gene suite and inverted repeat. *American Journal of Botany*, *102*, 1115–1127.

Scannell, D. R., Byrne, K. P., Gordon, J. L., Wong, S., & Wolfe, K. H. (2006). Multiple rounds of speciation associated with reciprocal gene loss in polyploid yeasts. *Nature*, *440*, 341–345.

Schumann, C. M., & Hancock, J. F. (1989). Paternal inheritance of plastids in *Medicago sativa*. *Theoretical and Applied Genetics*, *78*, 863–866.

Schwarz, E. N., Ruhlman, T. A., Sabir, J. S. M., Hajrah, N. H., Alharbi, N. S., Al-Malki,-A. L., et al. (2015). Plastid genome sequences of legumes reveal parallel inversions and multiple losses of *rps16* in papilionoids. *Journal of Systematics and Evolution*, *53*, 458–468.

Schwarz, E. N., Ruhlman, T. A., Weng, M.-L., Khiyami, M. A., Sabir, J. S. M., Hajrah, N. H., et al. (2017). Plastome-wide nucleotide substitution rates reveal accelerated rates in Papilionoideae and correlations with genome features across legume subfamilies. *Journal of Molecular Evolution*, *84*, 187–203.

Shi, C., Wang, S., Xia, E.-H., Jiang, J.-J., Zeng, F.-C., & Gao, L.-Z. (2016). Full transcription of the chloroplast genome in photosynthetic eukaryotes. *Scientific Reports*, *6*, 30135.

Sloan, D. B., Alverson, A. J., Wu, M., Palmer, J. D., & Taylor, D. R. (2012). Recent acceleration of plastid sequence and structural evolution coincides with extreme mitochondrial divergence in the angiosperm genus *Silene*. *Genome Biology and Evolution*, *4*, 294–306.

Sloan, D. B., Triant, D. A., Forrester, N. J., Bergner, L. M., Wu, M., & Taylor, D. R. (2014). A recurring syndrome of accelerated plastid genome evolution in the angiosperm tribe Sileneae (Caryophyllaceae). *Molecular Phylogenetics and Evolution*, *72*, 82–89.

Sloan, D. B., Triant, D. A., Wu, M., & Taylor, D. R. (2014). Cytonuclear interactions and relaxed selection accelerate sequence evolution in organelle ribosomes. *Molecular Biology and Evolution*, *31*, 673–682.

Smith, S. E. (1989). Influence of parental genotype on plastid inheritance in *Medicago sativa*. *The Journal of Heredity*, *80*, 214–217.
Stegemann, S., & Bock, R. (2006). Experimental reconstruction of functional gene transfer from the tobacco plastid genome to the nucleus. *The Plant Cell*, *18*, 2869–2878.
Stegemann, S., Hartmann, S., Ruf, S., & Bock, R. (2003). High-frequency gene transfer from the chloroplast genome to the nucleus. *Proceedings of the National Academy of Sciences of the United States of America*, *100*, 8828–8833.
Stern, D. B., & Gruissem, W. (1987). Control of plastid gene expression: 3' inverted repeats act as mRNA processing and stabilizing elements, but do not terminate transcription. *Cell*, *51*, 1145–1157.
Tilney-Bassett, R. A. (1973). The control of plastid inheritance in *Pelargonium* II. *Heredity*, *30*, 1–13.
Tilney-Bassett, R. A., & Abdel-Wahab, O. (1979). In D. R. Newth & M. Balls (Eds.), *Maternal effects and plastid inheritance. Maternal effects in development: The fourth symposium of the British society for developmental biology* (pp. 29–45). London: Cambridge University Press.
Tilney-Bassett, R. A., & Almouslem, A. (1989). Variation in plastid inheritance between *Pelargonium* cultivars and their hybrids. *Heredity*, *63*, 145–153.
Timmis, J. N., Ayliffe, M. A., Huang, C. Y., & Martin, W. (2004). Endosymbiotic gene transfer: Organelle genomes forge eukaryotic chromosomes. *Nature Reviews. Genetics*, *5*, 123–135.
Ueda, M., Nishikawa, T., Fujimoto, M., Takanashi, H., Arimura, S., Tsutsumi, N., et al. (2008). Substitution of the gene for chloroplast RPS16 was assisted by generation of a dual targeting signal. *Molecular Biology and Evolution*, *25*, 1566–1575.
Weihe, A., Apitz, J., Pohlheim, F., Salinas-Hartwig, A., & Börner, T. (2009). Biparental inheritance of plastidial and mitochondrial DNA and hybrid variegation in *Pelargonium*. *Molecular Genetics and Genomics*, *282*, 587–593.
Weng, M.-L., Blazier, J. C., Govindu, M., & Jansen, R. K. (2014). Reconstruction of the ancestral plastid genome in Geraniaceae reveals a correlation between genome rearrangements, repeats, and nucleotide substitution rates. *Molecular Biology and Evolution*, *31*, 645–659.
Weng, M.-L., Ruhlman, T. A., Gibby, M., & Jansen, R. K. (2012). Phylogeny, rate variation, and genome size evolution of *Pelargonium* (Geraniaceae). *Molecular Phylogenetics and Evolution*, *64*, 654–670.
Weng, M.-L., Ruhlman, T. A., & Jansen, R. K. (2016). Plastid–nuclear interaction and accelerated coevolution in plastid ribosomal genes in Geraniaceae. *Genome Biology and Evolution*, *8*, 1824–1838.
Weng, M.-L., Ruhlman, T. A., & Jansen, R. K. (2017). Expansion of inverted repeat does not decrease substitution rates in *Pelargonium* plastid genomes. *New Phytologist*, *214*, 842–851.
Williams, A. V., Boykin, L. M., Howell, K. A., Nevill, P. G., & Small, I. (2015). The complete sequence of the *Acacia ligulata* chloroplast genome reveals a highly divergent *clpP1* gene. *PLoS One*, *10*, e0125768.
Wojciechowski, M. F., Lavin, M., & Sanderson, M. J. (2004). A phylogeny of legumes (Leguminosae) based on analysis of the plastid *matK* gene resolves many well-supported subclades within the family. *American Journal of Botany*, *91*, 1846–1862.
Yadrikhinskiy, A. K., & Bogdanova, V. S. (2011). Nuclear-cytoplasm conflict in crosses of pea subspecies is controlled by alleles of a nuclear gene on linkage group III. *Doklady Biological Sciences*, *441*, 396–399.
Yamaguchi, K., & Subramanian, A. R. (2000). The plastid ribosomal proteins. Identification of all the proteins in the 50s subunit of an organelle ribosome (chloroplast). *Journal of Biological Chemistry*, *275*, 28466–28482.
Yamaguchi, K., & Subramanian, A. R. (2003). Proteomic identification of all plastid-specific ribosomal proteins in higher plant chloroplast 30S ribosomal subunit. *European Journal of Biochemistry*, *270*, 190–205.

Yao, J.-L., Cohen, D., & Rowland, R. E. (1994). Plastid DNA inheritance and plastome-genome incompatibility in interspecific hybrids of *Zantedeschia* (Araceae). *Theoretical and Applied Genetics*, *88*, 255–260.

Zhang, Q., Liu, Y., & Sodmergen (2003). Examination of the cytoplasmic DNA in male reproductive cells to determine the potential for cytoplasmic inheritance in 295 angiosperm species. *Plant and Cell Physiology*, *44*, 941–951.

Zhang, J., Ruhlman, T. A., Mower, J. P., & Jansen, R. K. (2013). Comparative analyses of two Geraniaceae transcriptomes using next-generation sequencing. *BMC Plant Biology*, *13*, 228.

Zhang, J., Ruhlman, T. A., Sabir, J. S. M., Blazier, J. C., & Jansen, R. K. (2015). Coordinated rates of evolution between interacting plastid and nuclear genes in Geraniaceae. *The Plant Cell*, *27*, 563–573.

Zhang, J., Ruhlman, T. A., Sabir, J. S. M., Blazier, J. C., Weng, M.-L., Park, S., et al. (2016). Coevolution between nuclear-encoded DNA replication, recombination, and repair genes and plastid genome complexity. *Genome Biology and Evolution*, *8*, 622–634.

Zhang, Q., & Sodmergen (2010). Why does biparental plastid inheritance revive in angiosperms? *Journal of Plant Research*, *123*, 201–206.

Zhang, H., Yang, Z., Shen, Y., & Tong, L. (2003). Crystal structure of the carboxyltransferase domain of acetyl-coenzyme A carboxylase. *Science*, *299*, 2064–2067.

CHAPTER NINE

Structural Diversity Among Plastid Genomes of Land Plants

Jeffrey P. Mower[1], Trisha L. Vickrey
University of Nebraska, Lincoln, NE, United States
[1]Corresponding author: e-mail address: jpmower@unl.edu

Contents

Abstract

The plastome of land plants is often considered to be highly conserved in sequence, structure, and content. This is particularly true for nonvascular land plants, for which few changes to the plastome have occurred throughout their evolutionary history. In vascular plants, however, the plastomic structure is more dynamic. Many lycophytes, most ferns, and particular lineages of seed plants have experienced extensive structural rearrangements, including inversions and modifications to the size and content of the IR. In this review, we describe the typical structural features of the land plant plastome, the major variations to this canonical structure that occur in various lineages, and the evolutionary implications of this structural variation.

Advances in Botanical Research, Volume 85
ISSN 0065-2296
https://doi.org/10.1016/bs.abr.2017.11.013

1. INTRODUCTION

Over the past 30 years, the plastome has been a favoured target of plant biologists for comparative genomics, starting from the first sequenced genomes in 1986 (Ohyama et al., 1986; Shinozaki et al., 1986) to the thousands of complete sequences available today. In particular, the small size, conserved sequence and structure, and high cellular copy number of the plastome of green plants make it an ideal candidate for high-throughput sequencing and assembly. This explosion of sequencing has enabled comparative evolutionary analysis on a massive scale, involving dozens to hundreds of individual plastomes (e.g. Bock, Kane, Ebert, & Rieseberg, 2014; Jansen et al., 2007; Knox, 2014; Parks, Cronn, & Liston, 2009; Ruhfel, Gitzendanner, Soltis, Soltis, & Burleigh, 2014; Vargas, Ortiz, & Simpson, 2017; Zhu, Guo, Gupta, Fan, & Mower, 2016).

Yet, despite the ease of sequencing new plastomes, there is an extreme bias in the organismal diversity represented by these data. The vast majority of plastome sequences available in the public sequence databases are derived specifically from angiosperms. Gymnosperms and green algae have also been extensively sampled from over 100 species in each group, and ferns to a somewhat lesser extent with just over 50 sampled species. In contrast, there is very poor representation of lycophytes (five sequenced species), hornworts (two sequenced species), and liverworts (five sequenced species), which is surprising in consideration of the glut of complete plastome sequences that are increasing at an exponential pace.

In spite of the shortcomings of this biased taxonomic sampling, much has been gleaned about the evolutionary diversity of the plastome over the past 30 years. In this review, we focus on describing the structural diversity of the plastome among photosynthetic members of land plants. We also discuss what this diversity tells us about the evolution of plastome structure over time as well as the evolutionary effects that structural changes cause.

2. TYPICAL STRUCTURE OF A LAND PLANT PLASTOME

The plastome from most land plants is 120–160 kb in length and organized into two single-copy regions (termed LSC and SSC) separated by two copies of the IR (termed IR_A and IR_B). Each genome tends to contain approximately 80 protein-coding genes, 4 rRNAs, and 30 tRNAs (Jansen & Ruhlman, 2012; Wicke, Schneeweiss, dePamphilis, Muller, &

Quandt, 2011). Across land plants, the content of the IR nearly universally includes all 4 rRNAs and 5 tRNAs, and in some lineages (especially seed plants), a small number of protein genes and additional tRNAs are also present (Zhu et al., 2016). There are several explanations for the variation in plastome size observed across land plants: expansions and contractions of the IR, gene loss, intron loss, variation in the size of intergenic spacer regions, and variation in abundance of smaller repetitive sequences. For example, the additional genes in the IR of seed plants result in an IR that is generally larger (20–30 kb) compared with other land plant groups (10–15 kb). Other common variants on the conserved structure include loss or pseudogenization of the entire suite of *ndh* genes (e.g. Blazier, Guisinger, & Jansen, 2011; Wakasugi et al., 1994; Wickett et al., 2008) or loss of one copy of the IR (e.g. Cai et al., 2008; Guisinger, Kuehl, Boore, & Jansen, 2011; Wu, Wang, Hsu, Lin, & Chaw, 2011), both of which have occurred repeatedly during the evolution of land plants.

Even prior to complete sequencing, it was already well established that plastomes map as circular molecules (Rochaix, 1978) and exist as two isomeric forms via homologous recombination between the two IR copies (Kolodner & Tewari, 1979; Palmer, 1983). What is less well established is how often these genomes exist as circular chromosomes in vivo. Early electron microscopy observations of plastids recovered circular molecules consistent in size with data obtained from analyses of DNA reassociation kinetics (Bedbrook & Kolodner, 1979). This, coupled with mapping evidence, led to a tractable depiction of the topology of plastid DNA molecules as circular. However, these early experiments may have either disregarded larger, more complex forms of DNA as contaminants or removed them through fractionation (Bendich, 2004).

More recent electrophoretic and microscopic analyses have generally recovered linear molecules and more complex multibranched conglomerations. For example, linear and multibranched linear structures have been repeatedly observed in different angiosperms (Bendich & Smith, 1990; Lilly, Havey, Jackson, & Jiang, 2001; Oldenburg & Bendich, 2004; Rowan, Oldenburg, & Bendich, 2004; Scharff & Koop, 2006; Shaver, Oldenburg, & Bendich, 2008). In all cases, however, some circular DNA was still recovered, the abundance of which was dependent on species, tissue type, and experimental design. For example, the amount of circular ptDNA observed in *Nicotiana tabacum* ranged from 27% using pulsed-field gel electrophoresis (PFGE; Shaver et al., 2008) to 45% using fibre-based fluorescence in situ hybridization (Lilly et al., 2001), while it was estimated to be only 3%–4% in *Zea mays* using PFGE (Oldenburg & Bendich, 2004). Further investigations of linear structures through restriction fragment mapping have demonstrated specific end sites near putative origins of replication (Oldenburg & Bendich, 2004, 2016; Scharff & Koop, 2006; Shaver et al., 2008), rather than random end sites as would be expected if the linear pieces resulted from breakage of circular molecules. Finally, linear arrangements of plastid DNA, such as head-to-tail concatemers, are consistent with circular mapping. Together, these data support a linear chromosome as the major structural conformation in vivo.

3. CONSERVED PLASTOMES AND INFERENCE OF ANCESTRAL STRUCTURES

At the structural level, the plastome of most land plant lineages has evolved in a very conservative manner. Across the diversity of land plants, plastomes are largely collinear, requiring just a small number of inversions and IR expansions to explain the large-scale structural rearrangements among major lineages (Fig. 1). In fact, plastomes from most nonvascular land plants (represented by the hornwort *Nothoceros*, the mosses *Sphagnum* and *Tetraphis*, and the liverworts *Marchantia* and *Pellia*) exhibit no rearrangements relative to one another (Fig. 1), and this shared gene order was likely established in a green algal ancestor prior to the colonization of land (Turmel, Otis, & Lemieux, 2006).

The content of the IR has also remained remarkably constant among nearly all nonvascular land plant plastomes and several of their closest green algal relatives (Fig. 2). By examining the IR boundaries in nonvascular land plants (Bell, Boore, Mishler, & Hyvonen, 2014; Forrest, Wickett, Cox, & Goffinet, 2011; Grosche, Funk, Maier, & Zauner, 2012; Ohyama et al., 1986; Oliver et al., 2010; Sugiura, Kobayashi, Aoki, Sugita, & Sugita, 2003; Villarreal, Forrest, Wickett, & Goffinet, 2013), it can be inferred that (1) the ancestral land plant IR included 5 tRNAs and all 4 rRNAs in the order *trnN-GUU–trnR-ACG–rrn5–rrn4.5–rrn23–trnA-UGC–trnI-GAU–rrn16–trnV-GAC*; (2) at the IR/SSC borders, *trnN-GUU* is adjacent to either *ndhF* or *chlL*; and (3) at the IR/LSC borders, *trnV-GAC* lies next to either *trnI-CAU* or the 3′-part of the *trans*-spliced *rps12* gene. Together, these results provide strong evidence that the plastome arrangement of most nonvascular land plants is the same as that in the common ancestor of all land plants.

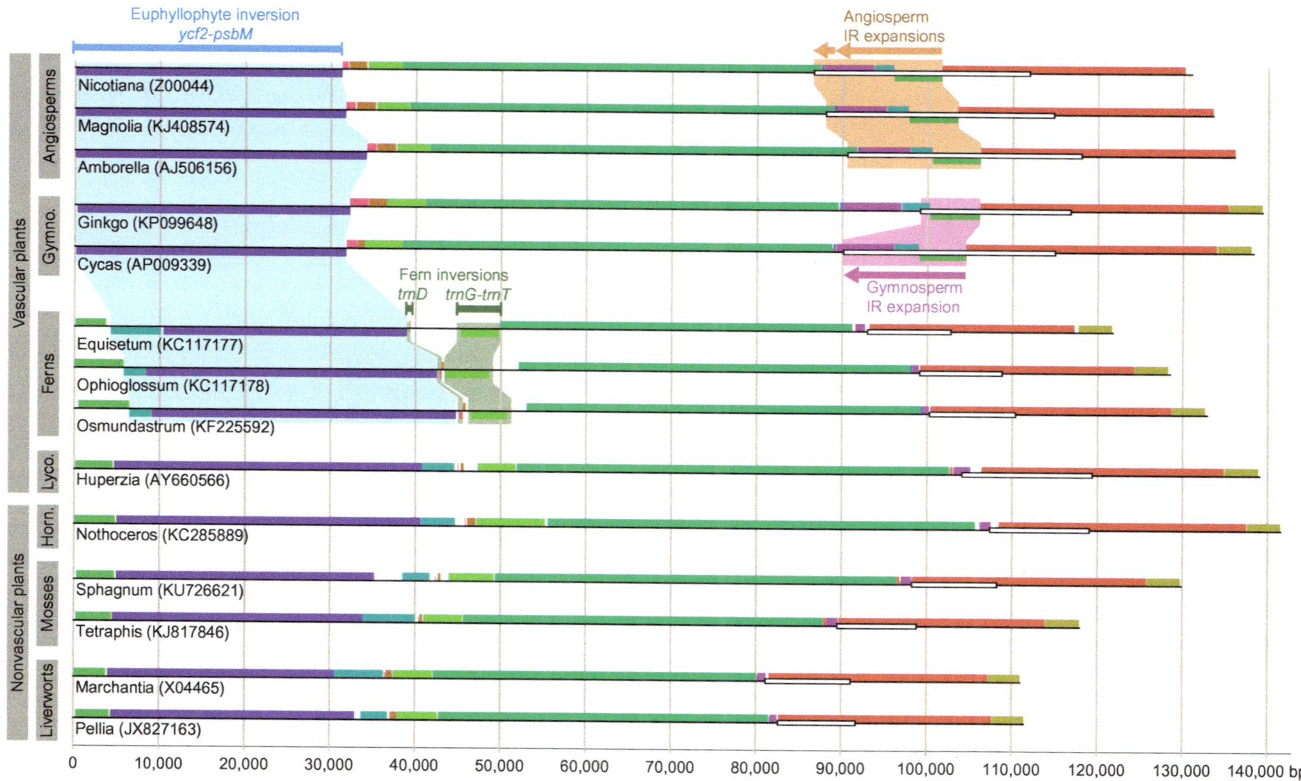

Fig. 1 Genome alignment highlighting diagnostic changes among land plant plastomes. A MAUVE plot is shown that denotes segments of genome homology among species with *coloured boxes*. Only one copy of the IR for each species was included, and its location is marked by a *horizontal white box*. The length of the genome is given by the scale at the bottom. Diagnostic changes are highlighted in *blue* for euphyllophytes, in *green* for ferns, in *pink* for gymnosperms, and in *orange* for angiosperms.

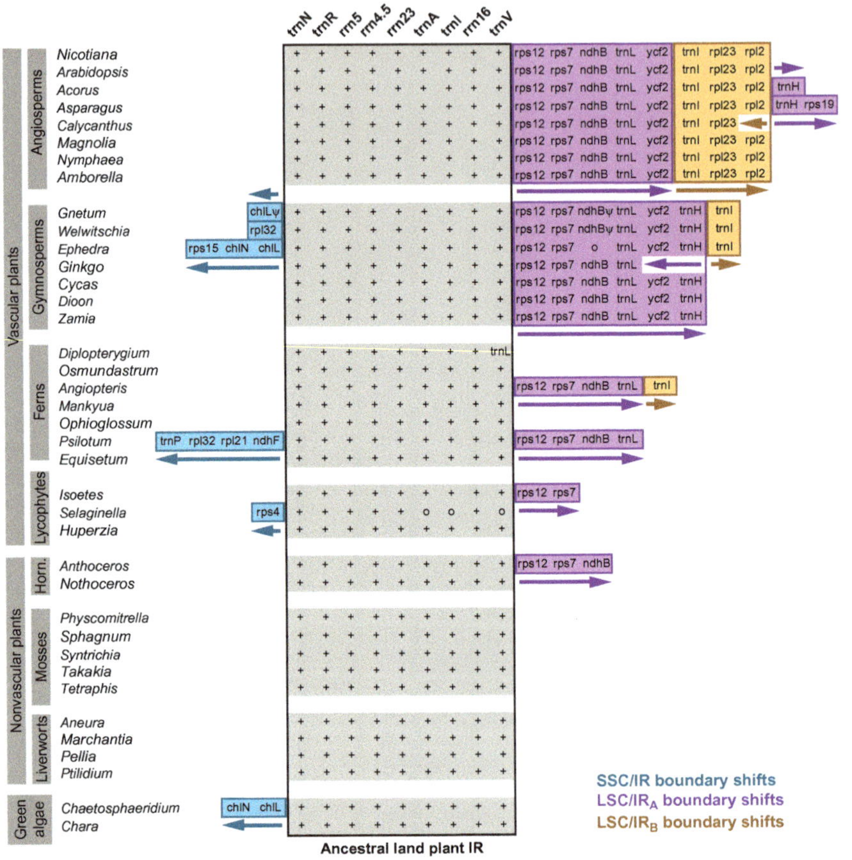

Fig. 2 Comparison of gene content in the IR of land plants. The extent of the ancestral land plant IR is shaded in *grey*, and the gene names are listed at the top. Presence of the gene is shown by a *plus*, loss of the gene is marked with a *circle*, and when the ancestral gene is replaced it is labelled with the name of the nonhomologous gene. Boundary shifts that have occurred in particular lineages are shown in *blue* for IR/SSC shifts, in *purple* for LSC/IR_A shifts, and in *yellow* for LSC/IR_B shifts. The direction of each boundary shift is marked with a *horizontal arrow*.

Plastomic structure is less well conserved among vascular plants. Nevertheless, the plastome from the lycophyte *Huperzia* is almost fully collinear with nonvascular land plant plastomes (Fig. 1; Wolf et al., 2005), suggesting that the common ancestor of vascular plants had a plastome that was very similar to those of nonvascular plants. For euphyllophytes (angiosperms, gymnosperms, and ferns), a ~35 kb inversion was discovered that is diagnostic for this group (Fig. 1; Raubeson & Jansen, 1992a), indicating that this large inversion, which spans the *psbM* to *ycf2* region in the LSC

(Grewe, Guo, Gubbels, Hansen, & Mower, 2013), most likely occurred in the euphyllophyte ancestor. Although the structure and gene content of the IR have changed in some ferns and lycophytes, the majority of lineages have retained the ancestral land plant arrangement at one or both IR boundaries (Fig. 2). This implies that the ancestral land plant arrangement for the IR, in which *trnN-GUU* and *trnV-GAC* are located at the IR/SSC and IR/LSC boundaries, respectively (Grewe et al., 2013; Zhu et al., 2016), was retained in the common ancestor of vascular plants and the common ancestor of euphyllophytes.

Each of the three descendant lineages of euphyllophytes has experienced their own set of diagnostic structural changes (Fig. 1). All ferns uniquely share a small ~3 kb inversion from *trnG-GCC* to *trnT-GGU* (Grewe et al., 2013; Karol et al., 2010; Wolf, Rowe, Sinclair, & Hasebe, 2003) and a smaller inversion involving *trnD-GUC* only (Gao, Yi, Yang, Su, & Wang, 2009; Gao, Zhou, Wang, Su, & Wang, 2011). In seed plants, the IR has expanded substantially relative to other land plants. However, the precise series of events that led to this change is difficult to reconstruct unambiguously. Gymnosperms, cycads, and gnetophytes have an IR that has extended into the LSC to include 3′-*rps12*, *rps7*, *ndhB*, *trnL-CAA*, *ycf2*, and *trnH-GUG* (McCoy, Kuehl, Boore, & Raubeson, 2008; Wu & Chaw, 2015; Wu, Lai, Lin, Wang, & Chaw, 2009; Wu, Wang, Liu, & Chaw, 2007). The IR in the *Ginkgo* plastome extends only to *trnL-CAA*; however, a *trnH-GUG* gene is still retained in duplicate just outside of the IR/LSC boundary, suggesting that the IR of the *Ginkgo* ancestor also extended as far as *trnH-GUG* before contracting to its present position (Lin, Wu, Huang, & Chaw, 2012). Collectively, these observations demonstrate that the IR in the common ancestor of gymnosperms underwent an expansion to *trnH-GUG* (Fig. 2).

Most angiosperms have an even larger IR, but this was the result of two separate expansions in the angiosperm ancestor (Fig. 2). The first expansion moved 3′-*rps12*, *rps7*, *ndhB*, *trnL-CAA*, and *ycf2* (but not *trnH-GUG*) from one end of the LSC into the IR. The absence of *trnH-GUG* from this initial angiosperm IR expansion is a small but important difference that distinguishes this event from the gymnosperm IR expansion. Subsequently, a second angiosperm IR expansion moved *trnI-CAU*, *rpl23*, and *rpl2* from the other end of the LSC into the IR. To put these expansions of the IR in angiosperms and gymnosperms in evolutionary context, three equally parsimonious scenarios arise: (1) IR_A expanded to *ycf2* in the seed plant ancestor, and then IR_A subsequently expanded to *trnH-GUG* in gymnosperms, while IR_B expanded to *rpl2* in angiosperms, (2) IR_A expanded to *trnH-GUG* in the seed plant ancestor, and then in angiosperms the IR contracted to *ycf2* before IR_B expanded to *rpl2*, or (3) IR_A expanded independently to *trnH-GUG* in gymnosperms and to *ycf2* in angiosperms, followed by a second angiosperm expansion of IR_B to *rpl2*.

4. STRUCTURAL DIVERSITY OF THE PLASTOME AMONG PLANTS

While the overall pattern of structural change has been slow during land plant evolution, not all plant lineages have maintained this conservative evolutionary trajectory. Many species have experienced their own lineage-specific changes, some of which are dramatic. Below, we describe the diversity of plastome structures among the major land plant lineages and their closest green algal relatives.

4.1 Streptophytic Green Algae

In contrast to the generally conserved plastomes of most land plants, the plastomes of green algae exhibit remarkable variation in size, structure, and content. For a comprehensive description of the evolutionary diversity of all green algal plastomes, the reader is directed to the chapter "Evolution of the plastid genome in green algae" by Turmel and Lemieux. Here, a brief overview of this variation is presented to provide context to the origins of the plastomic diversity of land plants.

The plastomes of streptophytic green algae exhibit nearly twofold variation in size, from 107 kb for *Coleochaete* to more than 200 kb for some species in Desmidiales and Klebsormidiales (Lemieux, Otis, & Turmel, 2016). The small *Coleochaete* plastome is the result of loss of eight genes and one copy of the IR, while the largest genomes have expanded due to either a greatly enlarged IR in Klebsormidiales or an increase in intergenic, often repetitive DNA in Desmidiales (Civan, Foster, Embley, Seneca, & Cox, 2014; Lemieux et al., 2016). In addition to *Coleochaete*, the IR has been lost from several species of Zygnematophyceae (Civan et al., 2014; Lemieux et al., 2016; Turmel, Otis, & Lemieux, 2005). The Zygnematophyceae plastomes are also distinctive because they are highly rearranged relative to one another and to other green algae, making it difficult to reconstruct the numbers of independent losses and possible regains of the IR that have occurred in this group (Civan et al., 2014; Lemieux et al., 2016; Turmel et al., 2005).

Plastomes from other streptophytic algal lineages contain an IR and are less rearranged, yet at least 10 inversions must still be inferred for each genome relative to the ancestral land plant structure (Civan et al., 2014; Lemieux et al., 2016; Lemieux, Otis, & Turmel, 2007; Turmel et al., 2006). The IRs in *Chara* and *Chaetosphaeridium* are very similar to those of nonvascular land plants, except for a short IR expansion in *Chaetosphaeridium* (Fig. 2). The IRs of *Mesostigma* and *Chlorokybus* are more reduced (Lemieux et al., 2007; Lemieux, Otis, & Turmel, 2000), whereas the IR of Klebsormidiales has expanded substantially to include between 8 and 26 additional genes on each flank of this reduced core (Civan et al., 2014; Lemieux et al., 2016).

4.2 Nonvascular Land Plants

Among nonvascular land plants, the liverworts have the smallest plastomes, ranging from 119 to 121 kb in size (Forrest et al., 2011; Grosche et al., 2012; Myszczyński et al., 2017; Ohyama et al., 1986), while the parasitic liverwort *Aneura mirabilis* (Metzgeriales) has an aberrantly small genome of only 108 kb due to substantial gene loss (Wickett et al., 2008). The plastomes of mosses are slightly larger and more variable, ranging in size from 149 kb in *Takakia* (Takakiales) to 136–140 kb among *Eosphagnum*, *Flatbergium*, and *Sphagnum* (Sphagnales), and to only 123–124 kb among Bryopsida, which includes the majority of moss species (Sugiura et al., 2003; Oliver et al., 2010; Bell et al., 2014; Shaw et al., 2016; Myszczyński et al., 2017; GenBank accession number AP014702 for the unpublished Takakia plastome). This taxonomically broad sampling among mosses suggests that this size range represents most, if not all, species. The two sequenced hornwort plastomes (Kugita et al., 2003; Villarreal et al., 2013), at 153 kb in *Nothoceros* (Dendrocerotales) and 161 kb in *Anthoceros* (Anthocerotales), are substantially larger than moss and liverwort plastomes, although further sequencing is needed to assess the full range of plastome sizes among hornworts.

Despite the rather large and lineage-specific differences in size among nonvascular land plant plastomes, their genomic structures are very static (Fig. 1). In fact, most of the genomes are fully collinear. There are just two inversions that have been identified among the 17 sequenced plastomes. The first involves a 71 kb inversion in the LSC of the *Physcomitrella patens* plastome (Sugiura et al., 2003). It was subsequently shown to be diagnostic of three families in Funariidae (Funariaceae, Disceliaceae, and Encalyptaceae) but not Gigaspermaceae or Timmiaceae (Goffinet et al., 2007). The second inversion, which is unique to the *A. mirabilis* plastome, affects a small 1 kb region containing the LSC genes *psbE* and *petL* (Wickett et al., 2008). This inversion must have occurred quite recently, as it was not detected in any of the sequenced plastomes from the close relative *Aneura pinguis* (Myszczyński et al., 2017).

As already mentioned, the IR boundaries are also very similar among nearly all nonvascular land plant plastomes (Fig. 2). The lone exception among sequenced nonvascular land plant plastomes is found in *Anthoceros*, in which the 3′-*rps12*, *rps7*, and *ndhB* genes are located within the IR rather than the LSC (Kugita et al., 2003). This change appears to be the result of a unique IR expansion because of its restricted distribution among members of Anthocerotaceae (Villarreal et al., 2013) and its unique IR/LSC boundary compared with other land plants (Fig. 2; Grewe et al., 2013; Zhu et al., 2016).

4.3 Lycophytes

Plastomes from five species of lycophytes, representing just three distinct genera, are available (Guo, Zhang, Shrestha, & Zhang, 2016; Karol et al., 2010; Smith, 2009; Tsuji et al., 2007; Wolf et al., 2005). All are quite similar in size: 144 kb in *Selaginella* (Selaginellales), 145 kb in *Isoetes* (Isoetales), and 154 kb in *Huperzia* (Lycopodiales). The two *Huperzia* plastomes are 99.8% identical in sequence and their gene orders are collinear (Guo et al., 2016). The structures of the two *Selaginella* plastomes are less well conserved, including a 20 kb inversion, two relocations, and some differences in gene content in *Selaginella uncinata* (Smith, 2009; Tsuji et al., 2007). The *Isoetes* plastome has a few unique changes, such as a relocation of *ycf2* from the LSC into the SSC and an inversion of the *chlL*–*chlN* gene cluster (Karol et al., 2010). There is a clear need for additional sequencing in this group, first to determine whether *Huperzia* is representative of other genera in Lycopodiales, and second to assess the extent of structural diversity among species in Isoetales and Selaginellales.

Relative to the conserved IR organization among nonvascular land plant plastomes, the lycophyte IR is more variable (Fig. 2). The *Huperzia* IR is very slightly modified, in which it has expanded to incorporate portions of *ndhF* on the IR/SSC border and 3′-*rps12* on the IR/LSC border (Karol et al., 2010; Wolf et al., 2005). The *Isoetes* IR has a different expansion at the IR/LSC border that moved the 3′-*rps12* and *rps7* genes from the LSC into the IR (Karol et al., 2010). The *Selaginella* IR has experienced several unique events, including the loss or pseudogenization of three tRNAs (*trnA-UGC*, *trnI-GAU*, and *trnV-GAC*), expansion of the IR/SSC border that moved *rps4* into the IR, and expansion of the IR/LSC border in *S. uncinata* that brought *rpl23* into the IR (Smith, 2009; Tsuji et al., 2007).

4.4 Ferns

Complete plastome sequences are available from over 50 diverse species of ferns (Gao et al., 2013, 2009; Grewe et al., 2013; Karol et al., 2010; Kim, Chung, & Kim, 2014; Labiak & Karol, 2017; Lu, Zhang, Du, Wen, & Li, 2015; Raman, Choi, & Park, 2016; Roper et al., 2007; Wei et al., 2017; Wolf et al., 2003; Zhong, Fong, Collins, McLenachan, & Penny, 2014; Zhu et al., 2016), representing 10 of the 11 fern orders, leaving Hymenophyllales as the sole order still lacking a complete sequence. Genome size is generally larger in the leptosporangiate ferns from Osmundales (143 kb), Gleicheniales (151 kb), Schizaeales (135–157 kb), Salviniales (152 kb), Cyatheales (157–168 kb), and Polypodiales (148–157 kb). Genome size is also large in Marattiales (154 kb), the closest relatives to leptosporangiate ferns, but smaller in the more distantly related lineages of Equisetales (132–133 kb), Ophioglossales (138–146 kb), and Psilotales (139–140 kb).

In addition to the *trnD-GUC* and *trnG-GCC* to *trnT-GGU* inversions shared by all ferns (Fig. 1), it has been recognized for many years that most, but not all, leptosporangiate ferns have experienced two overlapping inversions affecting most of the IR as well as an expansion of the IR into the LSC (Hasebe & Iwatsuki, 1992; Raubeson & Stein, 1995; Stein et al., 1992). These overlapping inversions specifically affect the core leptosporangiate ferns (Salviniales, Cyatheales, and Polypodiales) and Schizaeales (Gao et al., 2013, 2009; Kim et al., 2014; Roper et al., 2007; Wolf, Roper, & Duffy, 2010; Wolf et al., 2003). The core leptosporangiate ferns share another pair of overlapping inversions affecting ~3 kb of the LSC, resulting in the relocation of the gene set *trnD-GUC–trnY-GUA–trnE-UUC* and inversion of the gene set *trnC-GCA–petN–psbM* (Gao et al., 2013, 2009, 2011; Raman et al., 2016; Roper et al., 2007; Wolf et al., 2010). Subsequent to these changes, the Cyatheales and Polypodiales plastomes stabilized to remain collinear with one another (Gao et al., 2009; Grewe et al., 2013; Raman et al., 2016), while an IR contraction affected members of Salviniales that moved exon 2 of *ndhB* from the IR to the LSC (Gao et al., 2013). In Schizaeales, the IR has expanded in both *Schizaea* and *Actinostachys*, although in different ways, while *Actinostachys* additionally experienced a substantial number of gene losses (Labiak & Karol, 2017), perhaps related to its mycoheterotrophic gametophyte lifestyle (Merckx et al., 2013).

Outside of the core leptosporangiate ferns and Schizaeales, there are relatively fewer structural changes. The plastome of *Diplopterygium*, the only member of Gleicheniales that is fully sequenced, exhibits a unique *trnV-GCA* to *trnL-CAA* inversion that overlaps the IR/LSC boundary, which moved *trnL-CAA* into the IR and *trnV-GCA* into the LSC (Kim et al., 2014). *Angiopteris* (Marattiales) and *Psilotum* (Psilotales) share an apparent convergent expansion of the IR that moved 3′-*rps12*, *rps7*, *ndhB*, and *trnL-CAA* from the LSC into the IR, and then the IR further expanded independently in the two groups, incorporating *trnI-CAU* from the LSC in *Angiopteris* and *ndhF*, *rpl21*, *rpl32*, and *trnP-GGG* from the SSC in *Psilotum* (Fig. 2; Grewe et al., 2013; Roper et al., 2007).

In contrast to all of the plastomic changes detected in most ferns, the plastomes from *Equisetum* (Equisetales), *Ophioglossum* (Ophioglossales), and *Osmundastrum* (Osmundales) are collinear with one another (Fig. 1; Grewe et al., 2013; Kim et al., 2014). Furthermore, they are fully collinear with nonvascular land plants and the lycophyte *Huperzia*, except for the fern-specific *trnD-GUC* and *trnG-GCC* to *trnT-GGU* inversions already discussed (Fig. 1). Thus, it is most parsimonious to assume that the plastome arrangement in Equisetales, Ophioglossales, and Osmundales represents the gene order of the common ancestor of all ferns. However, the IR expansion to *trnL-CAA* in Marattiales and Psilotales is intriguing because a similar genomic expansion was also inferred to have occurred in the ancestor of core leptosporangiates plus Schizaeales (Grewe et al., 2013). These may represent convergent evolutionary events, although it is only slightly less parsimonious to assume that this expansion occurred in the fern common ancestor, followed by contractions in other fern lineages with smaller IRs. Denser sampling of early diverging leptosporangiate ferns is needed to more fully assess the early structural evolution of fern plastomes.

4.5 Gymnosperms

At least 110 plastomes have been completely sequenced to date from gymnosperms. A full description of the evolutionary diversity of these plastomes is provided in the chapter "Evolution of gymnosperm plastid genomes" by Chaw et al. Here, we provide a brief summary of the major structural variation observed among five distinct lineages that are distinguished primarily by the presence (cycads, *Ginkgo*, gnetophytes) or absence (cupressophytes and Pinaceae) of an IR.

Among the IR-containing clades of gymnosperms, cycads and *Ginkgo* are notable for their structural conservation of the plastome, whereas gnetophyte plastomes are more structurally diverse. For Cycadales, plastomes are available from 12 species, revealing an extreme level of conservation (Jiang, Hinsinger, & Strijk, 2016; Wu & Chaw, 2015; Wu et al., 2007). Genomes range from 162 kb (*Dioon*) to 166 kb (*Macrozamia*) in size and display the ancestral gymnosperm gene order, featuring the gymnosperm-specific IR expansion to *trnH-GUG* (Fig. 2). The plastome from *Ginkgo biloba*, the sole living member of Ginkgoales, is 157 kb long with a structure that is nearly identical to cycads, with the exception of the lineage-specific contraction of the IR that moved *ycf2* to the LSC while leaving two copies of *trnH-GUG* at both ends of the LSC (Lin et al., 2012). In contrast to cycads and *Ginkgo*, plastomes from nine species of gnetophytes (two *Ephedra*, six *Gnetum*, and one *Welwitschia*) are much smaller in size (Ephedrales: 110 kb; Gnetales: 113–115 kb, Welwitschiales: 119–120 kb) and more variable in structure, including at least three inversions, a relocation, and multiple gene losses (Hou, Wikström, Strijk, & Rydin, 2016; McCoy et al., 2008; Wu et al., 2009; Zhu et al., 2016).

Species of cupressophytes (including Araucariales and Cupressales) and Pinaceae are defined by the absence of a large IR found in most land plants (Raubeson & Jansen, 1992b; Tsudzuki et al., 1992). The IR was postulated to have been lost independently in the two groups (Wu & Chaw, 2014; Wu, Wang, et al., 2011); however, concerns over this conclusion have been raised due to the need to consider both isomeric forms of the plastome in the IR-containing ancestor (Yi, Gao, Wang, Su, & Wang, 2013). Although the large IR has been lost, sequenced plastomes from >80 different species (9 Araucariales, 133–147 kb; 41 Cupressales, 121–138 kb; 39 Pinaceae, 107–124 kb) are characterized by the presence of one or more pairs of shorter repeats that tend to promote infrequent inversions over evolutionary timescales, which leads to a high degree of structural variability (Guo et al., 2014; Hirao, Watanabe, Kurita, Kondo, & Takata, 2008; Lin, Huang, Wu, Hsu, & Chaw, 2010; Wu, Lin, Hsu, Wang, & Chaw, 2011; Yi et al., 2013). Some of this structural variation can be detected within single individuals, indicating that substoichiometric isomers of the plastome coexist with the major genomic arrangement (do Nascimento Vieira et al., 2016; Guo et al., 2014; Hsu, Wu, & Chaw, 2016; Qu, Wu, Chaw, & Yi, 2017; Wu & Chaw, 2016).

4.6 Angiosperms

As of October 2017, there were more than 2700 complete plastome sequences from over 2000 different species of angiosperms in GenBank.

The general features and evolutionary diversity of the angiosperm plastome have been the subject of several recent, extensive reviews (Daniell, Lin, Yu, & Chang, 2016; Jansen & Ruhlman, 2012; Ruhlman & Jansen, 2014), so in this chapter we highlight some of the major trends in structural diversity. However, we have excluded discussion of plastome structural variation of heterotrophic plants, which is covered in the chapter "Molecular evolution of plastid genomes in parasitic flowering plants" by Wicke and Naumann and elsewhere (Graham, Lam, & Merckx, 2017), as well as any detailed discussion of the structurally diverse members of Geraniaceae, covered in the chapter "Aberration or analogy? The atypical plastomes of Geraniaceae" by Ruhlman and Jansen.

Among eudicots, magnoliids, and "basal" angiosperms, the majority of plastomes have retained the ancestral angiosperm structure, as exemplified by the eudicots *Arabidopsis* and *Nicotiana*, the magnoliid *Magnolia*, and the early diverging species *Amborella* and *Nymphaea* (Fig. 1). Many monocots tend to have a slightly expanded IR, as seen in *Acrous* and *Asparagus* (Fig. 2), but are otherwise largely collinear with the ancestral angiosperm plastome. Small expansions or contractions of the IR were also found in many distinct angiosperm clades (e.g. Downie & Jansen, 2015; Goulding, Olmstead, Morden, & Wolfe, 1996; Wang et al., 2008; Wicke, Schaferhoff, dePamphilis, & Muller, 2014). Large IR expansions of at least several kilobases are less common, but examples exist in *Pelargonium* (Chumley et al., 2006; Weng, Ruhlman, & Jansen, 2017), *Berberis* (Ma et al., 2013), Trochodendraceae (Sun et al., 2013), *Annona* (Blazier, Ruhlman, et al., 2016), *Plantago* (Zhu et al., 2016), and *Trithuria* (Gruenstaeudl, Nauheimer, & Borsch, 2017). Large IR contractions are even rarer, with just a few examples in Austrobaileyales (Gruenstaeudl et al., 2017; Hansen et al., 2007) and Lauraceae (Song et al., 2017). Complete loss of the IR is also rare, but it has been observed in some species from Fabaceae (Cai et al., 2008; Palmer, Osorio, Aldrich, & Thompson, 1987), Geraniaceae (Guisinger et al., 2011; Blazier, Jansen, et al., 2016; Ruhlman, Zhang, Blazier, Sabir, & Jansen, 2017), Orobanchaceae (Wicke et al., 2013), and Cactaceae (Sanderson et al., 2015).

Given the high degree of structural conservation in the plastome across land plants, it was generally believed until quite recently that the plastome was resistant to the acquisition of foreign DNA, unlike plant mitochondrial and nuclear genomes which frequently acquire foreign DNA through intracellular and horizontal transfer (Mower, Jain, & Hepburn, 2012; Timmis, Ayliffe, Huang, & Martin, 2004). However, in the past 5 years, reports of foreign DNA in angiosperm plastomes have begun to accumulate. The first such case was detected in the carrot plastome (Goremykin, Salamini, Velasco, & Viola, 2009; Iorizzo et al., 2012), which involved a 1.5 kb region of plastid DNA of mitochondrial origin (termed PLMT; Mower et al., 2012). Shortly thereafter, a 2.4 kb PLMT was detected in milkweed (Ku, Chung, Chen, & Kuo, 2013; Straub, Cronn, Edwards, Fishbein, & Liston, 2013). Subsequent studies have provided evidence for PLMTs in a wide variety of angiosperms, including bamboo (Ma, Zhang, Guo, & Li, 2015), various members of Apiales (Downie & Jansen, 2015; Spooner, Ruess, Iorizzo, Senalik, & Simon, 2017), *Paspalum* (Burke et al., 2016), and cashew (Rabah et al., 2017). Yet more recently, the first potential case of plastid DNA derived from the nuclear genome (termed PLNC; Mower et al., 2012) was reported in *Caucalis platycarpos* (Spooner et al., 2017). However, it is unclear whether this is a true PLNC, as it is perhaps more likely that the nuclear gene was first transferred to the mitochondrial genome and then transferred to the plastome (Rabah et al., 2017). In Campanulaceae, there are many insertions, some of which contain ORFs with some degree of codon-level conservation consistent with protein functionality; these ORFs were postulated to have originated from the nucleus, although this hypothesis currently lacks evidence (Knox, 2014).

5. FUNCTIONAL EFFECT OF STRUCTURAL CHANGE

5.1 The IR and Copy-Dependent Repair

Easily the most noticeable evolutionary effect of the plastome's structure is the reduced rate of synonymous nucleotide substitutions in the IR. This observation was first made by Wolfe, Li, and Sharp (1987) using a small number of genes from several pairs of angiosperm species. Subsequent studies made the same observations using complete plastome sequences focused on particular families of angiosperms (Gaut, 1998; Kim, Park, & Kim, 2009; Maier, Neckermann, Igloi, & Kossel, 1995; Perry & Wolfe, 2002; Wicke et al., 2014; Yamane, Yano, & Kawahara, 2006; Yi & Kim, 2012; Yi, Lee, Sun, Chung, & Kim, 2012). More recent studies with greatly expanded taxonomic sampling, including dozens of angiosperm families (Zhu et al., 2016) as well as representatives from gymnosperms and ferns (Li, Kuo, Pryer, & Rothfels, 2016; Wu & Chaw, 2016; Zhu et al., 2016), again identified slower rates for IR-localized genes. Together, these studies have consistently reported an approximately two- to fourfold reduction in synonymous substitution rates for IR genes relative to SC genes in euphyllophytes, which is generally interpreted to be due to copy-dependent repair driven by biased gene conversion (Birky & Walsh, 1992).

An interesting corollary to this copy-dependent repair phenomenon is that genes transferred from the IR into the LSC or SSC should experience rate acceleration consistent with their new genomic position, while genes transferred from the LSC or SSC into the IR should experience rate deceleration like other IR genes. Indeed, in a pair of IR-lacking legumes, the transfer of ancestral IR genes into the SC region resulted in increased substitution rates (Perry & Wolfe, 2002). A broader follow-up study (Zhu et al., 2016) identified increased substitution rates for nine additional cases of IR-to-SC gene transitions, including not only angiosperms but also gymnosperms and ferns, providing strong confirmation of the increased rate effect for IR-to-SC transitions. Examination of synonymous rates during IR expansions, in five different groups of angiosperms and ferns, provided the first evidence for an SC-to-IR effect, whereby genes transferred from the SC regions to the IR experienced reduced rates of evolution, as expected given their new location in the IR (Li et al., 2016; Zhu et al., 2016). Together, these results provide strong evidence that localization of a gene in the IR confers a copy-correction benefit.

Some exceptions to this reduced IR rate effect do exist, however. In particular, a small number of IR genes have highly elevated synonymous rates in *Pelargonium*, *Plantago*, and *Silene*, which appears to be the result of locus-specific effects (Weng et al., 2017; Zhu et al., 2016). The underlying evolutionary basis for localized rate increases in plant organellar genomes has been postulated to be due to error-prone repair of double-strand breaks (Magee et al., 2010) or error-prone gene conversion using reverse-transcribed transcriptional templates (Zhu, Guo, Jain, & Mower, 2014). It must be also pointed out that these three genera are notorious for having extremely unusual mitochondrial genomes (Cho, Mower, Qiu, & Palmer, 2004; Mower, Touzet, Gummow, Delph, & Palmer, 2007; Parkinson et al., 2005; Sloan et al., 2012), so it is possible that the atypical evolutionary characteristics of both organellar genomes have an overlapping mechanistic basis. In *Ginkgo*, an accelerated substitution rate was not observed for the *ycf2* gene, despite its transfer from the IR into the SC (Lin et al., 2012). In this case, the IR-to-SC transfer was postulated to be a recent event, and the overall plastid substitution rate in *Ginkgo* is slow, which may have limited the accumulation of mutations consistent with the gene's new genomic position.

5.2 Repeats, Structural Rearrangements, and Substoichiometric Shifting

Early mapping studies of legume and conifer plastomes identified a connection between IR loss and increased rearrangements, suggesting that the IR may impose structural constraint on the plastome by impeding rearrangement events (Palmer et al., 1987; Palmer & Thompson, 1982). Additional support for this hypothesis was garnered after complete plastome sequencing from some of the IR-lacking lineages, which revealed extreme rearrangement after IR loss in plastomes from *Trifolium* (Cai et al., 2008), *Erodium* (Guisinger et al., 2011), and conifers (Hirao et al., 2008; Lin et al., 2010; Wu, Lin, et al., 2011). Notably, all of these IR-lacking plastomes have accumulated novel small repeats that associate with the rearrangement endpoints, suggesting a role for repeat-mediated recombination in generating plastome structural diversity.

The constraining effect of the IR on plastome structural evolution may not be as strong as originally suspected, however. More in-depth sequencing from IR-lacking species from legumes (Jansen, Wojciechowski, Sanniyasi, Lee, & Daniell, 2008; Sabir et al., 2014), *Erodium* (Blazier, Jansen, et al., 2016), and the Saguaro cactus (Sanderson et al., 2015) has uncovered plastomes that are much less rearranged despite the loss of the IR. In addition, there are quite a few plant lineages that are highly rearranged yet have retained the IR, including species from Geraniaceae (Chumley et al., 2006; Guisinger et al., 2011; Weng et al., 2017), Oleaceae (Lee, Jansen, Chumley, & Kim, 2007), Campanulaceae (Haberle, Fourcade, Boore, & Jansen, 2008; Knox, 2014), *Plantago* (Zhu et al., 2016), gnetophytes (Wu et al., 2009), leptosporangiate ferns (Gao et al., 2013, 2009; Kim et al., 2014; Wolf et al., 2003), and many lycophytes (Karol et al., 2010; Smith, 2009; Tsuji et al., 2007). The emerging consensus is that the presence of smaller repeats, rather than the loss of the IR, is the major driver of plastomic rearrangement.

In many seed plants with small repeats in their plastomes, the repeats are recombinationally active, leading to structural rearrangements over short evolutionary timescales in Pinaceae (Tsumura, Suyama, & Yoshimura, 2000; Wu, Lin, et al., 2011), cupressophytes (do Nascimento Vieira et al., 2016; Guo et al., 2014; Qu et al., 2017; Wu & Chaw, 2014), Fabaceae (Gurdon & Maliga, 2014), and Geraniaceae (Ruhlman et al., 2017). In fact, the different genomic isomers created by homologous recombination at repeats have been shown to coexist, but at different stoichiometry, within single individuals of several different species of cupressophytes (do Nascimento Vieira et al., 2016; Guo et al., 2014; Qu et al., 2017). Moreover, the major and minor stoichiometric forms have clearly shifted over time (Guo et al., 2014; Qu et al., 2017), which is reminiscent of the process known as substoichiometric shifting that affects plant mitochondrial genomes (Arrieta-Montiel & Mackenzie, 2011; Woloszynska, 2010). Substoichiometric shifting of plant mitochondrial genomes can have strong phenotypic effects on the organism (Arrieta-Montiel & Mackenzie, 2011); it remains to be seen whether a similar process affecting plant plastomes has any functional consequences.

5.3 A Dosage Effect for IR-Localized Genes

One less well-appreciated effect of the IR is the doubling of transcription that it affords through a gene dosage effect. From an evolutionary standpoint, it is perhaps unsurprising to observe that the genes nearly universally present within the IR of plants and algae—namely, the 4 rRNAs plus *trnI-GAU* and *trnA-UGC* (Fig. 2; Zhu et al., 2016)—are some of the most highly expressed genes in the plastomes of diverse plants, such as the model angiosperm *Arabidopsis* (Castandet, Hotto, Strickler, & Stern, 2016; Hotto, Schmitz, Fei, Ecker, & Stern, 2011), the monocot barley (Zhelyazkova et al., 2012), and the fern *Psilotum* (Guo, Grewe, & Mower, 2015). The need for high levels of rRNA expression is obvious, as they are crucial components for plastid ribosomes. High levels of *trnI-GAU* and *trnA-UGC* expression are also consistent with the abundance of codons in plant chloroplasts that are recognized by the anticodons of these tRNAs (e.g. Maier et al., 1995; Sato, Nakamura, Kaneko, Asamizu, & Tabata, 1999; Wolf et al., 2003).

Overall, it appears that the doubling of these particular genes in the IR is not a coincidence, but instead a concerted evolutionary outcome to maximize levels of expression through a gene dosage effect. This suggests that one of the primary functions of the IR may be for increased gene dosage. Indeed, it is well appreciated in the plant transformation community that the doubled gene copy afforded by the IR is a simple means to increase expression, and in fact the *trnA/trnI* region of the IR is the most commonly used target of insertion to achieve high levels of transgene expression in the plastid (Daniell et al., 2016; Verma & Daniell, 2007).

6. WHAT IS LEFT IN PLASTOME STRUCTURAL RESEARCH?

With the relative ease in obtaining complete plastome sequences from next-generation sequencing, it is no surprise that the number of sequenced plant plastomes has increased (and will continue to increase) at an exponential pace. This sequencing revolution has proven to be an incredible boon to understanding the tempo and pattern of evolutionary change in plastomes across land plants. Nevertheless, there is still a dearth of knowledge about plastomic diversity in some plant lineages. In particular, the lycophytes, hornworts, and liverworts remain poorly sampled, especially in comparison to the thousands of plastomes available from angiosperms. In ferns, sampling is such that there is now at least one representative from almost every order, which has revealed a large amount of structural change, most notably in core leptosporangiate ferns. However, our understanding of the origins of this diversity is hampered by the limited sampling from early-diverging leptosporangiate ferns (Gleicheniales, Hymenophyllales, Osmundales) and their closest eusporangiate relatives (Marattiales). There is the potential for discovering novel structural changes in these underrepresented groups, and by defining the true extent of this diversity, we can verify or refine our inferences about the earliest events of plastome evolution in land plants.

One often overlooked area of concern in plastome sequencing is the quality of the finished product. Next-generation sequencing approaches, whether they produce short reads using small library inserts or very long but highly error-prone reads, present difficult challenges for the assembly of an accurate genome sequence. There are now many approaches to assist in plastome assembly, including "black box" automated tools and strategies that rely on a reference genome to guide the assembly or the filtering of reads prior to assembly. There is little doubt that these automated tools and reference-guided approaches are likely to perform well on evolutionarily conserved plastomes. However, it is still unclear how they perform on species with plastomes that are highly divergent in structure or sequence, or that contain an abundance of recombinationally active repeats such as in cupressophytes and Geraniaceae. It is also unclear how users of these approaches will be able to identify incorrect assemblies when they arise. There is a dire need for the development of standard practices to verify the accuracy of an assembly. Evaluation of read-pair coverage across the genome is one simple approach to ensure that read pairs map in the proper orientation and expected distance at roughly equal depth across the genome (and at twice the depth in the IR). Performing multiple independent assemblies, from different subsets of the data or different assembly parameter settings, is another useful approach for verification. The highest level of quality can be achieved by using multiple sequencing platforms (such as a combination of short read and very long read technologies) to limit the propagation of platform-specific errors, although the additional cost required for such a multifaceted approach will be a deterrent for many projects.

Several authors have pointed out the need to move beyond the simple production of yet more plastome sequences and instead put more focus on hypothesis-driven analyses, functional genomics, and experimental biology (Sanita Lima, Woods, Cartwright, & Smith, 2016; Smith, 2017; Tonti-Filippini, Nevill, Dixon, & Small, 2017). With regard to plastome structure, there are several major questions to pursue. First of all, does the plastome ever exist as a functional circular chromosome? Detailed studies of plastome structure indicate that most genome copies are linear and multibranched molecules. While circular molecules do exist, it is unclear whether they represent functional replicating molecules or accidental by-products of replication of linear molecules. If the predominant structure is linear, when did this structure shift? Are there any evolutionary benefits to the expression, replication, repair, or inheritance of plastomes as linear molecules? Further, characterization of linear and branched structures, particularly their terminal sequences, will help elucidate the mechanisms of plastome replication and maintenance as well as the role that topology plays in these processes (Maréchal & Brisson, 2010; Oldenburg & Bendich, 2015).

Other outstanding questions relate to the evolution of gene order. There are many conserved operons within the plastome, but they are not universally conserved. Does gene order matter in plastomes? Is the prevalence of conserved plastome gene order across the many different lineages of land plant a consequence of selection against some reduced level of functionality introduced by inversions and relocations? Or is the evolutionary stasis of plastomes a by-product of the presence of the IR that inhibits many types of rearrangements? Or is there selection against the proliferation of smaller repeats, perhaps to maintain streamlined genomes, that then indirectly inhibits rearrangement as a side effect? The recent finding of substoichiometric shifting of plastome structures suggests that these changes could have phenotypic consequences (Guo et al., 2014), but this is speculative. Does substoichiometric shifting in plant plastomes have any functional relevance?

Finally, an emerging trend in plastid phylogenetics is that the resulting topologies can conflict with phylogenies based on data from the nuclear or mitochondrial genome. There is a need to better understand the sources of this phylogenetic incongruence. Is the incongruence due to technical or biological issues? Certainly, as the field of plastid phylogenomics emerged, an argument was made that the oversampling of sequence data coupled with an undersampling of taxa may lead to strong phylogenetic support for incorrect topologies (Leebens-Mack et al., 2005; Soltis et al., 2004). But with the exponential increase of plastomes (and nuclear genomes) becoming available, a reassessment of this explanation is overdue. Biological issues that could result in phylogenetic discord include incomplete lineage sorting of chloroplast haplotypes or hybridization and chloroplast capture (Folk, Mandel, & Freudenstein, 2017; Percy et al., 2014). These processes can provide a possible explanation for unusual phylogenetic results, but it can be difficult to distinguish between them, and it is unclear whether any other alternatives exist. For example, what is the effect of plastome sequence and assembly errors on these analyses?

Certainly, much has been learned over the past 30 years of plastome sequencing, but plenty of work remains. Targeted sequencing of taxonomically underrepresented taxa will increase our understanding of genomic diversity, while improved bioinformatics approaches will improve the accuracy of completed assemblies. Most importantly, it will be important to harness the massive number of genomic data sets to address the many issues in plastome biology that remain unresolved.

ACKNOWLEDGEMENTS

The Mower lab gratefully acknowledges the support for plastome research from the National Science Foundation and the University of Nebraska—Lincoln.

REFERENCES

Arrieta-Montiel, M. P., & Mackenzie, S. A. (2011). Plant mitochondrial genomes and recombination. In F. Kempken (Ed.), *Plant mitochondria* (pp. 65–82). New York: Springer.

Bedbrook, J. R., & Kolodner, R. (1979). The structure of chloroplast DNA. *Annual Review of Plant Physiology*, *30*, 593–620.

Bell, N. E., Boore, J. L., Mishler, B. D., & Hyvonen, J. (2014). Organellar genomes of the four-toothed moss, *Tetraphis pellucida*. *BMC Genomics*, *15*, 383.

Bendich, A. J. (2004). Circular chloroplast chromosomes: The grand illusion. *The Plant Cell*, *16*, 1661–1666.

Bendich, A. J., & Smith, S. B. (1990). Moving pictures and pulsed-field gel electrophoresis show linear DNA molecules from chloroplasts and mitochondria. *Current Genetics*, *17*, 421–425.

Birky, C. W., & Walsh, J. B. (1992). Biased gene conversion, copy number, and apparent mutation-rate differences within chloroplast and bacterial genomes. *Genetics*, *130*, 677–683.

Blazier, C. J., Guisinger, M. M., & Jansen, R. K. (2011). Recent loss of plastid-encoded *ndh* genes within *Erodium* (Geraniaceae). *Plant Molecular Biology*, *76*, 263–272.

Blazier, J. C., Jansen, R. K., Mower, J. P., Govindu, M., Zhang, J., Weng, M. L., et al. (2016). Variable presence of the inverted repeat and plastome stability in *Erodium*. *Annals of Botany*, *117*, 1209–1220.

Blazier, J. C., Ruhlman, T. A., Weng, M. L., Rehman, S. K., Sabir, J. S., & Jansen, R. K. (2016). Divergence of RNA polymerase alpha subunits in angiosperm plastid genomes is mediated by genomic rearrangement. *Scientific Reports*, *6*, 24595.

Bock, D. G., Kane, N. C., Ebert, D. P., & Rieseberg, L. H. (2014). Genome skimming reveals the origin of the Jerusalem Artichoke tuber crop species: Neither from Jerusalem nor an artichoke. *New Phytologist*, *201*, 1021–1030.

Burke, S. V., Wysocki, W. P., Zuloaga, F. O., Craine, J. M., Pires, J. C., Edger, P. P., et al. (2016). Evolutionary relationships in Panicoid grasses based on plastome phylogenomics (Panicoideae; Poaceae). *BMC Plant Biology*, *16*, 140.

Cai, Z., Guisinger, M., Kim, H. G., Ruck, E., Blazier, J. C., McMurtry, V., et al. (2008). Extensive reorganization of the plastid genome of *Trifolium subterraneum* (Fabaceae) is associated with numerous repeated sequences and novel DNA insertions. *Journal of Molecular Evolution*, *67*, 696–704.

Castandet, B., Hotto, A. M., Strickler, S. R., & Stern, D. B. (2016). ChloroSeq, an optimized chloroplast RNA-Seq bioinformatic pipeline, reveals remodeling of the organellar transcriptome under heat stress. *G3: Genes, Genomes, Genetics*, *6*, 2817–2827.

Cho, Y., Mower, J. P., Qiu, Y. L., & Palmer, J. D. (2004). Mitochondrial substitution rates are extraordinarily elevated and variable in a genus of flowering plants. *Proceedings of the National Academy of Sciences of the United States of America*, *101*, 17741–17746.

Chumley, T. W., Palmer, J. D., Mower, J. P., Fourcade, H. M., Calie, P. J., Boore, J. L., et al. (2006). The complete chloroplast genome sequence of *Pelargonium* x *hortorum*: Organization and evolution of the largest and most highly rearranged chloroplast genome of land plants. *Molecular Biology and Evolution*, *23*, 2175–2190.

Civan, P., Foster, P. G., Embley, M. T., Seneca, A., & Cox, C. J. (2014). Analyses of charophyte chloroplast genomes help characterize the ancestral chloroplast genome of land plants. *Genome Biology and Evolution*, *6*, 897–911.

Daniell, H., Lin, C. S., Yu, M., & Chang, W. J. (2016). Chloroplast genomes: Diversity, evolution, and applications in genetic engineering. *Genome Biology*, *17*, 134.

do Nascimento Vieira, L., Rogalski, M., Faoro, H., de Freitas Fraga, H. P., Dos Anjos, K. G., Picchi, G. F. A., et al. (2016). The plastome sequence of the endemic Amazonian conifer, *Retrophyllum piresii* (Silba) CN page, reveals different recombination events and plastome isoforms. *Tree Genetics & Genomes*, *12*, 10.

Downie, S. R., & Jansen, R. K. (2015). A comparative analysis of whole plastid genomes from the Apiales: Expansion and contraction of the inverted repeat, mitochondrial to plastid transfer of DNA, and identification of highly divergent noncoding regions. *Systematic Botany*, *40*, 336–351.

Folk, R. A., Mandel, J. R., & Freudenstein, J. V. (2017). Ancestral gene flow and parallel organellar genome capture result in extreme phylogenomic discord in a lineage of angiosperms. *Systematic Biology*, *66*, 320–337.

Forrest, L. L., Wickett, N. J., Cox, C. J., & Goffinet, B. (2011). Deep sequencing of *Ptilidium* (Ptilidiaceae) suggests evolutionary stasis in liverwort plastid genome structure. *Plant Ecology and Evolution*, *144*, 29–43.

Gao, L., Wang, B., Wang, Z. W., Zhou, Y., Su, Y. J., & Wang, T. (2013). Plastome sequences of *Lygodium japonicum* and *Marsilea crenata* reveal the genome organization transformation from basal ferns to core leptosporangiates. *Genome Biology and Evolution*, *5*, 1403–1407.

Gao, L., Yi, X., Yang, Y. X., Su, Y. J., & Wang, T. (2009). Complete chloroplast genome sequence of a tree fern *Alsophila spinulosa*: Insights into evolutionary changes in fern chloroplast genomes. *BMC Evolutionary Biology*, *9*, 130.

Gao, L., Zhou, Y., Wang, Z. W., Su, Y. J., & Wang, T. (2011). Evolution of the *rpoB-psbZ* region in fern plastid genomes: Notable structural rearrangements and highly variable intergenic spacers. *BMC Plant Biology*, *11*, 64.

Gaut, B. S. (1998). Molecular clocks and nucleotide substitution rates in higher plants. In M. K. Hecht, R. J. Macintyre, & M. T. Clegg (Eds.), *Evolutionary biology* (pp. 93–120). New York: Plenum Press.

Goffinet, B., Wickett, N. J., Werner, O., Ros, R. M., Shaw, A. J., & Cox, C. J. (2007). Distribution and phylogenetic significance of the 71-kb inversion in the plastid genome in Funariidae (Bryophyta). *Annals of Botany*, *99*, 747–753.

Goremykin, V. V., Salamini, F., Velasco, R., & Viola, R. (2009). Mitochondrial DNA of *Vitis vinifera* and the issue of rampant horizontal gene transfer. *Molecular Biology and Evolution*, *26*, 99–110.

Goulding, S. E., Olmstead, R. G., Morden, C. W., & Wolfe, K. H. (1996). Ebb and flow of the chloroplast inverted repeat. *Molecular and General Genetics*, *252*, 195–206.

Graham, S. W., Lam, V. K. Y., & Merckx, V. S. F. T. (2017). Plastomes on the edge: The evolutionary breakdown of mycoheterotroph plastid genomes. *New Phytologist*, *214*, 48–55.

Grewe, F., Guo, W., Gubbels, E. A., Hansen, A. K., & Mower, J. P. (2013). Complete plastid genomes from *Ophioglossum californicum*, *Psilotum nudum*, and *Equisetum hyemale* reveal an ancestral land plant genome structure and resolve the position of Equisetales among monilophytes. *BMC Evolutionary Biology*, *13*, 8.

Grosche, C., Funk, H. T., Maier, U. G., & Zauner, S. (2012). The chloroplast genome of *Pellia endiviifolia*: Gene content, RNA-editing pattern, and the origin of chloroplast editing. *Genome Biology and Evolution*, *4*, 1349–1357.

Gruenstaeudl, M., Nauheimer, L., & Borsch, T. (2017). Plastid genome structure and phylogenomics of Nymphaeales: Conserved gene order and new insights into relationships. *Plant Systematics and Evolution*, *303*, 1251–1270.

Guisinger, M. M., Kuehl, J. V., Boore, J. L., & Jansen, R. K. (2011). Extreme reconfiguration of plastid genomes in the angiosperm family Geraniaceae: Rearrangements, repeats, and codon usage. *Molecular Biology and Evolution*, *28*, 583–600.

Guo, W., Grewe, F., Cobo-Clark, A., Fan, W., Duan, Z., Adams, R. P., et al. (2014). Predominant and substoichiometric isomers of the plastid genome coexist within *Juniperus* plants and have shifted multiple times during cupressophyte evolution. *Genome Biology and Evolution*, *6*, 580–590.

Guo, W., Grewe, F., & Mower, J. P. (2015). Variable frequency of plastid RNA editing among ferns and repeated loss of uridine-to-cytidine editing from vascular plants. *PLoS One*, *10*, e0117075.

Guo, Z. Y., Zhang, H. R., Shrestha, N., & Zhang, X. C. (2016). Complete chloroplast genome of a valuable medicinal plant, *Huperzia serrata* (Lycopodiaceae), and comparison with its congener. *Applications in Plant Sciences*, *4*, 1600071.

Gurdon, C., & Maliga, P. (2014). Two distinct plastid genome configurations and unprecedented intraspecies length variation in the *accD* coding region in *Medicago truncatula*. *DNA Research*, *21*, 417–427.

Haberle, R. C., Fourcade, H. M., Boore, J. L., & Jansen, R. K. (2008). Extensive rearrangements in the chloroplast genome of *Trachelium caeruleum* are associated with repeats and tRNA genes. *Journal of Molecular Evolution*, *66*, 350–361.

Hansen, D. R., Dastidar, S. G., Cai, Z., Penaflor, C., Kuehl, J. V., Boore, J. L., et al. (2007). Phylogenetic and evolutionary implications of complete chloroplast genome sequences of four early-diverging angiosperms: *Buxus* (Buxaceae), *Chloranthus* (Chloranthaceae),

Dioscorea (Dioscoreaceae), and *Illicium* (Schisandraceae). *Molecular Phylogenetics and Evolution, 45*, 547–563.

Hasebe, M., & Iwatsuki, K. (1992). Gene localization on the chloroplast DNA of the maiden hair fern; *Adiantum capillus-veneris*. *Botanical Magazine Tokyo, 105*, 413–419.

Hirao, T., Watanabe, A., Kurita, M., Kondo, T., & Takata, K. (2008). Complete nucleotide sequence of the *Cryptomeria japonica* D. Don. chloroplast genome and comparative chloroplast genomics: Diversified genomic structure of coniferous species. *BMC Plant Biology, 8*, 70.

Hotto, A. M., Schmitz, R. J., Fei, Z., Ecker, J. R., & Stern, D. B. (2011). Unexpected diversity of chloroplast noncoding RNAs as revealed by deep sequencing of the *Arabidopsis* transcriptome. *G3: Genes, Genomes, Genetics, 1*, 559–570.

Hou, C., Wikström, N., Strijk, J. S., & Rydin, C. (2016). Resolving phylogenetic relationships and species delimitations in closely related gymnosperms using high-throughput NGS, Sanger sequencing and morphology. *Plant Systematics and Evolution, 302*, 1345–1365.

Hsu, C. Y., Wu, C. S., & Chaw, S. M. (2016). Birth of four chimeric plastid gene clusters in Japanese umbrella pine. *Genome Biology and Evolution, 8*, 1776–1784.

Iorizzo, M., Senalik, D., Szklarczyk, M., Grzebelus, D., Spooner, D., & Simon, P. (2012). De novo assembly of the carrot mitochondrial genome using next generation sequencing of whole genomic DNA provides first evidence of DNA transfer into an angiosperm plastid genome. *BMC Plant Biology, 12*, 61.

Jansen, R. K., Cai, Z., Raubeson, L. A., Daniell, H., Depamphilis, C. W., Leebens-Mack, J., et al. (2007). Analysis of 81 genes from 64 plastid genomes resolves relationships in angiosperms and identifies genome-scale evolutionary patterns. *Proceedings of the National Academy of Sciences of the United States of America, 104*, 19369–19374.

Jansen, R. K., & Ruhlman, T. A. (2012). Plastid genomes of seed plants. In R. Bock & V. Knoop (Eds.), *Genomics of chloroplasts and mitochondria* (pp. 103–126). Netherlands: Springer.

Jansen, R. K., Wojciechowski, M. F., Sanniyasi, E., Lee, S. B., & Daniell, H. (2008). Complete plastid genome sequence of the chickpea (*Cicer arietinum*) and the phylogenetic distribution of *rps12* and *clpP* intron losses among legumes (Leguminosae). *Molecular Phylogenetics and Evolution, 48*, 1204–1217.

Jiang, G. F., Hinsinger, D. D., & Strijk, J. S. (2016). Comparison of intraspecific, interspecific and intergeneric chloroplast diversity in cycads. *Scientific Reports, 6*, 31473.

Karol, K. G., Arumuganathan, K., Boore, J. L., Duffy, A. M., Everett, K. D., Hall, J. D., et al. (2010). Complete plastome sequences of *Equisetum arvense* and *Isoetes flaccida*: Implications for phylogeny and plastid genome evolution of early land plant lineages. *BMC Evolutionary Biology, 10*, 321.

Kim, H. T., Chung, M. G., & Kim, K.-J. (2014). Chloroplast genome evolution in early diverged leptosporangiate ferns. *Molecules and Cells, 37*, 372.

Kim, Y. K., Park, C. W., & Kim, K. J. (2009). Complete chloroplast DNA sequence from a Korean endemic genus, *Megaleranthis saniculifolia*, and its evolutionary implications. *Molecules and Cells, 27*, 365–381.

Knox, E. B. (2014). The dynamic history of plastid genomes in the Campanulaceae *sensu lato* is unique among angiosperms. *Proceedings of the National Academy of Sciences of the United States of America, 111*, 11097–11102.

Kolodner, R., & Tewari, K. K. (1979). Inverted repeats in chloroplast DNA from higher plants. *Proceedings of the National Academy of Sciences of the United States of America, 76*, 41–45.

Ku, C., Chung, W. C., Chen, L. L., & Kuo, C. H. (2013). The complete plastid genome sequence of Madagascar periwinkle *Catharanthus roseus* (L.) G. Don: Plastid genome evolution, molecular marker identification, and phylogenetic implications in asterids. *PLoS One, 8*, e68518.

Kugita, M., Kaneko, A., Yamamoto, Y., Takeya, Y., Matsumoto, T., & Yoshinaga, K. (2003). The complete nucleotide sequence of the hornwort (*Anthoceros formosae*) chloroplast genome: Insight into the earliest land plants. *Nucleic Acids Research, 31*, 716–721.
Labiak, P. H., & Karol, K. G. (2017). Plastome sequences of an ancient fern lineage reveal remarkable changes in gene content and architecture. *American Journal of Botany, 104*, 1008–1018.
Lee, H. L., Jansen, R. K., Chumley, T. W., & Kim, K. J. (2007). Gene relocations within chloroplast genomes of *Jasminum* and *Menodora* (Oleaceae) are due to multiple, overlapping inversions. *Molecular Biology and Evolution, 24*, 1161–1180.
Leebens-Mack, J., Raubeson, L. A., Cui, L., Kuehl, J. V., Fourcade, H. M., Chumley, T. W., et al. (2005). Identifying the basal angiosperm node in chloroplast genome phylogenies: Sampling one's way out of the Felsenstein zone. *Molecular Biology and Evolution, 22*, 1948–1963.
Lemieux, C., Otis, C., & Turmel, M. (2000). Ancestral chloroplast genome in *Mesostigma viride* reveals an early branch of green plant evolution. *Nature, 403*, 649–652.
Lemieux, C., Otis, C., & Turmel, M. (2007). A clade uniting the green algae *Mesostigma viride* and *Chlorokybus atmophyticus* represents the deepest branch of the Streptophyta in chloroplast genome-based phylogenies. *BMC Biology, 5*, 2.
Lemieux, C., Otis, C., & Turmel, M. (2016). Comparative chloroplast genome analyses of streptophyte green algae uncover major structural alterations in the Klebsormidiophyceae, Coleochaetophyceae and Zygnematophyceae. *Frontiers in Plant Science, 7*, 697.
Li, F. W., Kuo, L. Y., Pryer, K. M., & Rothfels, C. J. (2016). Genes translocated into the plastid inverted repeat show decelerated substitution rates and elevated GC content. *Genome Biology and Evolution, 8*, 2452–2458.
Lilly, J. W., Havey, M. J., Jackson, S. A., & Jiang, J. (2001). Cytogenomic analyses reveal the structural plasticity of the chloroplast genome in higher plants. *The Plant Cell, 13*, 245–254.
Lin, C. P., Huang, J. P., Wu, C. S., Hsu, C. Y., & Chaw, S. M. (2010). Comparative chloroplast genomics reveals the evolution of Pinaceae genera and subfamilies. *Genome Biology and Evolution, 2*, 504–517.
Lin, C. P., Wu, C. S., Huang, Y. Y., & Chaw, S. M. (2012). The complete chloroplast genome of *Ginkgo biloba* reveals the mechanism of inverted repeat contraction. *Genome Biology and Evolution, 4*, 374–381.
Lu, J. M., Zhang, N., Du, X. Y., Wen, J., & Li, D. Z. (2015). Chloroplast phylogenomics resolves key relationships in ferns. *Journal of Systematics and Evolution, 53*, 448–457.
Ma, J., Yang, B., Zhu, W., Sun, L., Tian, J., & Wang, X. (2013). The complete chloroplast genome sequence of *Mahonia bealei* (Berberidaceae) reveals a significant expansion of the inverted repeat and phylogenetic relationship with other angiosperms. *Gene, 528*, 120–131.
Ma, P. F., Zhang, Y. X., Guo, Z. H., & Li, D. Z. (2015). Evidence for horizontal transfer of mitochondrial DNA to the plastid genome in a bamboo genus. *Scientific Reports, 5*, 11608.
Magee, A. M., Aspinall, S., Rice, D. W., Cusack, B. P., Semon, M., Perry, A. S., et al. (2010). Localized hypermutation and associated gene losses in legume chloroplast genomes. *Genome Research, 20*, 1700–1710.
Maier, R. M., Neckermann, K., Igloi, G. L., & Kossel, H. (1995). Complete sequence of the maize chloroplast genome: Gene content, hotspots of divergence and fine-tuning of genetic information by transcript editing. *Journal of Molecular Biology, 251*, 614–628.
Maréchal, A., & Brisson, N. (2010). Recombination and the maintenance of plant organelle genome stability. *New Phytologist, 186*, 299–317.
McCoy, S. R., Kuehl, J. V., Boore, J. L., & Raubeson, L. A. (2008). The complete plastid genome sequence of *Welwitschia mirabilis*: An unusually compact plastome with accelerated divergence rates. *BMC Evolutionary Biology, 8*, 130.

Merckx, V. S. F. T., Freudenstein, J. V., Kissling, J., Christenhusz, M. J. M., Stotler, R. E., Crandall-Stotler, B., et al. (2013). Taxonomy and classification. In V. Merckx (Ed.), *Mycoheterotrophy: The biology of plants living on fungi* (pp. 19–101). New York, NY: Springer New York.

Mower, J. P., Jain, K., & Hepburn, N. J. (2012). The role of horizontal transfer in shaping the plant mitochondrial genome. In M.-D. Laurence (Ed.), *Advances in botanical research* (pp. 41–69). Academic Press.

Mower, J. P., Touzet, P., Gummow, J. S., Delph, L. F., & Palmer, J. D. (2007). Extensive variation in synonymous substitution rates in mitochondrial genes of seed plants. *BMC Evolutionary Biology*, 7, 135.

Myszczyński, K., Bączkiewicz, A., Buczkowska, K., Ślipiko, M., Szczecińska, M., & Sawicki, J. (2017). The extraordinary variation of the organellar genomes of the *Aneura pinguis* revealed advanced cryptic speciation of the early land plants. *Scientific Reports*, 7, 9804.

Ohyama, K., Fukuzawa, H., Kohchi, T., Shirai, H., Sano, T., Sano, S., et al. (1986). Chloroplast gene organization deduced from complete sequence of liverwort *Marchantia polymorpha* chloroplast DNA. *Nature*, *322*, 572–574.

Oldenburg, D. J., & Bendich, A. J. (2004). Most chloroplast DNA of maize seedlings in linear molecules with defined ends and branched forms. *Journal of Molecular Biology*, *335*, 953–970.

Oldenburg, D. J., & Bendich, A. J. (2015). DNA maintenance in plastids and mitochondria of plants. *Frontiers in Plant Science*, *6*, 883.

Oldenburg, D. J., & Bendich, A. J. (2016). The linear plastid chromosomes of maize: Terminal sequences, structures, and implications for DNA replication. *Current Genetics*, *62*, 431–442.

Oliver, M. J., Murdock, A. G., Mishler, B. D., Kuehl, J. V., Boore, J. L., Mandoli, D. F., et al. (2010). Chloroplast genome sequence of the moss *Tortula ruralis*: Gene content, polymorphism, and structural arrangement relative to other green plant chloroplast genomes. *BMC Genomics*, *11*, 143.

Palmer, J. D. (1983). Chloroplast DNA exists in two orientations. *Nature*, *301*, 92–93.

Palmer, J. D., Osorio, B., Aldrich, J., & Thompson, W. F. (1987). Chloroplast DNA evolution among legumes—Loss of a large inverted repeat occurred prior to other sequence rearrangements. *Current Genetics*, *11*, 275–286.

Palmer, J. D., & Thompson, W. F. (1982). Chloroplast DNA rearrangements are more frequent when a large inverted repeat sequence is lost. *Cell*, *29*, 537–550.

Parkinson, C. L., Mower, J. P., Qiu, Y. L., Shirk, A. J., Song, K., Young, N. D., et al. (2005). Multiple major increases and decreases in mitochondrial substitution rates in the plant family Geraniaceae. *BMC Evolutionary Biology*, *5*, 73.

Parks, M., Cronn, R., & Liston, A. (2009). Increasing phylogenetic resolution at low taxonomic levels using massively parallel sequencing of chloroplast genomes. *BMC Biology*, 7, 84.

Percy, D. M., Argus, G. W., Cronk, Q. C., Fazekas, A. J., Kesanakurti, P. R., Burgess, K. S., et al. (2014). Understanding the spectacular failure of DNA barcoding in willows (*Salix*): Does this result from a trans-specific selective sweep? *Molecular Ecology*, *23*, 4737–4756.

Perry, A. S., & Wolfe, K. H. (2002). Nucleotide substitution rates in legume chloroplast DNA depend on the presence of the inverted repeat. *Journal of Molecular Evolution*, *55*, 501–508.

Qu, X. J., Wu, C. S., Chaw, S. M., & Yi, T. S. (2017). Insights into the existence of isomeric plastomes in Cupressoideae (Cupressaceae). *Genome Biology and Evolution*, *9*, 1110–1119.

Rabah, S. O., Lee, C., Hajrah, N. H., Makki, R. M., Alharby, H. F., Alhebshi, A. M., et al. (2017). Plastome sequencing of ten nonmodel crop species uncovers a large insertion of mitochondrial DNA in cashew. *The Plant Genome*, 7, 20.

Raman, G., Choi, K. S., & Park, S. (2016). Phylogenetic relationships of the fern *Cyrtomium falcatum* (Dryopteridaceae) from Dokdo Island based on chloroplast genome sequencing. *Genes*, *7*, 115.

Raubeson, L. A., & Jansen, R. K. (1992a). Chloroplast DNA evidence on the ancient evolutionary split in vascular land plants. *Science*, *255*, 1697–1699.

Raubeson, L. A., & Jansen, R. K. (1992b). A rare chloroplast-DNA structural mutation is shared by all conifers. *Biochemical Systematics and Ecology*, *20*, 17–24.

Raubeson, L. A., & Stein, D. B. (1995). Insights into fern evolution from mapping chloroplast genomes. *American Fern Journal*, *85*, 193–204.

Rochaix, J. D. (1978). Restriction endonuclease map of the chloroplast DNA of *Chlamydomonas reinhardii*. *Journal of Molecular Biology*, *126*, 597–617.

Roper, J. M., Kellon Hansen, S., Wolf, P. G., Karol, K. G., Mandoli, D. F., Everett, K. D., et al. (2007). The complete plastid genome sequence of *Angiopteris evecta* (G. Forst.) Hoffm. (Marattiaceae). *American Fern Journal*, *97*, 95–106.

Rowan, B. A., Oldenburg, D. J., & Bendich, A. J. (2004). The demise of chloroplast DNA in *Arabidopsis*. *Current Genetics*, *46*, 176–181.

Ruhfel, B. R., Gitzendanner, M. A., Soltis, P. S., Soltis, D. E., & Burleigh, J. G. (2014). From algae to angiosperms-inferring the phylogeny of green plants (Viridiplantae) from 360 plastid genomes. *BMC Evolutionary Biology*, *14*, 23.

Ruhlman, T. A., & Jansen, R. K. (2014). The plastid genomes of flowering plants. *Methods in Molecular Biology*, *1132*, 3–38.

Ruhlman, T. A., Zhang, J., Blazier, J. C., Sabir, J. S. M., & Jansen, R. K. (2017). Recombination-dependent replication and gene conversion homogenize repeat sequences and diversify plastid genome structure. *American Journal of Botany*, *104*, 559–572.

Sabir, J., Schwarz, E., Ellison, N., Zhang, J., Baeshen, N. A., Mutwakil, M., et al. (2014). Evolutionary and biotechnology implications of plastid genome variation in the inverted-repeat-lacking clade of legumes. *Plant Biotechnology Journal*, *12*, 743–754.

Sanderson, M. J., Copetti, D., Burquez, A., Bustamante, E., Charboneau, J. L., Eguiarte, L. E., et al. (2015). Exceptional reduction of the plastid genome of saguaro cactus (*Carnegiea gigantea*): Loss of the *ndh* gene suite and inverted repeat. *American Journal of Botany*, *102*, 1115–1127.

Sanita Lima, M., Woods, L. C., Cartwright, M. W., & Smith, D. R. (2016). The (in)complete organelle genome: Exploring the use and nonuse of available technologies for characterizing mitochondrial and plastid chromosomes. *Molecular Ecology Resources*, *16*, 1279–1286.

Sato, S., Nakamura, Y., Kaneko, T., Asamizu, E., & Tabata, S. (1999). Complete structure of the chloroplast genome of *Arabidopsis thaliana*. *DNA Research*, *6*, 283–290.

Scharff, L. B., & Koop, H. U. (2006). Linear molecules of tobacco ptDNA end at known replication origins and additional loci. *Plant Molecular Biology*, *62*, 611–621.

Shaver, J. M., Oldenburg, D. J., & Bendich, A. J. (2008). The structure of chloroplast DNA molecules and the effects of light on the amount of chloroplast DNA during development in Medicago truncatula. *Plant Physiology*, *146*, 1064–1074.

Shaw, A. J., Devos, N., Liu, Y., Cox, C. J., Goffinet, B., Flatberg, K. I., et al. (2016). Organellar phylogenomics of an emerging model system: *Sphagnum* (peatmoss). *Annals of Botany*, *118*, 185–196.

Shinozaki, K., Ohme, M., Tanaka, M., Wakasugi, T., Hayashida, N., Matsubayashi, T., et al. (1986). The complete nucleotide sequence of the tobacco chloroplast genome: Its gene organization and expression. *The EMBO Journal*, *5*, 2043–2049.

Sloan, D. B., Alverson, A. J., Chuckalovcak, J. P., Wu, M., McCauley, D. E., Palmer, J. D., et al. (2012). Rapid evolution of enormous, multichromosomal genomes in flowering plant mitochondria with exceptionally high mutation rates. *PLoS Biology*, *10*, e1001241.

Smith, D. R. (2009). Unparalleled GC content in the plastid DNA of *Selaginella*. *Plant Molecular Biology*, *71*, 627–639.
Smith, D. R. (2017). Goodbye genome paper, hello genome report: The increasing popularity of "genome announcements" and their impact on science. *Briefings in Functional Genomics*, *16*, 156–162.
Soltis, D. E., Albert, V. A., Savolainen, V., Hilu, K., Qiu, Y. L., Chase, M. W., et al. (2004). Genome-scale data, angiosperm relationships, and "ending incongruence": A cautionary tale in phylogenetics. *Trends in Plant Science*, *9*, 477–483.
Song, Y., Yu, W. B., Tan, Y., Liu, B., Yao, X., Jin, J., et al. (2017). Evolutionary comparisons of the chloroplast genome in Lauraceae and insights into loss events in the magnoliids. *Genome Biology and Evolution*, *9*, 2354–2364.
Spooner, D. M., Ruess, H., Iorizzo, M., Senalik, D., & Simon, P. (2017). Entire plastid phylogeny of the carrot genus (*Daucus*, Apiaceae): Concordance with nuclear data and mitochondrial and nuclear DNA insertions to the plastid. *American Journal of Botany*, *104*, 296–312.
Stein, D. B., Conant, D. S., Ahearn, M. E., Jordan, E. T., Kirch, S. A., Hasebe, M., et al. (1992). Structural rearrangements of the chloroplast genome provide an important phylogenetic link in ferns. *Proceedings of the National Academy of Sciences of the United States of America*, *89*, 1856–1860.
Straub, S. C., Cronn, R. C., Edwards, C., Fishbein, M., & Liston, A. (2013). Horizontal transfer of DNA from the mitochondrial to the plastid genome and its subsequent evolution in milkweeds (Apocynaceae). *Genome Biology and Evolution*, *5*, 1872–1885.
Sugiura, C., Kobayashi, Y., Aoki, S., Sugita, C., & Sugita, M. (2003). Complete chloroplast DNA sequence of the moss *Physcomitrella patens*: Evidence for the loss and relocation of *rpoA* from the chloroplast to the nucleus. *Nucleic Acids Research*, *31*, 5324–5331.
Sun, Y. X., Moore, M. J., Meng, A. P., Soltis, P. S., Soltis, D. E., Li, J. Q., et al. (2013). Complete plastid genome sequencing of Trochodendraceae reveals a significant expansion of the inverted repeat and suggests a Paleogene divergence between the two extant species. *PLoS One*, *8*, e60429.
Timmis, J. N., Ayliffe, M. A., Huang, C. Y., & Martin, W. (2004). Endosymbiotic gene transfer: Organelle genomes forge eukaryotic chromosomes. *Nature Reviews Genetics*, *5*, 123–135.
Tonti-Filippini, J., Nevill, P. G., Dixon, K., & Small, I. (2017). What can we do with 1000 plastid genomes? *The Plant Journal*, *90*, 808–818.
Tsudzuki, J., Nakashima, K., Tsudzuki, T., Hiratsuka, J., Shibata, M., Wakasugi, T., et al. (1992). Chloroplast DNA of black pine retains a residual inverted repeat lacking rRNA genes: Nucleotide sequences of *trnQ*, *trnK*, *psbA*, *trnI* and *trnH* and the absence of *rps16*. *Molecular and General Genetics*, *232*, 206–214.
Tsuji, S., Ueda, K., Nishiyama, T., Hasebe, M., Yoshikawa, S., Konagaya, A., et al. (2007). The chloroplast genome from a lycophyte (microphyllophyte), *Selaginella uncinata*, has a unique inversion, transpositions and many gene losses. *Journal of Plant Research*, *120*, 281–290.
Tsumura, Y., Suyama, Y., & Yoshimura, K. (2000). Chloroplast DNA inversion polymorphism in populations of *Abies* and *Tsuga*. *Molecular Biology and Evolution*, *17*, 1302–1312.
Turmel, M., Otis, C., & Lemieux, C. (2005). The complete chloroplast DNA sequences of the charophycean green algae *Staurastrum* and *Zygnema* reveal that the chloroplast genome underwent extensive changes during the evolution of the Zygnematales. *BMC Biology*, *3*, 22.
Turmel, M., Otis, C., & Lemieux, C. (2006). The chloroplast genome sequence of *Chara vulgaris* sheds new light into the closest green algal relatives of land plants. *Molecular Biology and Evolution*, *23*, 1324–1338.

Vargas, O. M., Ortiz, E. M., & Simpson, B. B. (2017). Conflicting phylogenomic signals reveal a pattern of reticulate evolution in a recent high-Andean diversification (Asteraceae: Astereae: *Diplostephium*). *New Phytologist, 214*, 1736–1750.
Verma, D., & Daniell, H. (2007). Chloroplast vector systems for biotechnology applications. *Plant Physiology, 145*, 1129–1143.
Villarreal, J. C., Forrest, L. L., Wickett, N., & Goffinet, B. (2013). The plastid genome of the hornwort *Nothoceros aenigmaticus* (Dendrocerotaceae): Phylogenetic signal in inverted repeat expansion, pseudogenization, and intron gain. *American Journal of Botany, 100*, 467–477.
Wakasugi, T., Tsudzuki, J., Ito, S., Nakashima, K., Tsudzuki, T., & Sugiura, M. (1994). Loss of all *ndh* genes as determined by sequencing the entire chloroplast genome of the black pine *Pinus thunbergii*. *Proceedings of the National Academy of Sciences of the United States of America, 91*, 9794–9798.
Wang, R. J., Cheng, C. L., Chang, C. C., Wu, C. L., Su, T. M., & Chaw, S. M. (2008). Dynamics and evolution of the inverted repeat-large single copy junctions in the chloroplast genomes of monocots. *BMC Evolutionary Biology, 8*, 36.
Wei, R., Yan, Y.-H., Harris, A. J., Kang, J.-S., Shen, H., Xiang, Q.-P., et al. (2017). Plastid phylogenomics resolve deep relationships among eupolypod II ferns with rapid radiation and rate heterogeneity. *Genome Biology and Evolution, 9*, 1646–1657.
Weng, M. L., Ruhlman, T. A., & Jansen, R. K. (2017). Expansion of inverted repeat does not decrease substitution rates in *Pelargonium* plastid genomes. *New Phytologist, 214*, 842–851.
Wicke, S., Muller, K. F., de Pamphilis, C. W., Quandt, D., Wickett, N. J., Zhang, Y., et al. (2013). Mechanisms of functional and physical genome reduction in photosynthetic and nonphotosynthetic parasitic plants of the broomrape family. *The Plant Cell, 25*, 3711–3725.
Wicke, S., Schaferhoff, B., dePamphilis, C. W., & Muller, K. F. (2014). Disproportional plastome-wide increase of substitution rates and relaxed purifying selection in genes of carnivorous Lentibulariaceae. *Molecular Biology and Evolution, 31*, 529–545.
Wicke, S., Schneeweiss, G. M., dePamphilis, C. W., Muller, K. F., & Quandt, D. (2011). The evolution of the plastid chromosome in land plants: Gene content, gene order, gene function. *Plant Molecular Biology, 76*, 273–297.
Wickett, N. J., Zhang, Y., Hansen, S. K., Roper, J. M., Kuehl, J. V., Plock, S. A., et al. (2008). Functional gene losses occur with minimal size reduction in the plastid genome of the parasitic liverwort *Aneura mirabilis*. *Molecular Biology and Evolution, 25*, 393–401.
Wolf, P. G., Karol, K. G., Mandoli, D. F., Kuehl, J., Arumuganathan, K., Ellis, M. W., et al. (2005). The first complete chloroplast genome sequence of a lycophyte, *Huperzia lucidula* (Lycopodiaceae). *Gene, 350*, 117–128.
Wolf, P. G., Roper, J. M., & Duffy, A. M. (2010). The evolution of chloroplast genome structure in ferns. *Genome, 53*, 731–738.
Wolf, P. G., Rowe, C. A., Sinclair, R. B., & Hasebe, M. (2003). Complete nucleotide sequence of the chloroplast genome from a leptosporangiate fern, *Adiantum capillus-veneris* L. *DNA Research, 10*, 59–65.
Wolfe, K. H., Li, W. H., & Sharp, P. M. (1987). Rates of nucleotide substitution vary greatly among plant mitochondrial, chloroplast, and nuclear DNAs. *Proceedings of the National Academy of Sciences of the United States of America, 84*, 9054–9058.
Woloszynska, M. (2010). Heteroplasmy and stoichiometric complexity of plant mitochondrial genomes—Though this be madness, yet there's method in't. *Journal of Experimental Botany, 61*, 657–671.
Wu, C. S., & Chaw, S. M. (2014). Highly rearranged and size-variable chloroplast genomes in conifers II clade (cupressophytes): Evolution towards shorter intergenic spacers. *Plant Biotechnology Journal, 12*, 344–353.

Wu, C. S., & Chaw, S. M. (2015). Evolutionary stasis in cycad plastomes and the first case of plastome GC-biased gene conversion. *Genome Biology and Evolution*, *7*, 2000–2009.
Wu, C. S., & Chaw, S. M. (2016). Large-scale comparative analysis reveals the mechanisms driving plastomic compaction, reduction, and inversions in conifers II (cupressophytes). *Genome Biology and Evolution*, *8*, 3740–3750.
Wu, C. S., Lai, Y. T., Lin, C. P., Wang, Y. N., & Chaw, S. M. (2009). Evolution of reduced and compact chloroplast genomes (cpDNAs) in gnetophytes: Selection toward a lower-cost strategy. *Molecular Phylogenetics and Evolution*, *52*, 115–124.
Wu, C. S., Lin, C. P., Hsu, C. Y., Wang, R. J., & Chaw, S. M. (2011). Comparative chloroplast genomes of Pinaceae: Insights into the mechanism of diversified genomic organizations. *Genome Biology and Evolution*, *3*, 309–319.
Wu, C. S., Wang, Y. N., Hsu, C. Y., Lin, C. P., & Chaw, S. M. (2011). Loss of different inverted repeat copies from the chloroplast genomes of Pinaceae and cupressophytes and influence of heterotachy on the evaluation of gymnosperm phylogeny. *Genome Biology and Evolution*, *3*, 1284–1295.
Wu, C. S., Wang, Y. N., Liu, S. M., & Chaw, S. M. (2007). Chloroplast genome (cpDNA) of *Cycas taitungensis* and 56 cp protein-coding genes of *Gnetum parvifolium*: Insights into cpDNA evolution and phylogeny of extant seed plants. *Molecular Biology and Evolution*, *24*, 1366–1379.
Yamane, K., Yano, K., & Kawahara, T. (2006). Pattern and rate of indel evolution inferred from whole chloroplast intergenic regions in sugarcane, maize and rice. *DNA Research*, *13*, 197–204.
Yi, X., Gao, L., Wang, B., Su, Y.-J., & Wang, T. (2013). The complete chloroplast genome sequence of *Cephalotaxus oliveri* (Cephalotaxaceae): Evolutionary comparison of *Cephalotaxus* chloroplast DNAs and insights into the loss of inverted repeat copies in gymnosperms. *Genome Biology and Evolution*, *5*, 688–698.
Yi, D. K., & Kim, K. J. (2012). Complete chloroplast genome sequences of important oilseed crop *Sesamum indicum* L. *PLoS One*, *7*, e35872.
Yi, D. K., Lee, H. L., Sun, B. Y., Chung, M. Y., & Kim, K. J. (2012). The complete chloroplast DNA sequence of *Eleutherococcus senticosus* (Araliaceae); comparative evolutionary analyses with other three asterids. *Molecules and Cells*, *33*, 497–508.
Zhelyazkova, P., Sharma, C. M., Forstner, K. U., Liere, K., Vogel, J., & Borner, T. (2012). The primary transcriptome of barley chloroplasts: Numerous noncoding RNAs and the dominating role of the plastid-encoded RNA polymerase. *The Plant Cell*, *24*, 123–136.
Zhong, B., Fong, R., Collins, L. J., McLenachan, P. A., & Penny, D. (2014). Two new fern chloroplasts and decelerated evolution linked to the long generation time in tree ferns. *Genome Biology and Evolution*, *6*, 1166–1173.
Zhu, A., Guo, W., Gupta, S., Fan, W., & Mower, J. P. (2016). Evolutionary dynamics of the plastid inverted repeat: The effects of expansion, contraction, and loss on substitution rates. *New Phytologist*, *209*, 1747–1756.
Zhu, A., Guo, W., Jain, K., & Mower, J. P. (2014). Unprecedented heterogeneity in the synonymous substitution rate within a plant genome. *Molecular Biology and Evolution*, *31*, 1228–1236.

CHAPTER TEN

Plastome Phylogenetics: 30 Years of Inferences Into Plant Evolution

Matthew A. Gitzendanner*, Pamela S. Soltis†,‡,§, Ting-Shuang Yi¶, De-Zhu Li¶, Douglas E. Soltis*,†,‡,§,1

*Department of Biology, University of Florida, Gainesville, FL, United States
†Florida Museum of Natural History, University of Florida, Gainesville, FL, United States
‡Genetics Institute, University of Florida, Gainesville, FL, United States
§Biodiversity Institute, University of Florida, Gainesville, FL, United States
¶Germplasm Bank of Wild Species, Kunming Institute of Botany, Chinese Academy of Sciences, Kunming, PR China
[1]Corresponding author: e-mail address: dsoltis@ufl.edu

Contents

Abstract

From restriction site analyses to whole plastid genome sequences, our understanding of green plant (*Viridiplantae*; ~500,000 extant species) evolutionary relationships over the past three decades has largely been informed by analyses of the plastid genome. The plastid genome has informed studies ranging from population genetics to phylogenetics, the latter ranging from the intraspecific level to studies of all green plants. Diverse portions of the genome ranging from plastid spacers to entire genomes provide valuable data for plant evolutionary biologists. Recent phylogenetic analyses using whole plastid genomes sampled from over 2000 species representing all major groups of green plants have both solidified our understanding of relationships and highlighted the few key nodes in plant evolutionary history that remain unresolved. Likewise, detailed large-scale analyses of plastomes across angiosperms reinforce firmly supported nodes but fail to resolve a handful of remaining questionable relationships. The long history of plastid phylogenetics will serve as a reference point as scientists

Advances in Botanical Research, Volume 85
ISSN 0065-2296
https://doi.org/10.1016/bs.abr.2017.11.016

continue to expand beyond the plastid genome and include more nuclear and mitochondrial data in their analyses. These comparisons are crucial in that recent studies indicate some discordance between nuclear and plastid gene trees both across green plants as a whole and within angiosperms. Rather than being a source of concern, these discordances point to the complex and intriguing one-billion-year evolutionary history of the green plant clade, a clade that is foundational to life on Earth.

1. INTRODUCTION

It has been over 30 years since the first phylogenetic analyses based on plastid DNA markers (e.g. Palmer, Jorgensen, & Thompson, 1985; Palmer & Zamir, 1982). Since that time, plastid DNA has provided a variety of data sources, from restriction site variants to large-scale inversions to nucleotide substitutions in single genes, spacers, and ultimately entire genomes. This variation has been harnessed to address questions of phylogeny, population structure, and a host of other topics in evolutionary biology. As data from the plastid genome (i.e. the plastome) have accumulated, they have provided the framework for studies of plastome structure, sequence, and evolution, further promoting research in plastid engineering (e.g. Daniell, 2006; Maliga, 2001), among other topics. Despite the breadth of research facilitated by analyses of plastome variation, the greatest impact has been in revolutionizing our understanding of plant phylogeny. In this chapter, we trace the development of the use of the plastome in phylogenetics, summarize current understanding of green plant phylogeny in general and angiosperm phylogeny in particular based on plastid-based analyses, and consider both the challenges and future prospects of plastid phylogenetics.

2. THE PLASTID GENOME AND PLANT SYSTEMATICS

2.1 Attributes of the Plastome

The plastid genome has long been the workhorse of plant molecular systematics. Typically 120–150 kb in size, the plastome has numerous advantages for phylogeny reconstruction, including uniparental inheritance and conservation of structure and rate of sequence evolution (Fig. 1). Moreover, it encodes key photosynthetic genes, such as the large subunit of RuBisCo (*rbcL*, ribulose-1,5-bisphosphate carboxylase/oxygenase), and therefore

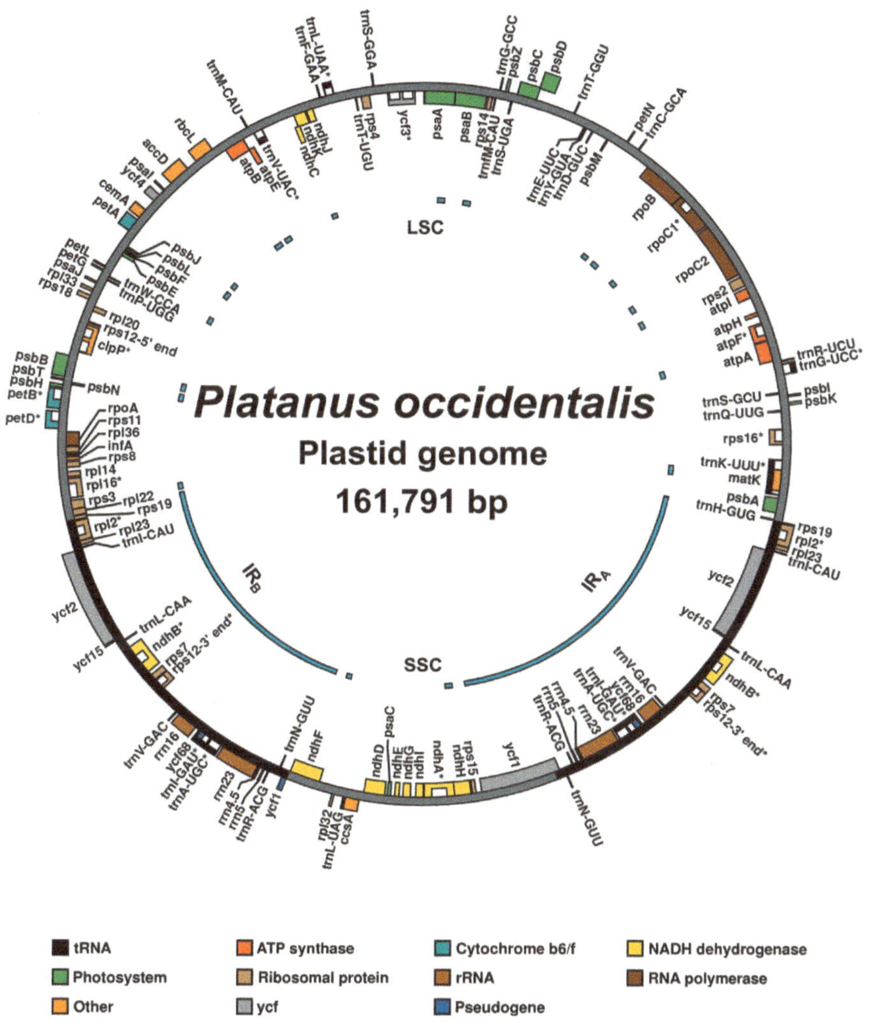

Fig. 1 Annotated plastid genome of *Platanus occidentalis* showing locations of genes and introns. *Asterisks* (*) following the gene names denote the presence of introns in the gene, with *white boxes* showing approximate locations of the introns. *Blue bars* inside the genome map indicate portions of the genome determined through Sanger sequencing. *Reprinted from fig. 2 of Moore, M. J., Dhingra, A., Soltis, P. S., Shaw, R., Farmerie, W. G., Folta, K. M., et al. (2006). Rapid and accurate pyrosequencing of angiosperm plastid genomes.* BMC Plant Biology, 6, *17.*

has been well characterized at the genetic and protein levels for decades. These useful features have been well reviewed elsewhere (e.g. Clegg & Zurawski, 1992; Palmer, 1985, 1987; Zurawski & Clegg, 1987). For our purposes, we will simply note that these characteristics made the plastid genome an ideal tool in the early history of plant molecular systematics. These features facilitated PCR primer design and use of the same suite of primers (see below) over broad portions of the plant branch of the tree of life. In contrast, the nuclear genome is large and complex with genes undergoing biparental inheritance. Moreover, most nuclear genes are part of gene families, and all plant genomes have undergone multiple rounds of whole-genome duplication (i.e. polyploidy; Green Plant Consortium, 2018). Thus, identifying homologous copies of genes across broad phylogenetic distances is often difficult. As reviewed elsewhere, the mitochondrial genome also has utility in reconstructing green plant phylogeny (e.g. Duff & Nickrent, 1999; Qiu et al., 2007, 2006) but exhibits instances of horizontal gene transfer, which are fascinating and informative about plant evolution but may be problematic for phylogeny reconstruction (reviewed in Davis & Xi, 2015; Keeling & Palmer, 2008; Won & Renner, 2003; Xi et al., 2013).

Importantly, the conserved rate of evolution of the plastid genome has resulted in numerous applications at diverse levels across the breadth of the green plant tree of life. Via *rbcL* sequencing, gene space could be used at deep phylogenetic levels, including studies of all green plants, land plants, ferns, seed plants, and angiosperms, as well as within families in these major clades (e.g. Chase et al., 1993; Conran et al., 2000; Källersjö et al., 1998; Morgan, Soltis, & Robertson, 1994; Wolf, Soltis, & Soltis, 1994 reviewed in Soltis & Soltis, 1998 see below). Importantly, faster evolving genes and spacer regions make it possible to investigate phylogenetic relationships at shallow levels—among genera and sometimes even among species. However, levels of variation detected with a small number of targeted loci are often insufficient for resolving interspecific relationships, requiring the use of many plastid loci and/or the inclusion of nuclear loci, such as ITS (Internal Transcribed Spacer of the nuclear ribosomal cistron). Ironically, following early studies demonstrating sufficient restriction site variation within species to identify groups of populations (e.g. Soltis, Soltis, Ranker, & Ness, 1989), plastid data (both restriction site variants and intergenic spacer sequences) have been applied extensively in phylogeographic studies. Perhaps the popularity of plastid data for intraspecific studies, even though plastid data cannot always distinguish among closely related species, can be attributed to the fact that most phylogeographic analyses do not require complete resolution among samples; instead, they search for groups of haplotypes, and plastid data are well suited for this task. Many analyses that have relied on plastid microsatellites, restriction site variation, and/or intergenic spacer sequences have helped shape our understanding of plant population structure and of migrations associated with glaciation (e.g. Deguilloux, Dumolin-Lapègue, Gielly, Grivet, & Petit, 2003; Ferris, Oliver, Davy, & Hewitt, 1995; Marsico, Hellmann, & Romero-Severson, 2009; Petit et al., 1997; Soltis, Gitzendanner, Strenge, & Soltis, 1997; Soltis, Morris, McLachlan, Manos, & Soltis, 2006). However, in this chapter, we will focus on the many contributions of plastid data to our understanding of deep phylogenetic relationships in green plants.

2.2 A Brief History of Plastid Phylogenetics

The plastid genome era in systematics really began in earnest with the publication of a series of papers showing the great potential of plastid DNA variation based on the limited knowledge of the genome at that time (Palmer, Jansen, Michaels, Chase, & Manhart, 1988; Ritland & Clegg, 1987; Zurawski & Clegg, 1987). Ritland and Clegg (1987) showed that several plastid genes were ideal in terms of rate of evolution for resolving phylogeny in green plants. Early papers based on restriction site variation demonstrated the value of molecular data for phylogenetics (e.g. Palmer & Zamir, 1982) and helped convince a community sceptical of cladistic methods that explicit phylogenetic analyses following the logic of Hennig (1950, 1966) were superior to the practices of 'evolutionary systematics' sensu Mayr (1969). Without the clean restriction site data provided by a burgeoning cohort of plant molecular systematists, phylogenetic methods might have been the topic of continued controversy for years into the future, but molecular systematists quickly embraced the new technologies and the new analytical methods that were emerging simultaneously. The acquisition and analysis of sequence data, even before the dawn of PCR, further supported the view that rates of plastid genome evolution were appropriate for addressing questions of green plant phylogeny (e.g. Doebley, Durbin, Golenberg, Clegg, & Ma, 1990; Palmer et al., 1988; Soltis, Soltis, Clegg, & Durbin, 1990), and other early papers reinforced the power of *rbcL* as a phylogenetic marker (e.g. Les, Garvin, & Wimpee, 1991; Giannasi, Zurawski, Learn, & Clegg, 1992; reviewed in Chase et al., 1993; Soltis & Soltis, 1995).

With the advent of PCR, the floodgates were open in terms of the sequencing of *rbcL* and other genes. But the wide early usage of *rbcL* was greatly enhanced by Zurawski and Clegg, who made aliquots of PCR and sequencing primers readily available for free to all researchers. This gesture greatly facilitated the rapid and widespread sequencing of *rbcL* by many members of the botanical community, each working on his/her particular group of interest. The use of other plastid genes soon followed, including the widespread sequencing of *atpB* (e.g. Savolainen, Chase, et al., 2000; Savolainen, Fay, et al., 2000) and *matK* (e.g. Johnson & Soltis, 1995; Les et al., 1999).

A key element of the successful and broad use of plastid gene sequences was the willingness of botanists to readily exchange unpublished DNA sequences and to work collaboratively to achieve major goals. The initial result was the now classic Chase et al. (1993) *rbcL* paper for 500 terminals (499 species) by 47 authors. Also of note were the accompanying papers in the same volume of *Annals of the Missouri Botanical Garden* (volume 80, number 3) focused on major subclades of angiosperms (e.g. Conti, Fischbach, & Sytsma, 1993; Morgan & Soltis, 1993; Qiu, Chase, Les, & Parks, 1993; Smith, Kress, & Zimmer, 1993; see overview by Chase et al., 1993). Not only did these papers collectively provide the first broad phylogenetic molecular hypotheses for seed plant relationships, they also provided a standard of community input and collaboration that transformed not only plant systematics but the field of systematics in general. Later studies involving additional plastid genes followed this same collaborative approach. Chase et al. (1993) was followed by a series of papers, including the use of *atpB* (Savolainen, Chase, et al., 2000; Savolainen, Fay, et al., 2000), as well as studies that combined 3 genes, 2 of which were plastid-encoded (*rbcL, atpB*, and 18S rDNA; Soltis, Soltis, & Chase, 1999; Soltis et al., 2000), 5 genes, 3 of which were plastid (Burleigh, Hilu, & Soltis, 2009), and 17 genes, 11 of which were plastid (Soltis et al., 2011).

In addition to broad analyses of major clades of green life, other studies focused on major subclades, including the monocots (e.g. Chase et al., 2006, 2000; Chase, Stevenson, Wilkin, & Rudall, 1995; Duvall et al., 1993; Givnish et al., 2006). There were also numerous foundational studies at finer scales—within orders and families of flowering plants (e.g. Conti et al., 1993; Kron & Chase, 1993; Michaels et al., 1993; Morgan & Soltis, 1993; Smith et al., 1993). Ultimately, these early studies prompted collaborative classification of the angiosperms by the Angiosperm Phylogeny Group (APG, 1998), a broad collaboration to revise angiosperm classification. The APG classifications continue to be updated to the present (APG II, 2003; APG III, 2009; APG IV, 2016), relying primarily on plastid gene sequence data although nuclear gene sequence data are now playing more of a role and will continue to do so (e.g. Wickett et al., 2014). Although we have focused this discussion on angiosperms, and to a lesser extent seed plants, a new collaborative classification for lycophytes and monilophytes has also recently been published, following this same model (Pteridophyte Phylogeny Group, 2016). Beyond plant systematics, the entire, broader systematics community was impacted by the collaborative approach employed in the early days of plastid gene phylogenetics, and systematists investigating most major lineages of life adopted a similar model—a team approach to tackling big phylogenetic questions in lineages including fungi and many animal clades. It is fair to say that the early plastid phylogenetic era transformed systematics in general.

2.3 Methodological Advances in Acquiring Plastid Genome Data

The use of plastid gene sequences via a PCR approach continued for 15 years with, as noted, papers using more and more plastid genes, although 18S rDNA and mitochondrial genes were also sometimes included. The next transformational event in plastid gene sequencing history was the use of the entire plastid genome. Although early efforts made use of conserved regions throughout the plastid genome to amplify the entire genome via standard PCR and Sanger sequencing (e.g. Dhingra & Folta, 2005), the sequencing of the entire plastome in green plant phylogenetic studies became feasible on a broad scale with the advent of the first wave of next-generation sequencing technology, specifically 454 sequencing technology (Moore et al., 2006). Several key, early papers revealed the power of the complete plastid genome (with an emphasis on gene space) via 454 technology to resolve problematic areas in the tree of life (e.g. Jansen et al., 2007; Moore, Bell, Soltis, & Soltis, 2007; Moore, Soltis, Bell, Burleigh, & Soltis, 2010).

Other methods and technologies have also been employed to acquire the plastid genome. The plastid genome can be recovered from transcriptomes (Leister, 2003), the most massive example of this approach being the plastid tree for over 1100 plants (Gitzendanner, Soltis, Wong, Ruhfel, & Soltis, 2018) resulting from the One Thousand Plant Transcriptome Project (1KP; onekp.com). Because plant cells contain many copies of the plastid genome, its sheer abundance has made it possible to sequence many plastid genomes at once via a single lane of Illumina and sample barcoding (Straub et al., 2012). Meanwhile, long PCR (Cronn et al., 2008) and hybrid (target) capture methods (Cronn et al., 2012; Stull et al., 2013) focused on plastid genomes while making use of novel methods. Most recently, the current widespread use of target capture for nuclear genes also typically yields the complete plastid genome without the need for the design of plastid-specific baits—there is sufficient coverage of the plastid genome in the off-target reads for complete, or near-complete, plastid genome assembly (e.g. Weitemier et al., 2014).

3. PLASTOME PHYLOGENY: STATE OF THE TREE

Recent studies of complete plastome sequences by Ruhfel, Gitzendanner, Soltis, Soltis, and Burleigh (2014) and Gitzendanner et al. (2018) largely agree in the deep-level patterns of relationships recovered for green plants. A recent investigation of most angiosperm families (D.-Z. Li et al., unpublished data) provides the most comprehensive look at deep-level angiosperm relationships based on the plastome. We will focus our overviews of green plant and angiosperm relationships following Gitzendanner et al. (2018) and D.-Z. Li et al. (unpublished data), respectively.

3.1 Summary of Green Plant Phylogeny

Complete plastome gene data sets recover two well-supported clades of *Viridiplantae* (green plants), *Chlorophyta* and *Streptophyta*, consistent with all major studies of green plant phylogeny in the past 20 years (reviewed in Gitzendanner et al., 2018; Wickett et al., 2014) (Fig. 2). (Throughout this chapter, italicized names of larger groups correspond to clade names for which phylogenetic definitions have been provided [e.g. Cantino et al., 2007; Podani, 2015]; nonitalicized names reflect traditional Linnaean taxonomy.) Within *Streptophyta*, Coleochaetales, Charales, and Zygnematales are successive sisters to land plants (*Embryophyta*). The position of Zygnematales as the immediate sister to land plants was unexpected based on morphology and recent DNA studies based on a few genes, but agrees with Timme, Bachvaroff, and Delwiche (2012), Ruhfel et al. (2014), and recent nuclear gene analyses (Wickett et al., 2014). In contrast, other recent studies using fewer genes had favoured Charales or Coleochaetales as sister to land plants. This phylogenetic placement of Zygnematales as sister to land plants is a major finding with important evolutionary implications. That is, Charales, Coleochaetales, and embryophytes are characterized by complex morphological characters, including apical growth with branching, parental retention of the egg, and plasmodesmata in the gametophyte stage of the life cycle. In contrast, Zygnematales lack these features and are unicellular or filamentous and reproduce by conjugation and not by motile cells with flagella. Phylogenies based on plastome and large nuclear data sets therefore

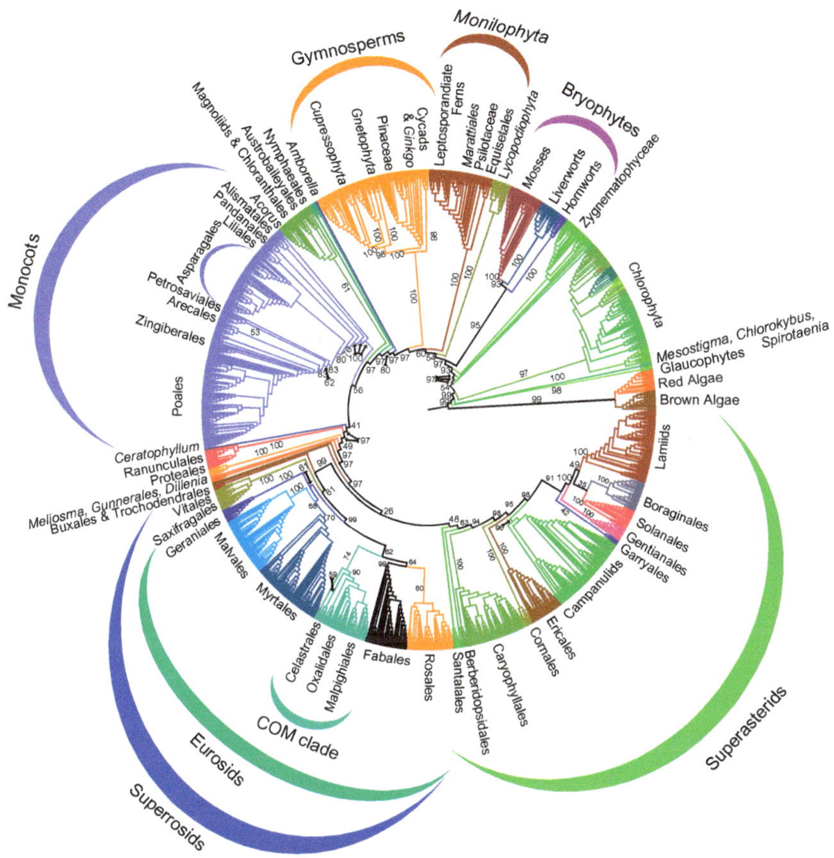

Fig. 2 Summary of the plastid phylogeny of 1827 *Viridiplantae* taxa and 52 outgroups using 78 protein-coding genes. Bootstrap values (from 100 bootstrap replicates) of many larger clades are noted; see Gitzendanner et al. (2018) for all bootstrap values. *Reprinted from Gitzendanner, M. A., Soltis, P. S., Wong, G. S.-K., Ruhfel, B. R., & Soltis, S. E. (2018). Plastid phylogenomic analysis of green plants: A billion years of evolutionary history. American Journal of Botany. In press.*

indicate that the complex features noted for Coleochaetales, Charales, and land plants actually originated in the common ancestor of these groups and Zygnematales and that there was a reduction in complexity on the branch to Zygnematales (reviewed in Gitzendanner et al., 2018; Wickett et al., 2014).

Within land plants, another noteworthy result is the recent recovery of a bryophyte clade (hornworts sister to a clade of mosses + liverworts) by Gitzendanner et al. (2018). Previous studies have recovered nearly all possible relationships among the three bryophyte lineages (reviewed in Ruhfel et al., 2014; Wickett et al., 2014). Significantly, plastid-based reconstructions, such as that of Gitzendanner et al. (2018), recover a bryophyte clade and not a grade. However, nuclear-based data sets show a more complicated story, with a coalescent-based species tree derived from 410 nuclear gene trees supporting a bryophyte clade, while, in contrast, a concatenated analysis of these genes yields support for hornworts sister to the remaining land plants (Green Plant Consortium, 2018). The possible monophyly of bryophytes has important evolutionary implications. Liverworts were sometimes recovered in earlier phylogenetic analyses as sister to all other land plants. Liverworts lack stomata and also lack a columella, a columnar mass of sterile tissue in the sporangium present in other land plants. With a placement of liverworts sister to other land plants, stomata and a columella were inferred to have evolved on the branch to all remaining land plants. However, the well-supported recovery of a bryophyte clade indicates that stomata and a columella may have been ancestral in extant land plants and then lost in liverworts.

The bryophyte clade is in turn sister to vascular plants (*Tracheophyta*). Within vascular plants, lycophytes are sister to the remaining land plants or *Euphyllophyta* (*Monilophyta* + *Spermatophyta*). Within *Monilophyta* (ferns in the broad sense), complete plastid genome data recover *Equisetales* + *Psilotales* as sister to *Marattiales* + leptosporangiate ferns. Within *Spermatophyta* (seed plants), extant gymnosperms form a clade sister to flowering plants (angiosperms). Within extant gymnosperms, complete plastomes place a clade of cycads plus *Ginkgo* as sister to the remaining gymnosperms (conifers and Gnetales). These plastid data also place Gnetales within conifers as sister to non-Pinaceae (i.e. Cupressaceae) with strong support (Gitzendanner et al., 2018); this position of Gnetales has been referred to as the Gne-Cup placement. In contrast to large plastid data sets, large data sets of nuclear genes place Gnetales within Pinaceae (Gne-Pine; Wickett et al., 2014).

3.2 Summary of Angiosperm Phylogeny

Within angiosperms, plastid phylogenomic analyses reveal *Amborella* as sister to all remaining flowering plants, in agreement with most recent analyses (see review by Drew et al., 2014). *Amborella* is then followed successively by Nymphaeales (water lilies) and then Austrobaileyales as sisters to all other extant flowering plants (D.-Z. Li et al., unpublished data), a huge clade referred to as *Mesangiospermae* (Fig. 3). Although some studies identified a clade of *Amborella* + Nymphaeales as the sister of all other living angiosperms (e.g. Goremykin et al., 2012; Qiu et al., 2010; but see Simmons, 2017;

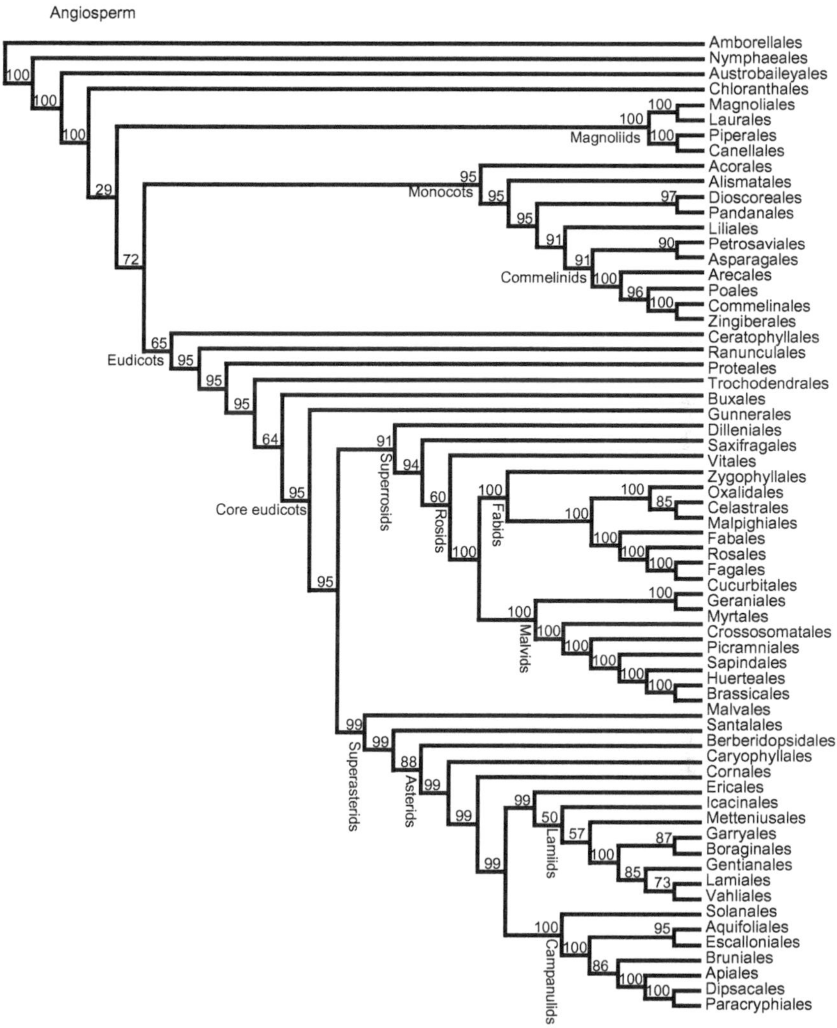

Fig. 3 Summary of the plastid phylogeny of 2881 seed plants (187 gymnosperms and 2694 angiosperms). Bootstrap values are from 1000 replicates. *Reprinted from D.-Z. Li et al. (unpublished data).*

Simmons & Gatesy, 2015), most recent phylogenetic studies applying plastid and/or nuclear data have identified *Amborella* alone as sister to the remaining extant angiosperms (e.g. Drew et al., 2014; Gitzendanner et al., 2018; Ruhfel et al., 2014; Soltis et al., 2011; Wickett et al., 2014).

Mesangiospermae comprise well-supported subclades of magnoliids, Chloranthales, monocots, *Ceratophyllum*, and eudicots. However, relationships among these subclades have been difficult to resolve, presumably due to rapid radiation (see Moore et al., 2007). Even the use of complete plastome data sets does not resolve with confidence relationships among Chloranthales, magnoliids, and the moderately supported monocots—*Ceratophyllum*—eudicots clade (D.-Z. Li et al., unpublished data; see also Gitzendanner et al., 2018) (Fig. 3). Most previous plastid gene analyses suggested a clade of Chloranthales + magnoliids sister to the remaining *Mesangiospermae* (e.g. Moore et al., 2007; Soltis et al., 2011), and, in fact, a tree of 1879 green plants based on nearly complete plastomes (Gitzendanner et al., 2018) finds this same relationship. However, other studies employing nuclear, mitochondrial, and plastid inverted repeat genes have suggested a Chloranthales + *Ceratophyllum* clade (Zeng et al., 2014 and references cited). Resolution of phylogenetic relationships among these five lineages of *Mesangiospermae* remains one of the most difficult problems in angiosperm phylogeny reconstruction (Davis, Xi, & Mathews, 2014), with as many as 15 poorly to moderately supported topologies having been proposed for mesangiosperms (Zeng et al., 2014). Rapid diversification within just a few million years likely hampers the reconstruction of relationships among these five clades (Moore et al., 2007). A relationship of monocots sister to (magnoliids (eudicots (Chloranthales + *Ceratophyllum*))) is strongly supported by some nuclear data sets (Zeng et al., 2014). Mesangiosperm relationships could be an area of possible conflict between nuclear and plastid topologies (see below for further discussion).

Relationships within monocots inferred from plastomes (or at least the protein-coding genes) (Gitzendanner et al., 2018; D.-Z. Li et al., unpublished data) are mostly consistent with those reported previously (Chase et al., 2006; Givnish et al., 2010; Graham et al., 2006; Hertweck et al., 2015; Soltis et al., 2011) and represented by APG IV (2016). The basic phylogenetic backbone of Acorales and then Alismatales as subsequent sisters to all other monocots is well supported in trees presented by both D.-Z. Li et al. (unpublished data) and Gitzendanner et al. (2018). All other major clades—Petrosaviales, Pandanales, Dioscoreales, Liliales, Asparagales, Arecales, Poales, Commelinales, Zingiberales, all recognized as orders by APG IV (2016), as well as the commelinid clade comprising the latter four orders—were recovered in trees from full plastome analyses. However, the placement of Petrosaviales differs between the studies of D.-Z. Li et al. (unpublished data) and Gitzendanner et al. (2018), and both studies disagree with the consensus placement of Petrosaviales as sister to all monocots except Acorales and Alismatales (APG IV, 2016). In D.-Z. Li et al. (unpublished data), Petrosaviales are nested within Asparagales, rendering the latter paraphyletic, whereas in Gitzendanner et al. (2018), they are weakly supported as sister to the commelinids. These conflicting placements deserve additional attention.

Within a well-supported eudicot clade, Ranunculales, Proteales, Trochodendrales, and Buxales are successive sisters to a well-supported clade of core eudicots (*Gunneridae*). These results are similar to those obtained with increasing support in recent plastid-based studies of eudicot phylogeny (e.g. Moore et al., 2010; Soltis et al., 2011) and are represented in APG IV (2016). As in most recent phylogenetic studies (see Gitzendanner et al., 2018; Soltis et al., 2011), Gunnerales are sister to the remaining core eudicots (*Pentapetalae*), which in turn comprise two major clades, the superasterids and superrosids. Although the placement of Dilleniales has been highly problematic (see Soltis et al., 2011), with complete plastid data, Dilleniales are strongly supported as sister to superrosids, consistent with previous analyses of plastome data (e.g. Moore et al., 2010).

Within superrosids, Saxifragales are sister to rosids, within which Vitales are sisters to the core rosids (eurosids of Chase et al., 1993). The eurosids in turn form two subclades, fabids and malvids. These results are in agreement with most previous analyses, although the relationships of Saxifragales, Vitales, and eurosids have varied (Gitzendanner et al., 2018; Soltis et al., 2011).

Within superasterids, Santalales are sister to other members of the clade; Berberidopsidales and Caryophyllales are then successive sisters to a well-supported clade of asterids. Within asterids, Cornales, followed by Ericales, are sisters to the remaining asterids (euasterids sensu Chase et al., 1993). Within the euasterids are well-supported clades of campanulids and lamiids. The backbone relationships recovered in recent analyses of plastid genome-scale data are generally the same as those revealed by previous plastome data (e.g. Moore et al., 2010); however, the most recent plastome-based study for angiosperms (D.-Z. Li et al., unpublished data) provides greater resolution and support among clades of asterids in particular than evident in previous studies.

Although many of the relationships described above seem to appear consistently in trees based on analyses of hundreds, and in one case, nearly 2000 species, some well-established relationships based on plastid data conflict with those inferred in recent analyses of hundreds of nuclear genes, also for hundreds of species. While vexing for those focused on resolution of a bifurcating phylogenetic tree for angiosperms and for green plants as a whole, such conflict is an entree into potentially significant evolutionary events, such as ancient hybridization, polyploidy, and incomplete lineage sorting (ILS) (see below).

4. PLASTOME PHYLOGENETICS: ONGOING CHALLENGES

Conflict between plastid trees and inferences from the nuclear genome—sometimes represented by taxonomy or morphology as well as by nuclear-based trees—goes back to the earliest studies of plastid phylogenetics (e.g. Palmer et al., 1985). In fact, such conflict appeared in a surprising number of early plastid restriction site analyses, in which plastids of one or a few populations of one species were nested within plastid-based clades of another species. These results demonstrated widespread hybridization and interspecific transfer of plastomes, a process described as chloroplast capture (Rieseberg & Soltis, 1991). Comparisons of plastid and nuclear trees have continued to suggest instances of chloroplast capture and provide some of the strongest evidence to date for extensive hybridization (e.g. Folk, Mandel, & Freudenstein, 2017).

Although conflict between plastid and nuclear trees is typically attributed to hybridization, other processes may yield discordance between nuclear and organellar trees (e.g. Doyle, 1992; Wendel & Doyle, 1998). In fact, patterns of discordance due to hybridization may be indistinguishable from those due to ILS. In many cases, aspects of the biology of the species—geographic distributions, sympatry, mating systems, etc.—may tip the scales in favour of hybridization vs ILS, but other cases may remain equivocal. The recent development of coalescent methods of analysis, particularly those aimed at multispecies coalescence (e.g. ASTRAL, Mirarab et al., 2014; ASTRAL-II, Mirarab & Warnow, 2015), enables tests of ancient hybridization vs ILS. However, deep coalescence may resemble ancient reticulation.

In plants, deep discordance between plastid and nuclear trees might be interpreted as ancient hybridization, given the propensity for interspecific hybridization among extant species. However, nuclear trees based on genes other than ribosomal genes (ITS, 18S rDNA) are still relatively uncommon (but increasing rapidly; e.g. Green Plant Consortium, 2018; Rothfels et al., 2015; Zeng et al., 2014; Zhao et al., 2016), and deep incongruence between nuclear and plastid trees is typically not strongly supported. Thus, possible cases of ancient reticulation are few. A likely example of ancient hybridization is found in the rosid clade of angiosperms (Sun et al., 2015). Plastid trees have long supported the placement of the large 'COM' clade (composed of Celastrales, Oxalidales, and Malpighiales) in the fabid clade; however, analyses based on both mitochondrial genes (Qiu et al., 2010) and morphology (Endress & Matthews, 2006) place the COM clade instead in the malvid clade. Our analyses of multiple data sets (Sun et al., 2015) likewise found these placements and support the hypothesis of ancient hybridization that resulted in essentially a chloroplast capture event early in the evolutionary history of the rosids. Given the high incidence of recent hybridization revealed by comparison of plastid vs nuclear trees, it is likely that such processes have been ongoing for many millions of years, leading to additional deep events of reticulation. However, such ancient events may not be discernible because the hybridization may have occurred between lineages in the early stages of divergence such that all evidence of hybridization may have been wiped out. As strongly supported nuclear trees are generated for comparison with plastid trees, we are likely to see additional areas of conflict, requiring resolution and interpretation of hybridization vs ILS as possible sources of the conflict. We see this prospect as both a challenge to clarifying phylogeny and as an opportunity to learn more about the evolutionary history of plants and the processes that have generated extant plant diversity.

Finally, how do we interpret what is essentially a one-locus tree? Even if based on 150 kb or more, a plastome-based tree represents the history of a single locus (e.g. Doyle, 1992). We argue that this perspective remains powerful, providing a uniparental, typically maternal view of plant phylogeny. However, the role of this maternal phylogeny, given the size and complexity of the nuclear genome, with each gene exhibiting its own evolutionary history, is unclear: it is just one of many thousands of gene phylogenies that form the basis of species relationships.

5. CONCLUSIONS

Plastids play a range of roles in the cells of green plants, from photosynthesis to terpene synthesis to dismantling the photosynthetic machinery of the chloroplast to storage of pigments, starch, fats, and proteins. Despite this diversity of function and the ability of plastids to differentiate and take on specialized roles, all plastids within an individual plant share the same plastid genome. As described elsewhere in this book, the functions of plastids are controlled by a complex interplay between nuclear and plastid genomes. Evolutionary dynamics of the plastome are therefore governed in part by selective factors that maintain cytonuclear interactions. These dynamics may vary among clades of photosynthetic eukaryotes, such as glaucophytes, diatoms, and green plants, and clades in which plastids have been lost (e.g. alveolates), and the signatures of evolution are carried in the structure and sequence of the plastomes.

Sequence data from the plastid genome have been the workhorse of plant systematics for roughly 30 years, and the plastome has to date played the dominant role in shaping our current understanding of phylogenetic relationships in plants at both deep and shallow levels, the latter including phylogeographic inference. However, although plastid genomes can now be routinely sequenced, at deep levels we may have largely realized the extent to which the plastome can resolve relationships. Additional plastid genomes will likely not improve our understanding of the relationships among bryophyte lineages or the placement of Gnetales nor resolve enigmatic relationships among major clades of angiosperms. However, the plastid genome will continue to be of value within many clades recognized as orders, families, and even genera. Complete plastid genome sequences hold remarkable untapped potential at those levels. Another major avenue of future research will involve rigorous comparisons of phylogenies based on the plastid genome with large topologies based on many nuclear genes. With the growing availability of larger and larger nuclear data sets, an increasing number of examples of discordance between plastid and nuclear-based topologies has emerged. Thus, the next generation of studies in plant molecular systematics will also involve analyses not only to clarify the well-supported examples of discordance but also to explore the causes of these discordances and the relative roles of ILS and ancient reticulation in green plant evolutionary history.

REFERENCES

Angiosperm Phylogeny Group (APG). (1998). An ordinal classification for the families of flowering plants. *Annals of the Missouri Botanical Garden, 85*, 531–553.

Angiosperm Phylogeny Group (APG) II. (2003). An update of the Angiosperm Phylogeny Group classification for the orders and families of flowering plants: APG II. *Botanical Journal of the Linnean Society, 141*, 399–436.

Angiosperm Phylogeny Group (APG) III. (2009). An update of the Angiosperm Phylogeny Group classification for the orders and families of flowering plants: APG III. *Botanical Journal of the Linnean Society, 161*, 105–121.

Angiosperm Phylogeny Group (APG) IV. (2016). An update of the Angiosperm Phylogeny Group classification for the orders and families of flowering plants: APG IV. *Botanical Journal of the Linnean Society, 181*, 1–20.

Burleigh, J. G., Hilu, K. W., & Soltis, D. E. (2009). Inferring phylogenies with incomplete data sets: A 5-gene, 567-taxon analysis of angiosperms. *BMC Evolutionary Biology, 9*, 61. https://doi.org/10.1186/1471-2148-9-61.

Cantino, P. D., Doyle, J. A., Graham, S. W., Judd, W. S., Olmstead, R. G., Soltis, D. E., et al. (2007). Towards a phylogenetic nomenclature of Tracheophyta. *Taxon, 56*, 1E–44E.

Chase, M., Fay, M., Devey, D., Maurin, O., Rønsted, N., Davies, T., et al. (2006). Multigene analyses of monocot relationships. *Aliso: A Journal of Systematic and Evolutionary Botany*, *22*, 63–75.

Chase, M. W., Soltis, D. E., Olmstead, R. G., Morgan, D., Les, D. H., Mishler, B. D., et al. (1993). Phylogenetic relationships among seed plants based on *rbcL* sequence data. *Annals of the Missouri Botanical Garden*, *80*, 528–580.

Chase, M., Soltis, D. E., Soltis, P., Rudall, P., Fay, M., Hahn, W., et al. (2000). Higher-level systematics of the monocotyledons: An assessment of current knowledge and a new classification. In K. L. Wilson & D. A. Morrison (Eds.), *Monocots: Systematics and evolution* (pp. 1–17). Collingwood, Australia: CSIRO.

Chase, M. W., Stevenson, D. W., Wilkin, P., Rudall, P., Rudall, P. J., Cribb, P. J., et al. (1995). Monocot systematics: A combined analysis. *Monocots: Systematics and evolution Vol. 7* (pp. 685–730). Kew, London, UK: Royal Botanic Gardens, Kew.

Clegg, M. T., & Zurawski, G. (1992). Chloroplast DNA and the study of plant phylogeny: Present status and future prospects. In P. S. Soltis, D. E. Soltis, & J. J. Doyle (Eds.), *Molecular systematics of plants* (pp. 1–14). New York: Chapman and Hall.

Conran, J. G., Wood, G. M., Martin, P. G., Dowd, J. M., Quinn, C. J., Gadek, P. A., et al. (2000). Generic relationships within and between the gymnosperm families Podocarpaceae and Phyllocladaceae based on an analysis of the chloroplast gene *rbcL*. *Australian Journal of Botany*, *48*, 715–724.

Conti, E., Fischbach, A., & Sytsma, K. J. (1993). Tribal relationships in Onagraceae—Implications from *rbcL* sequence data. *Annals of the Missouri Botanical Garden*, *80*, 672–685.

Cronn, R., Knaus, B. J., Liston, A., Maughan, P. J., Parks, M., Syring, J. V., et al. (2012). Targeted enrichment strategies for next-generation plant biology. *American Journal of Botany*, *99*, 291–311.

Cronn, R., Liston, A., Parks, M., Gernandt, D. S., Shen, R., & Mockler, T. (2008). Multiplex sequencing of plant chloroplast genomes using Solexa sequencing-by-synthesis technology. *Nucleic Acids Research*, *36*, e122.

Daniell, H. (2006). Production of biopharmaceuticals and vaccines in plants via the chloroplast genome. *Biotechnology Journal*, *1*, 1071–1079.

Davis, C. C., & Xi, X. (2015). Horizontal gene transfer in parasitic plants. *Current Opinion in Plant Biology*, *26*, 14–19.

Davis, C. C., Xi, Z., & Mathews, S. (2014). Plastid phylogenomics and green plant phylogeny: Almost full circle but not quite there. *BMC Biology*, *12*, 11.

Deguilloux, M.-F., Dumolin-Lapègue, S., Gielly, L., Grivet, D., & Petit, R. J. (2003). A set of primers for the amplification of chloroplast microsatellites in Quercus. *Molecular Ecology Notes*, *3*, 24–27.

Dhingra, A., & Folta, K. M. (2005). ASAP: Amplification, sequencing & annotation of plastomes. *BMC Genomics*, *6*, 176.

Doebley, J., Durbin, M., Golenberg, E. M., Clegg, M. T., & Ma, D. P. (1990). Evolutionary analysis of the large subunit of carboxylase (*rbcL*) nucleotide-sequence among the grasses (Gramineae). *Evolution*, *44*, 1097–1108.

Doyle, J. J. (1992). Gene trees and species trees: Molecular systematics as one-character taxonomy. *Systematic Botany*, *17*, 144–163.

Drew, B. T., Ruhfel, B. R., Smith, S. A., Moore, M. J., Briggs, B. G., Gitzendanner, M. A., et al. (2014). Another look at the root of the angiosperms reveals a familiar tale. *Systematic Biology*, *63*, 368–382.

Duff, R. J., & Nickrent, D. L. (1999). Phylogenetic relationships of land plants using mitochondrial small-subunit rDNA sequences. *American Journal of Botany*, *86*, 372–386.

Duvall, M. R., Clegg, M. T., Chase, M. W., Clark, W. D., Kress, W. J., Hills, H. G., et al. (1993). Phylogenetic hypotheses for the monocotyledons constructed from *rbcL* sequence data. *Annals of the Missouri Botanical Garden*, *80*, 607–619.

Endress, P. K., & Matthews, M. L. (2006). First steps towards a floral structural characterization of the major rosid subclades. *Plant Systematics and Evolution*, *260*, 223–251.

Ferris, C., Oliver, R. P., Davy, A. J., & Hewitt, G. M. (1995). Using chloroplast DNA to trace postglacial migration routes of oaks into Britain. *Molecular Ecology*, *4*, 731–738.

Folk, R. A., Mandel, J. R., & Freudenstein, J. V. (2017). Ancestral gene flow and parallel organellar genome capture result in extreme phylogenomic discord in a lineage of angiosperms. *Systematic Biology*, *66*(3), 320–337.

Giannasi, D. E., Zurawski, G., Learn, G., & Clegg, M. T. (1992). Evolutionary relationships of the Caryophyllidae based on comparative *rbcL* sequences. *Systematic Botany*, *17*(1), 15.

Gitzendanner, M. A., Soltis, P. S., Wong, G. S.-K., Ruhfel, B. R., & Soltis, S. E. (2018). Plastid phylogenomic analysis of green plants: A billion years of evolutionary history. *American Journal of Botany*. In press.

Givnish, T. J., Ames, M., McNeal, J. R., McKain, M. R., Steele, P. R., dePamphilis, C. W., et al. (2010). Assembling the tree of the monocotyledons: Plastome sequence phylogeny and evolution of Poales. *Annals of the Missouri Botanical Garden*, *97*, 584–616.

Givnish, T., Pires, J., Graham, S., McPherson, M., Prince, L., Patterson, T., et al. (2006). Phylogenetic relationships of monocots based on the highly informative plastid gene *ndhF*. *Aliso: A Journal of Systematic and Evolutionary Botany*, *22*, 28–51.

Goremykin, V. V., Nikiforova, S. V., Biggs, P. J., Zhong, B., DeLange, P., Martin, W., et al. (2012). The evolutionary root of flowering plants. *Systematic Biology*, *62*, 50–61.

Graham, S., Zgurski, J., McPherson, M., Cherniawsky, D., Saarela, J., Horne, E., et al. (2006). Robust inference of monocot deep phylogeny using an expanded multigene plastid data set. *Aliso: A Journal of Systematic and Evolutionary Botany*, *22*, 3–21.

Green Plant Consortium. (2018). *A phylogenomic view of evolutionary complexity in green plants.* Submitted.

Hennig, W. (1950). *Grundzüge einer Theorie der Phylogenetischen Systematik.* Berlin: Deutscher Zentralverlag.

Hennig, W. (1966). *Phylogenetic systematics.* Urbana: University of Illinois Press.

Hertweck, K. L., Kinney, M. S., Stuart, S. A., Maurin, O., Mathews, S., Chase, M. W., et al. (2015). Phylogenetics, divergence times and diversification from three genomic partitions in monocots. *Botanical Journal of the Linnean Society*, *178*, 375–393.

Jansen, R. K., Cai, Z., Raubeson, L. A., Daniell, H., dePamphilis, C. W., Leebens-Mack, J., et al. (2007). Analysis of 81 genes from 64 plastid genomes resolves relationships in angiosperms and identifies genome-scale evolutionary patterns. *Proceedings of the National Academy of Sciences of the United States of America*, *104*, 19369–19374.

Johnson, L. A., & Soltis, D. E. (1995). Phylogenetic inference using *matK* sequences. *Annals of the Missouri Botanical Garden*, *82*, 149–175.

Källersjö, M., Farris, J. S., Chase, M. W., Bremer, B., Fay, M. F., Humphries, C. J., et al. (1998). Simultaneous parsimony jackknife analysis of 2538rbcL DNA sequences reveals support for major clades of green plants, land plants, seed plants and flowering plants. *Plant Systematics and Evolution*, *213*, 259–287.

Keeling, P. J., & Palmer, J. D. (2008). Horizontal gene transfer in eukaryotic evolution. *Nature Reviews. Genetics*, *9*(8), 605–618.

Kron, K. A., & Chase, M. W. (1993). Systematics of the Ericaceae, Empetraceae, Epacridaceae and related taxa based upon *rbcL* sequence data. *Annals of the Missouri Botanical Garden*, *80*, 735–741.

Leister, D. (2003). Chloroplast research in the genomic age. *Trends in Genetics*, *19*, 47–56.

Les, D. H., Garvin, D. K., & Wimpee, C. F. (1991). Molecular evolutionary history of ancient aquatic angiosperms. *Proceedings of the National Academy of Sciences of the United States of America*, *88*, 10119–10123.

Les, D. H., Schneider, E. L., Padgett, D. J., Soltis, P. S., Soltis, D. E., & Zanis, M. (1999). Phylogeny, classification and floral evolution of water lilies (Nymphaeaceae; Nymphaeales):

A synthesis of non-molecular, *rbcL*, *matK*, and 18S rDNA data. *Systematic Botany*, *24*, 28–46.
Maliga, P. (2001). Plastid engineering bears fruit. *Nature Biotechnology*, *19*, 826–827.
Marsico, T. D., Hellmann, J., & Romero-Severson, J. (2009). Patterns of seed dispersal and pollen flow in *Quercus garryana* (Fagaceae) following post-glacial climatic changes. *Journal of Biogeography*, *36*, 929–941.
Mayr, E. (1969). *Principles of systematic zoology*. New York: McGraw-Hill.
Michaels, H. J., Scott, K. M., Olmstead, R. G., Szaro, T., Jansen, R. K., & Palmer, J. D. (1993). Interfamilial relationships of the Asteraceae: Insights from *rbcL* sequence variation. *Annals of the Missouri Botanical Garden*, *80*, 742–751.
Mirarab, S., Reaz, R., Bayzid, M. S., Zimmermann, T., Swenson, M. S., & Warnow, T. (2014). ASTRAL: Genome-scale coalescent-based species tree estimation. *Bioinformatics (Oxford, England)*, *30*, i541–i548.
Mirarab, S., & Warnow, T. (2015). ASTRAL-II: Coalescent-based species tree estimation with many hundreds of taxa and thousands of genes. *Bioinformatics (Oxford, England)*, *31*, i44–52.
Moore, M. J., Bell, C. D., Soltis, P. S., & Soltis, D. E. (2007). Using plastid genome-scale data to resolve enigmatic relationships among basal angiosperms. *Proceedings of the National Academy of Sciences of the United States of America*, *104*, 19363–19368.
Moore, M. J., Dhingra, A., Soltis, P. S., Shaw, R., Farmerie, W. G., Folta, K. M., et al. (2006). Rapid and accurate pyrosequencing of angiosperm plastid genomes. *BMC Plant Biology*, *6*, 17.
Moore, M. J., Soltis, P. S., Bell, C. D., Burleigh, J. G., & Soltis, D. E. (2010). Phylogenetic analysis of 83 plastid genes further resolves the early diversification of eudicots. *Proceedings of the National Academy of Sciences of the United States of America*, *107*, 4623–4628.
Morgan, D. R., & Soltis, D. E. (1993). Phylogenetic-relationships among members of *Saxifragaceae sensu lato* based on rbcL sequence data. *Annals of the Missouri Botanical Garden*, *80*, 631–660.
Morgan, D. R., Soltis, D. E., & Robertson, K. R. (1994). Systematic and evolutionary implications of *rbc*L sequence variation in Rosaceae. *American Journal of Botany*, *81*, 890–903.
Palmer, J. D. (1985). Comparative organization of chloroplast genomes. *Annual Review of Genetics*, *19*, 335–354.
Palmer, J. D. (1987). Chloroplast DNA evolution and biosystematic uses of chloroplast DNA variation. *American Naturalist*, *130*, S6–S29.
Palmer, J. D., Jansen, R. K., Michaels, H. J., Chase, M. W., & Manhart, J. M. (1988). Analysis of chloroplast DNA variation. *Annals of the Missouri Botanical Garden*, *75*, 1180–1206.
Palmer, J. D., Jorgensen, R. A., & Thompson, W. F. (1985). Chloroplast DNA variation and evolution in *Pisum*: Patterns of change and phylogenetic analysis. *Genetics*, *109*, 195–213.
Palmer, J. D., & Zamir, D. (1982). Chloroplast DNA evolution and phylogenetic relationships in Lycopersicon. *Proceedings of the National Academy of Sciences of the United States of America*, *79*, 5006–5010.
Petit, R. J., Pineau, E., Demesure, B., Bailieri, R., Ducousso, A., & Kremer, A. (1997). Chloroplast DNA footprints of postglacial recolonization by oaks. *Proceedings of the National Academy of Sciences of the United States of America*, *94*, 9996–10001.
Podani, J. (2015). *A növények evolúciója és osztályozása*. Evolution and systematics of plantsBudapest: Eötvös Kiadó.
Pteridophyte Phylogeny Group. (2016). A community-derived classification for extant lycophytes and ferns. *Journal of Systematics and Evolution*, *54*, 563–603.
Qiu, Y. L., Chase, M. W., Les, D. H., & Parks, C. R. (1993). Molecular phylogenetics of the Magnoliidae cladistic analyses of nucleotide-sequences of the plastid gene *rbcL*. *Annals of the Missouri Botanical Garden*, *80*, 587–606.

Qiu, Y., Li, L., Wang, B., Chen, Z., Dombrovska, O., Lee, J., et al. (2007). A nonflowering land plant phylogeny inferred from nucleotide sequences of seven chloroplast, mitochondrial, and nuclear genes. *International Journal of Plant Sciences*, *168*, 691–708.

Qiu, Y.-L., Li, L., Wang, B., Chen, Z., Knoop, V., Groth-Malonek, M., et al. (2006). The deepest divergences in land plants inferred from phylogenomic evidence. *Proceedings of the National Academy of Sciences of the United States of America*, *103*, 15511–15516.

Qiu, Y.-L., Li, L., Wang, B., Xue, J.-Y., Hendry, T. A., Li, R.-Q., et al. (2010). Angiosperm phylogeny inferred from sequences of four mitochondrial genes. *Journal of Systematics and Evolution*, *48*, 391–425.

Rieseberg, L. H., & Soltis, D. E. (1991). Phylogenetic consequences of cytoplasmic gene flow in plants. *Evolutionary Trends in Plants*, *5*, 65–84.

Ritland, K., & Clegg, M. T. (1987). Evolutionary analysis of plant DNA sequences. *American Naturalist*, *130*, S74–S100.

Rothfels, C. J., Li, F.-W., Sigel, E. M., Huiet, L., Larsson, A., Burge, D. O., et al. (2015). The evolutionary history of ferns inferred from 25 low-copy nuclear genes. *American Journal of Botany*, *102*, 1089–1107.

Ruhfel, B. R., Gitzendanner, M. A., Soltis, P. S., Soltis, D. E., & Burleigh, J. G. (2014). From algae to angiosperms—Inferring the phylogeny of green plants (Viridiplantae) from 360 plastid genomes. *BMC Evolutionary Biology*, *4*, 23.

Savolainen, V., Chase, M. W., Hoot, S. B., Morton, C. M., Soltis, D. E., Bayer, C., et al. (2000). Phylogenetics of flowering plants based on combined analysis of plastid a*tpB* and r*bcL* gene sequences. *Systematic Biology*, *49*, 306–362.

Savolainen, V., Fay, M. F., Albach, D. C., Backlund, A., van der Bank, M., Cameron, K. M., et al. (2000). Phylogeny of the eudicots: A nearly complete familial analysis based on *rbcL* gene sequences. *Kew Bulletin*, *55*, 257–309.

Simmons, M. P. (2017). Mutually exclusive phylogenomic inferences at the root of the angiosperms: *Amborella* is supported as sister and observed variability is biased. *Cladistics*, *33*, 488–512.

Simmons, M. P., & Gatesy, J. (2015). Coalescence vs. concatenation: Sophisticated analyses vs. first principles applied to rooting the angiosperms. *Molecular Phylogenetics and Evolution*, *91*, 98–122.

Smith, J. F., Kress, W. J., & Zimmer, E. A. (1993). Phylogenetic analysis of the *Zingiberales* based on r*bcL* sequences. *Annals of the Missouri Botanical Garden*, *80*, 620–630.

Soltis, D. E., Gitzendanner, M. A., Strenge, D. D., & Soltis, P. S. (1997). Chloroplast DNA intraspecific phylogeography of plants from the Pacific Northwest of North America. *Plant Systematics and Evolution*, *206*, 353–373.

Soltis, D. E., Morris, A. B., McLachlan, J. S., Manos, P. S., & Soltis, P. S. (2006). Comparative phylogeography of unglaciated eastern North America. *Molecular Ecology*, *15*, 4261–4293.

Soltis, D. E., Smith, S. A., Cellinese, N., Wurdack, K. J., Tank, D. C., Brockington, S. F., et al. (2011). Angiosperm phylogeny: 17 genes, 640 taxa. *American Journal of Botany*, *98*, 704–730.

Soltis, P. S., & Soltis, D. E. (1995). Plant molecular systematics: Inferences of phylogeny and evolutionary processes. *Evolutionary Biology*, *28*, 139–194.

Soltis, D. E., & Soltis, P. S. (1998). Choosing an approach and an appropriate gene for phylogenetic analysis. In D. E. Soltis, P. S. Soltis, & J. Doyle (Eds.), *Molecular systematics of plants II* (pp. 1–42). Boston: Kluwer.

Soltis, P. S., Soltis, D. E., & Chase, M. W. (1999). Angiosperm phylogeny inferred from multiple genes as a tool for comparative biology. *Nature*, *402*, 402–404.

Soltis, D. E., Soltis, P. S., Chase, M. W., Mort, M. E., Albach, D. C., Zanis, M., et al. (2000). Angiosperm phylogeny inferred from 18S rDNA, *rbcL*, and *atpB* sequences. *Botanical Journal of the Linnean Society*, *133*, 381–461.

Soltis, D. E., Soltis, P. S., Clegg, M. T., & Durbin, M. (1990). *RbcL* sequence divergence and phylogenetic-relationships in *Saxifragaceae* sensu-lato. *Proceedings of the National Academy of Sciences of the United States of America, 87*, 4640–4644.

Soltis, D. E., Soltis, P. S., Ranker, T. A., & Ness, B. D. (1989). Chloroplast DNA variation in a wild plant, *Tolmiea menziesii*. *Genetics, 121*, 819–826.

Straub, S. C. K., Parks, M., Weitemier, K., Fishbein, M., Cronn, R. C., & Liston, A. (2012). Navigating the tip of the genomic iceberg: Next-generation sequencing for plant systematics. *American Journal of Botany, 99*, 349–364.

Stull, G. W., Moore, M. J., Douglas, N., Mandala, V., Kates, H.-R., Qi, X., et al. (2013). A targeted enrichment strategy for massively parallel sequencing of angiosperm plastid genomes. *Applications in Plant Sciences, 1*, 1200497.

Sun, M., Soltis, D. E., Soltis, P. S., Zhu, X., Burleigh, J. G., & Chen, Z. (2015). Deep phylogenetic incongruence in the angiosperm clade Rosidae. *Molecular Phylogenetics and Evolution, 83*, 156–166.

Timme, R. E., Bachvaroff, T. R., & Delwiche, C. F. (2012). Broad phylogenomic sampling and the sister lineage of land plants. *PLoS One, 7*, e29696.

Weitemier, K., Straub, S. C. K., Cronn, R. C., Fishbein, M., Schmickl, R., McDonnell, A., et al. (2014). Hyb-Seq: Combining target enrichment and genome skimming for plant phylogenomics. *Applications in Plant Sciences, 2*, 1400042.

Wendel, J. F., & Doyle, J. J. (1998). Phylogenetic incongruence: Window into genome history and molecular evolution. In D. Soltis, P. Soltis, & J. Doyle (Eds.), *Molecular systematics of plants II* (pp. 265–296). Boston, MA: Springer.

Wickett, N. J., Mirarab, S., Nguyen, N., Warnow, T., Carpenter, E., Matasci, N., et al. (2014). A phylotranscriptomics analysis of the origin and early diversification of land plants. *Proceedings of the National Academy of Sciences of the United States of America, 111*, E4859–E4868. www.pnas.org/cgi/doi/10.1073/pnas.1323926111.

Wolf, P. G., Soltis, P. S., & Soltis, D. E. (1994). Phylogenetic relationships of dennstaedtioid ferns: Evidence from *rbc*L sequences. *Molecular Phylogenetics and Evolution, 3*, 383–392.

Won, H., & Renner, S. S. (2003). Horizontal gene transfer from flowering plants to Gnetum. *Proceedings of the National Academy of Sciences of the United States of America, 100*, 10824–10829.

Xi, Z., Wang, Y., Bradley, R. K., Sugumaran, M., Marx, C. J., Rest, J. S., et al. (2013). Massive mitochondrial gene transfer in a parasitic flowering plant clade. *PLoS Genetics, 9*, e1003265.

Zeng, L., Zhang, Q., Sun, R., Kong, H., Zhang, N., & Ma, H. (2014). Resolution of deep angiosperm phylogeny using conserved nuclear genes and estimates of early divergence times. *Nature Communications, 5*, 4956.

Zhao, L., Li, X., Zhang, N., Zhang, S.-D., Yi, T.-S., Ma, H., et al. (2016). Phylogenomic analyses of large-scale nuclear genes provide new insights into the evolutionary relationships within the rosids. *Molecular Phylogenetics and Evolution, 105*, 166–176.

Zurawski, G., & Clegg, M. T. (1987). Evolution of higher-plant chloroplast DNA-encoded genes implications for structure function and phylogenetic studies. *Annual Review of Plant Physiology and Plant Molecular Biology, 38*, 391–418.

CHAPTER ELEVEN

Molecular Evolution of Plastid Genomes in Parasitic Flowering Plants

Susann Wicke*[,1], Julia Naumann[†]
*Institute for Evolution and Biodiversity, University of Münster, Münster, Germany
[†]Institute for Botany, Technische Universität Dresden, Dresden, Germany
[1]Corresponding author: e-mail address: susann.wicke@uni-muenster.de

Contents

Abstract

Heterotrophic carbon acquisition is the most unusual lifestyle in plants, whereby the heterotrophs obtain water, nutrients, and macromolecules from either another plant or a fungus. Besides numerous morphological changes that accompany the transition to a fully heterotrophic lifestyle in plants, the so-called parasitic reduction syndrome manifests at the molecular level, especially in the plastid genome. Here, we provide an overview of the sizes, architectures, and coding capacity of plastid genomes in heterotrophic land plants, with a major focus on flowering plants. Our compilation of plastomes of over 75 taxa covering 15 lineages of haustorial parasites and mycoheterotrophs reveals novel insights into the order of housekeeping gene losses, where apparently several plastid tRNA gene deletions precede the loss of ribosomal subunits. A comparison of the three major conceptual models of plastome degradation en route to heterotrophy in plants shows that plastid evolutionary trajectories are essentially convergent across lineages—independent of the feeding type. However, several

Advances in Botanical Research, Volume 85
ISSN 0065-2296
https://doi.org/10.1016/bs.abr.2017.11.014

questions regarding the series and timing of functional and physical gene losses remain unclear, in part because functional data are widely lacking. Nevertheless, the currently available evolutionary models of reductive plastome evolution provide excellent starting points for leaving the paths of descriptive science towards hypothesis-driven research.

1. INTRODUCTION

Parasitism represents the most extreme interaction between plants, where the parasite steals water and nutrients from another plant or fungus. The heterotrophic plant may acquire its carbon by connecting physically to another plant's vascular tissue (parasitic plants or haustorial parasites) or by means of a fungal association (mycoheterotrophic plants). Most parasites are partial heterotrophs and carry out photosynthesis to some extent, thereby ideally producing a positive net balance of organic carbon in addition to host-derived nutrients. Of these partial heterotrophs, some can fulfil their life cycle without ever connecting to a host plant (facultative heterotrophs or parasites), whereas others depend on nutrient supply by a host during at least certain developmental stages (obligate heterotrophs/parasites). At the end of the spectrum of obligate parasites are those that have lost the ability to convert light into chemical energy by CO_2 assimilation. As holoparasites or holo-heterotrophs, these plants retrieve organic and inorganic nutrients, organic carbon, macromolecules, and water mostly exclusively from their hosts (see Glossary for disambiguation of terms).

Numerous morphological changes accompany the transition to a parasitic lifestyle in plants. The "parasitic reduction syndrome" describes the traits that emerge convergently in all parasitic lineages as trophic specialization unfolds. This trait set includes the reduction of roots to stumpy, root-like structures called haustoria (in parasitic plants), an overall decrease in plant heights and of photosynthetic tissue, as well as the loss of light-harvesting pigments like chlorophyll (e.g. Barrett, Freudenstein, et al., 2014), so that, eventually, mostly achlorophyllous reproductive structures remain. On the genetic level, the parasitic reduction syndrome includes a dramatic functional and physical reduction of the heterotrophs' plastid genomes, where rampant gene loss and an acceleration of molecular evolutionary rates occur. Over 20 years after the publication of the first plastome of a parasitic plant, *Epifagus virginiana* (Orobanchaceae) (dePamphilis & Palmer, 1990; Wolfe, Morden, & Palmer, 1992), the reductive evolution of plastomes has become the best-characterized genomic modification that directly relates to the heterotrophic lifestyle in plants. Of the known at least 30 families that contain initially, partially, or fully heterotrophic taxa, comprehensive data for 15 lineages across the range of land plants are currently available (Fig. 1). Some of these, like Orchidaceae or Orobanchaceae, have sequence data for species covering independent transitions to parasitism within these groups. Most other lineages of heterotrophic seed plants have already been sequenced as of writing this contribution and are at different stages of analysis and publication (own data and personal communication with various authors). Moreover, we now have begun moving towards understanding the evolutionary trajectories of plastid genome reduction in nonphotosynthetic angiosperms rather than pursuing descriptive science.

Here, we critically review the current state of knowledge of heterotrophic plant plastid genomics. This chapter aims at providing an overview of the diversity of plastome sizes, architectures, and coding capacities in heterotrophic land plants. By comparing three major models of plastome degradation in heterotrophs, we clearly show that the plastid evolutionary trajectories are essentially convergent across lineages and independent of the feeding type of the various heterotrophic lineages. Therefore, another major focus of our review lies in identifying gaps of knowledge, which may help to improve the experimental designs of future studies and highlight the need for more function-based studies targeting the role and function of plastomes and plastids of heterotrophic plants in general.

2. PLASTOME SIZE DIVERSITY IN HETEROTROPHIC LAND PLANTS

The published plastomes of over 75 species of heterotrophic plants, regardless of the mode of carbon acquisition (haustorial parasitism, mycoheterotrophy), cover the entire range of facultatively to fully heterotrophic species. Ranging from 11 to 161 kb in size, heterotrophs cover about three times the range of autotrophic plants (Fig. 2). On the partial heterotrophs side of the spectrum are species like *Cassytha* (Lauraceae) (Song et al., 2017; Wu, Wang, Wu, Wang, & Chaw, 2017), Orchidaceae (Barrett, Freudenstein, et al., 2014; Feng et al., 2016), some Ericaceae (Logacheva, Schelkunov, Shtratnikova, Matveeva, & Penin, 2016; Yu, Wang, & Gong, 2017), Orobanchaceae (Fan et al., 2016; Uribe-Convers, Duke, Moore, & Tank, 2014; Wicke et al., 2016, 2013), as well as Santalales (Petersen, Cuenca, & Seberg, 2015; Su & Hu, 2016) that are all in a very

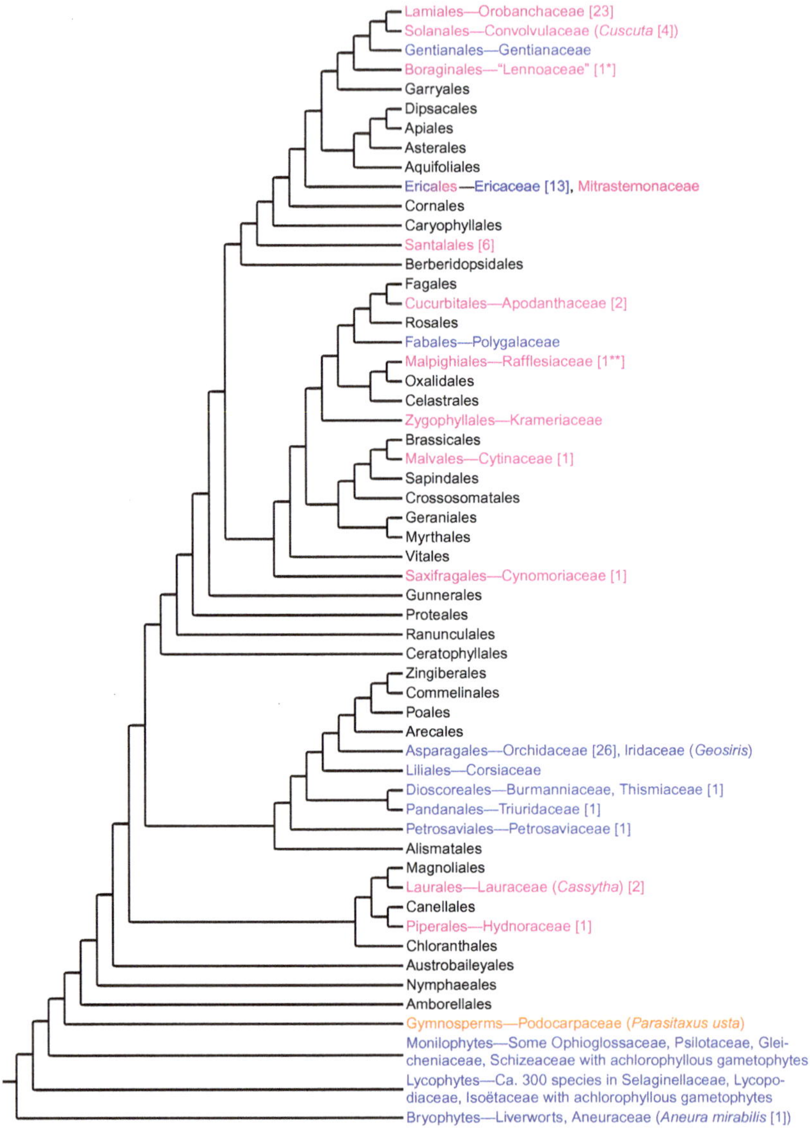

Fig. 1 Evolution of heterotrophy in land plants. The heterotrophic lifestyle has evolved multiple times independently during the evolution of land plants, giving rise to mycoheterotrophs (*blue*) and haustorial parasites (*pink*). The heterotrophic gymnosperm *Parasitaxus* (*orange*) has a unique physiology reminiscent of an intermediate between both feeding types. Ericales is the only lineage in which both feeding forms have arisen independently. Note that despite nonparasitic members in Santalales, we provide no information of the various parasitic lineages here as there are too many families (Su, Hu, Anderson, Der, & Nickrent, 2015 for details). Lineages for which plastid genomes have been analysed as of October 2017 are given with numbers in square brackets to indicate the number of studied species. The main text details information and references for all plastome data shown here. Relationships among angiosperms are according to APG IV (The Angiosperm Phylogeny Group, 2016). [*, Own data, unpublished; **, No plastome detected].

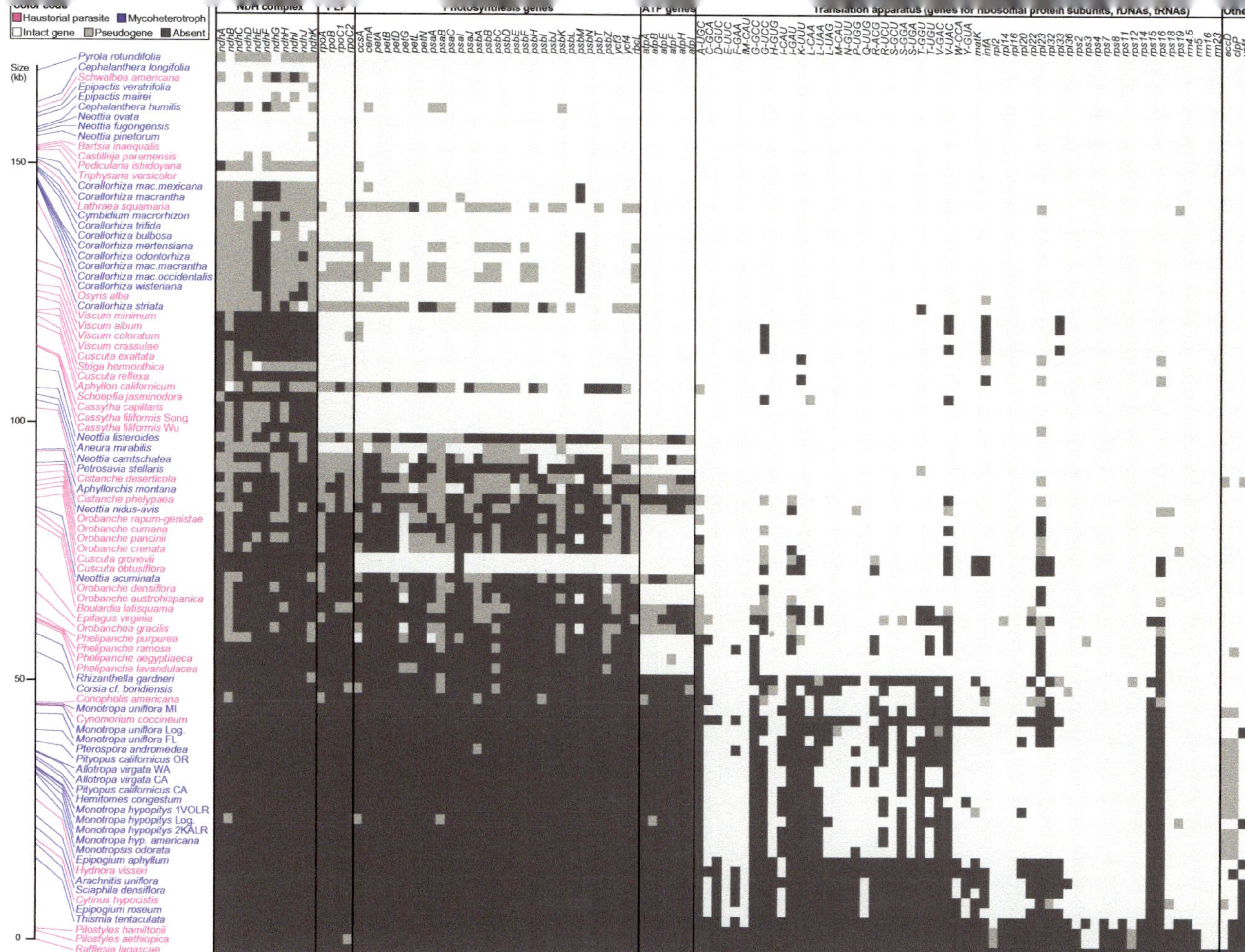

Fig. 2 See legend on next page.

early stage of reductive plastome evolution. Their plastomes are often within the size range of regular angiosperm chloroplast genomes and have a nearly complete plastid gene set. However, several genes are repeatedly lost or reside in the plastome as nonfunctional (pseudogene) copies, especially genes for the NAD(P)H dehydrogenase complex (*ndh* genes).

As the specialization on the heterotrophic lifestyle proceeds, plastid genomes undergo rapid changes regarding both gene content and structure. A great diversity of genome sizes thus are known from nonphotosynthetic Orobanchaceae (Cusimano & Wicke, 2016; Li et al., 2013; Samigullin, Logacheva, Penin, & Vallejo-Roman, 2016; Wicke et al., 2016, 2013) and Orchidaceae (Barrett & Davis, 2012; Barrett, Freudenstein, et al., 2014; Feng et al., 2016). *Lathraea squamaria* (Orobanchaceae; Samigullin et al., 2016) has the largest plastome of an achlorophyllous plant with 150.5 kb in length and 112 retained genes, of which 32 are pseudogenes. Another species, *Lathraea clandestina*, shows more physical and functional reductions with 138 kb and 41 gene losses, suggesting a rapid progression of plastome degeneration in this genus (Delavault, Russo, Lusson, & Thalouarn, 1996; own data, unpublished). Six partial heterotrophs have smaller plastomes than *L. squamaria*, with *Schoepfia jasminodora* (Santalales) being the smallest with 119 kb in length and 103 retained genes, including only three pseudogenes (Su & Hu, 2016). At the end of the genome size spectrum are holoparasites with extremely reduced plastomes as in the "endoparasite" *Pilostyles hamiltonii* (Apodanthaceae) with a little over 11 kb in size and possibly no more than five functional genes, if at all (Bellot & Renner, 2015), and in the mycoheterotroph *Thismia tentaculata* (Burmanniaceae), whose plastid genome is 16 kb in size and harbours seven genes (Lim, Barrett, Pang, & Davis, 2016). The variation in size, structure, and gene content between (and within) lineages can be speculated to reflect either different modes and paces of functional and physical reductions or time since the transition to (holo)heterotrophy, or both.

Fig. 2 Plastid-coding capacities of heterotrophic plants. Sorted by plastome size, the presence or absence of genes in currently sequenced plastid genomes of mycoheterotrophs (*blue*) and haustorial parasites (*pink*) is depicted for all plastid gene classes. Intact genes are highlighted in *off-white*, *light grey* highlights the presence of a gene as a pseudogene, and the absence of a gene is marked in *dark grey*. Note that the categorization of genes as intact, pseudogene, or absent is shown according to the scoring of the original publications or published updates thereof.

3. RECONSTRUCTING AND ANNOTATING DIVERGENT PLASTOMES OF HETEROTROPHS

3.1 Plastome Assembly

Plastomes of heterotrophic plants currently available in GenBank (last accessed: October 2017) have been generated by several different approaches. While the first plastomes of heterotrophs used plastome mapping and hierarchical shotgun sequencing of genome libraries (e.g. dePamphilis & Palmer, 1990; Funk, Berg, Krupinska, Maier, & Krause, 2007; McNeal, Kuehl, Boore, & dePamphilis, 2007; Wolfe, Morden, Palmer, et al., 1992), the majority of published plastomes were reconstructed from high-throughput data of whole-genome shotgun-sequencing projects (e.g. Braukmann, Broe, Stefanović, & Freudenstein, 2017; Schelkunov et al., 2015; Wicke et al., 2013). Long-range PCR in combination with high-throughput sequencing has been employed for a few taxa, where reference plastomes of close relatives were available (Funk et al., 2007; Li et al., 2013; Uribe-Convers et al., 2014). However, target enrichment via long-PCR or solution-based hybridization is not common, perhaps because the extent of plastome degradation in the species of interest is rarely known beforehand. Whole plastome capture thus might fail if the species to be sequenced is too divergent from those used for designing capture baits.

Oversampling read data through massively parallel sequencing in the so-called genome skimming, where many genomic regions are recovered by sequencing randomly fragmented DNA at minimal genome coverage, has become the commonly preferred method. This approach takes advantage of the normally high copy number of plastid genomes in a plant cell and is especially useful in lineages with little or no reliable reference data. However, reconstructing plastomes from these data can still be considerably more challenging in heterotrophs than for ordinary green plants. In several cases, the ratio of plastid DNA can be lower than in related green plants (e.g. Feng et al., 2016; Wicke et al., 2013). This increases the risk of accidently extracting and including divergent plastid-like reads in plastome assemblies and downstream analyses. As inserts of plastid DNA into the nuclear genome (nupts) and/or the mitochondrial genome (mipts) are still much less abundant than authentic plastid fragments, experimental and bioinformatic quantitative methods including coverage plots, quantitative PCR, or DNA gel plots are a gold standard for plastome assembly (see Wicke & Schneeweiss, 2015 for a detailed review). For plastome assembly, contigs should be retained only if they had already exceeded a certain coverage threshold during primary assembly (e.g. Straub et al., 2012).

Relying exclusively on divergence between nupts, mipts, and genuine plastid DNA copies can be misleading, because, first, it is not normally known which authentic plastid genes are pseudogenes, and, second, inserts of very recent transfers of plastid DNAs might not have accumulated sufficient amounts of mutations to distinguish them from the original. Also, abundance and preservation of plastid-like DNAs often are highly lineage-specific and may include insertions of near-complete copies of the entire plastome (Ayliffe, Scott, & Timmis, 1998; Bock & Timmis, 2008; Cusimano & Wicke, 2016; Lloyd, Rousseau-Gueutin, Timmis, Sheppard, & Ayliffe, 2012; Richly & Leister, 2004; Rousseau-Gueutin, Ayliffe, & Timmis, 2011). Although always a concern for the reconstruction of plastomes, assembly and reconstruction of plastid genomes of heterotrophs should take into account an adequate combination of (at least two of the following) coverage, read length, and read quality. Any a priori knowledge of the expected length in relation to overall genome size and the expected gene content will allow approximating the number of required read data (see Cronn et al., 2012 for details) and can give information on the completeness of an assembly.

Although many plastid contigs can be readily identified using BLAST-based approaches with well-curated plastome reference data, additional means are sometimes required to identify all plastome contigs and to distinguish those from mitochondrial DNA. For example, the plastome assembly of *Hydnora visseri* (Aristolochiaceae) required the use of stoichiometry plots of read depths relative to scaffold length and GC content (Naumann et al., 2016). Low GC contents and more low-complexity regions including higher amounts of repetitive DNA than in photosynthetic plants often characterize the plastome contigs of nonphotosynthetic plants and may hamper a reference-based assembly. These genomic features can result in contig breaks that may not be resolved correctly (Cronn et al., 2008; Moore et al., 2006; Straub et al., 2012). Therefore, combining reference-based assemblies with data from de novo approaches represent a reliable option for closing gaps in silico (Barrett, Freudenstein, et al., 2014; Barrett, Specht, et al., 2014; Eserman, Tiley, Jarret, Leebens-Mack, & Miller, 2014; Henriquez, Arias, Pires, Croat, & Schaal, 2014; Straub, Cronn, Edwards, Fishbein, & Liston, 2013; Wicke et al., 2013). As plastomes of heterotrophs are also often structurally aberrant, not only regarding their gene content, a confident assembly can be achieved using de novo assemblies with *k*-mer ranges, manual curation, and read mapping-based error correction, or PCR finishing.

3.2 Plastome Annotation

Chloroplast genes of autotrophic plants are mostly highly conserved. In heterotrophic plants, functional genes tend to become more divergent as selective constraints relax. This makes identifying genes in plastomes of heterotrophs by means of similarity-based methods particularly challenging (Bellot & Renner, 2015; Braukmann et al., 2017; Lim et al., 2016; Naumann et al., 2016; Roquet et al., 2016). A great number of those similarity-based annotation tools are now available specifically for plastome data (e.g. CGAP—Cheng, Zeng, Ren, & Liu, 2013; PLANN—Huang & Cronk, 2015; Verdant—McKain, Hartsock, Wohl, & Kellogg, 2017), but many underperform on highly divergent plastomes of heterotrophs. Annotation transfer tools as implemented in the widely accepted software *Geneious* (Biomatters, Inc.) are popular, but standardly assume little divergence between reference and target, and often fail to recognize intron/exon boundaries correctly. These tools are often not designed to include protein translations either, and thus, might annotate frame-shifted genes, those with stop codons or other mutations that might render the gene product nonfunctional. The *Dual Organellar Genome Annotator* (DOGMA, Wyman, Boore, & Jansen, 2004) web tool utilizes BLASTX and BLASTN for automated annotation as well as an automated tRNA prediction. It performs well on plastomes of heterotrophs when run with low stringency, i.e., assuming low similarity (<40%) to the reference plastomes, enabling the detection of very short and highly divergent genes. However, inaccurate annotation of intron/exon boundaries and gene start/stops still necessitates manual curation. While the ribosomal genes tend to be the most conserved elements of the plastid gene sets in parasites, many of the (retained) protein-coding genes can have shorter open reading frames (ORFs) than their orthologues in autotrophs. The local alignment to the queries retrieved by BLAST might often still be shorter than the corresponding ORF. Therefore, and ideally, ORF-finder-assisted annotations and prediction of RNA-editing sites should be applied for plastid protein-coding genes. Similarity alone can lead to incorrect (incomplete) annotation of gene boundaries since only the most conserved gene region may be identified. In addition, chances are that highly divergent genes are missed completely when not considering ORFs at all.

Although high-throughput sequencing has allowed us to gain insights into the extent of physical reduction and the structure of plastid genomes in many heterotrophic plant lineages, we still know little about the functionality of the retained genes. Gene content is commonly compared between lineages, as we do herein, but these comparisons are circumstantial, based on DNA evidence only. In fact, plastid gene expression data only exist for *E. virginiana* (Ems et al., 1995; Wolfe, Morden, Palmer, et al., 1992), *H. visseri* (Naumann et al., 2016), *Rhizanthella gardneri* (Orchidaceae) (Delannoy, Fujii, des Francs, Brundrett, & Small, 2011), *Cuscuta reflexa*, and *Cuscuta gronovii* (Convolvulaceae) (Funk et al., 2007), as well as a few Orobanchaceae species whose *accD*, *clpP*, *ycf1*, and *ycf2* genes are notoriously difficult to annotate (Wicke et al., 2016).

A yet almost inevitable problem arising from similarity-based annotations rather than expression data-based procedures is the classification of genes as intact or pseudogenes. While sequences with similarity to previously characterized genes of conserved function in most green plants may have an intact ORF, these genes are not necessarily functional. Similarly, genes with similarity to previously characterized proteins of conserved function may have many indels or substitutions, be it of both a synonymous and nonsynonymous nature, without rendering the gene nonfunctional. Various posttranscriptional mechanisms that correct or enhance transcript diversity are known in plants, including RNA editing, stop codon readthrough, and transcriptional slippage (e.g. Castandet & Araya, 2011; Lin et al., 2015; Meurer et al., 2002).

4. ARE WE ALWAYS DEALING WITH (PSEUDO)GENES?

Most sequence data of heterotrophs are obtained from genomic surveys, but additional experimental data are urgently needed to obtain evidence for the functionality of ORFs and annotated genes. Basing judgement exclusively on DNA similarity can be misleading. For instance, the *accD* gene varies drastically in length across heterotrophic plants: annotated as an intact gene, it ranges from 954 bp in *Phelipanche aegyptiaca* (Orobanchaceae) (Wicke et al., 2016) to 2094 bp in *Monotropa uniflora* (Ericaceae) (Braukmann et al., 2017); the median length of *accD* in heterotrophic plants is 1482 bp. Presumably, all of these *accD*-like ORFs are functional, but experimental proof is evidently needed. Plastid gene models thus are hypothetical until validated by species-specific expression or protein data. Studies of gene expression deliver important evidence and are powerful in finding the correct coding region. However, some caution should be used with the interpretation of these data. Gene expression does not necessarily mean that a gene product will also be active on the protein level, which, ultimately, represents the level of function. For example, a case study centring around several recent holoparasitic species of Orobanchaceae showed that *rbcL* is expressed but not translated into a functional peptide in some parasites (Randle & Wolfe, 2005).

Often, variations of "*the gene … is highly diverged and probably non-functional*" can be read in research reports, but, to our knowledge, the actual functional space of plastid genes, i.e., the extent to which nucleotide substitutions and indels can be tolerated on the functional (peptide) level, has not yet been determined—neither for photosynthesis genes nor for housekeeping genes. In the absence of clear criteria as to when a gene should be annotated as a pseudogene based on DNA evidence, it is the responsibility of the individual researcher to decide the category into which a gene in question belongs. There is as much unawareness of the functional realm of plastid proteins as there is on the extent of putative researcher bias in annotating plastid genes of unusual divergence. For example, assuming that a gene of a parasite has an intact ORF that is 35% shorter and 96% divergent in sequence compared with its equivalent in a phylogenetically closely related autotroph. How many researchers would classify this gene as "functional" or as "pseudogene"? Some sure would ask for evidence of gene expression, but when no RNA-grade materials of this plant (at its various developmental stages) are available, should this genomic region then better be left unannotated? Certainly not—but perhaps we could add an annotation note pointing others to this form of uncertainty.

An inspection of available sequences in GenBank shows that differences in gene annotation most often indeed pertain to categorizing genes as "intact" or "pseudogenes". However, it also seems as if different views exist as to when a gene is "absent". While one researcher might classify contiguous stretches of less than 10 amino acids as insignificant evidence for the retention of a pseudogene fragment, another researcher would annotate this region as pseudogene. In consequence, downstream analyses, like the reconstruction of ancestral gene content, will carry over discrepancies, no matter their origin, with the potential to severely influence the direction of data interpretation. Determining the degree of researcher bias in annotating plastomes of heterotrophs is hard. Hence, peers should be commended for their candour to admit that sometimes their categorization of genes as intact or pseudogenes may be wrong in the absence of functional data.

Does annotation quality matter? We think so. Many aspects in the field of heterotrophic plant plastomics centre on questions like which genes are lost, when that loss occurred, and in which lineages and how quickly. These questions cannot be answered with confidence if there are reservations about the accuracy of the underlying data. Ideally, the community would work towards refining existing annotations by adding gene expression and protein data. Considering the scarcity of some material paired with the remoteness of habitats where some heterotrophs grow, broadly complementing existing plastome sequences with new experimental data seems unrealistic. Another measure would be to implement best-practice standards with recommendations for assembly, gene finding, and annotation procedures and to clarify criteria for categorizing pseudogenes. Although many researchers might welcome such standardized procedures, how should the community handle published data that may not comply with these recommended procedures? Devising methods or best-practice procedures with a battery of tested software and recommendations for stringency settings or manual curation may also contribute to overcoming annotation biases. Also, when taxon sampling is sufficiently dense, the error of the reconstructed events of pseudogenization or loss-of-function deletions can be minimized to some extent. Nonetheless, it remains the risk to infer events at deeper nodes in a phylogenetic tree and thus in a common ancestor when really these events were independent (or vice versa).

5. EVOLUTION OF PLASTOME STRUCTURE UNDER RELAXED SELECTIVE CONSTRAINTS

Not all plastome size variation in heterotrophs is attributed to functional reduction. While the sequenced plastomes vary considerably in size, especially in holo-heterotrophs (Fig. 2), even closely related species with the same coding capacity may retain plastomes of notably different sizes due to variation in the lengths of the two large inverted repeats (IRs). Regardless of the extent of functional reduction, many plastomes of heterotrophs exhibit a normal quadripartite architecture (Wicke, Schneeweiss, dePamphilis, Müller, & Quandt, 2011) with a large and small single-copy region (LSC, SSC) and two IRs. For example, the highly divergent plastid genome of *H. visseri* has retained an IR, and its gene order is also mostly collinear with that of an ordinary chloroplast genome, but its IR boundaries have shifted drastically (Naumann et al., 2016). However, there are also many lineages of heterotrophs whose plastomes have modified IRs, including *Cynomorium coccineum* (Cynomoriaceae) (Bellot et al., 2016), *Epipogium* species (Orchidaceae) (Schelkunov et al., 2015), *Conopholis americana* (Orobanchaceae) (Wicke et al., 2016), *Cytinus hypocistis* (Cytinaceae)

(Roquet et al., 2016), *Pilostyles* (Apodanthaceae) (Bellot & Renner, 2015), and many, yet not all, nonphotosynthetic Ericaceae examined to date (Braukmann et al., 2017; Logacheva et al., 2016).

The IR is crucial for plastome stability and conservation (Maréchal & Brisson, 2010), which leads to the hypothesis that a plastome becomes more prone to rearrangements and decay once it has lost an IR copy (Wicke et al., 2016, 2013). Several lines of evidence indicate that some parasites in the Orobanchaceae even have inflated IRs, resulting in plastome sizes of >170 kb (own data, unpublished). Other structural changes such as large inversions are often coinciding, but not exclusively found in lineages with IR modifications (*Cuscuta* sp.: Funk et al., 2007; *Petrosavia*: Logacheva, Schelkunov, Nuraliev, Samigullin, & Penin, 2014; *Viscum minimum*: Petersen et al., 2015; some *Orobanche* and *Phelipanche*: Cusimano & Wicke, 2016; Wicke et al., 2016; some Ericaceae: Braukmann et al., 2017; Logacheva et al., 2016). These inversions are considerably more rare than segmental DNA deletions, but in a few cases they coincide with functional or physical gene losses (Petersen et al., 2015; Wicke et al., 2013). Although changes in collinearity have also been reported in photosynthetic heterotrophs of various lineages (Barrett, Freudenstein, et al., 2014; Petersen et al., 2015; Wicke et al., 2016), the generally high degree of structural conservation reported for most autotrophic angiosperm plastomes (Wicke et al., 2011) appears to be upheld in parasites for a long period of time. However, structural maintenance appears to experience relaxed selection as genome reduction proceeds. For example, in heterotrophic orchids of *Corallorhiza* (Barrett, Freudenstein, et al., 2014) and the *Neottia* tribe (Feng et al., 2016), which are both in an early state of functional reduction (Fig. 2), only a single case of structural rearrangement is known (a 16-kb inversion in a variety of *Corallorhiza maculata*—Barrett, Freudenstein, et al., 2014). In contrast, the highly reduced orchids *Epipogium* (Schelkunov et al., 2015) and *Rhizanthella* (Delannoy et al., 2011) exhibit extreme structural modifications, including the loss of the IR. Rampant functional reduction also often coincides with a decreasing GC content (Fig. 3), which may trigger structural rearrangements (Wicke et al., 2016, 2013)—or vice versa? However, the data thus far still seem equivocal on whether taxa with IR modification or loss experience drastic rearrangements and changes of GC content as a result. An alternative hypothesis is that any stimulus that introduces instability like functional relaxation as in holo-heterotrophs might affect rearrangements, fluctuation as well as loss of IRs, or both simultaneously. Clearly, further study is needed to adequately

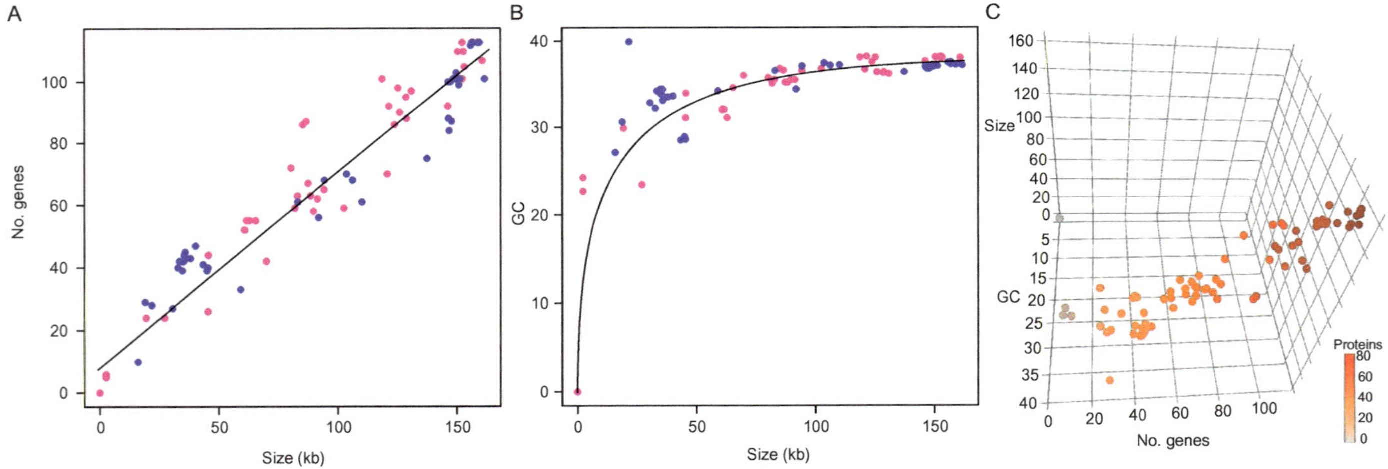

Fig. 3 Associations of plastid-coding capacity, GC content, and genome size. (A) Plastome size and coding capacity as inferred from the number of retained intact genes show a strong correlation across all heterotrophic plants. No differences appear to exist between mycoheterotrophs (*blue*) and haustorial parasites (*pink*). (B) GC content appears to be a nonlinear function of plastome size. Little size (and thus perhaps functional) reduction affects GC content less drastically while extensive physical deletions coincide with dramatically decreasing GC contents, whereby current data are still not sufficient to pinpoint the tipping point [*Colours* as in (A)]. (C) A three-dimensional scatter plot indicates that the nonlinear relationship between plastome size and GC content may be cocorrelated with other genomic traits such as functional reduction as evidenced by both the number of intact plastid genes (*x* axis) and the number of intact protein-coding genes (*colour-coded*), corroborating earlier predictions that AT richness in plastomes is characteristic for the progression of the parasitic reduction syndrome of heterotrophs (Wicke et al., 2016).

address the causation and interrelations of genomic trait changes and functional reduction in plastid genomes of heterotrophs.

6. FUNCTIONAL REDUCTION ALONG THE TRANSITION TO A NONPHOTOSYNTHETIC LIFESTYLE

Most of the genome size variation in heterotrophs clearly relates to the rampant gene losses in consequence of the transition to a nonphotosynthetic lifestyle. In general, there is a strong correlation of functional and physical reductions in heterotrophs (Fig. 3). Independent lineages often show a surprising convergence regarding the number and nature of retained genes (Fig. 2), although, of course, notable exceptions exist. It has emerged as a more general pattern that plastid *ndh* genes (see Martín & Sabater, 2010; Peltier & Cournac, 2002; Wicke et al., 2011 for a general review of plastid gene function) are the earliest functional losses in (most) partial heterotrophs. Although the functional and/or physical loss of *ndh* genes is exceptionally prevalent in photosynthetic parasites, the loss of *ndh* genes apparently is not linked to heterotrophy per se. These genes are also dispensable in many photosynthetic lineages, including photosynthetic, leaf-bearing orchids, which, at the seedling stage, are all so-called initial mycoheterotrophs, in carnivorous plants like Lentibulariaceae (Wicke, Schäferhoff, dePamphilis, & Müller, 2014), of which some are known to take up organic carbon from their prey, as well as in plant lineages that show no signs of heterotrophic carbon acquisition such as Geraniaceae (Blazier, Guisinger-Bellian, & Jansen, 2011) or gymnosperms like Pinaceae and Gnetales (Lin, Huang, Wu, Hsu, & Chaw, 2010; Wakasugi et al., 1994; Wu, Lai, Lin, Wang, & Chaw, 2009).

The series of functional reductions following the loss of *ndh* genes is less uniform. Several lines of evidence suggest that the plastid-encoded polymerase (PEP), which transcribes many plastid photosynthesis genes, may be dispensable at a rather early parasitic stage (Wicke et al., 2016). For example, in *Cuscuta*, a genus of stem parasites whose specialization is often unclear and likely to be in the range of physiological holoparasitism, PEP genes including the corresponding PEP promoters were already lost in spite of the retention of plastid photosynthesis genes and maintenance of their expression (Fig. 2) (Funk et al., 2007; McNeal et al., 2007). Pseudogenes of PEP subunits were also reported in some *Corallorhiza* species although only a few photosynthesis genes were lost (Barrett, Freudenstein, et al., 2014), and maximum likelihood-based reconstructions of the patterns of genes losses in Orobanchaceae also place PEP among the earliest functional reductions (Wicke et al., 2016, 2013).

Core photosynthesis genes show no clear pattern as to when and how they are lost from plastomes. Reconstruction of gene losses across a larger set of taxa within different heterotrophic orchid tribes and Orobanchaceae show that *psa/psb* (photosystem I and II) and *pet* genes (cytochrome b_6/f complex), as well as *ccsA* (haem attachment factor), *cemA* (inner membrane protein for CO_2 uptake), and *pafI/ycf3* and *pafII/ycf4* (both photosystem assembly factors) are functionally lost around the boundary to holoparasitism, but there is no indication of a specific order of losses (Barrett, Freudenstein, et al., 2014; Feng et al., 2016; Wicke et al., 2013). In other words, based on all currently available data, it remains uncertain as to whether photosystems and photosystem-associated genes are lost before plastid cytochromes and electron transport complexes. Location effects from the proximity to essential neighbouring genes and/or their localization in an operon apparently determine the survival time as genes become dispensable (Wicke et al., 2013). However, gene size also has an impact on gene retention, apparently allowing short genes, especially tRNAs, to escape deletion (Lohan & Wolfe, 1998). New clues to the series of photosynthesis gene losses might come from partial heterotrophs, such as some *Viscum* species (Petersen et al., 2015) or the mixotrophic orchids *Cephalanthera humilis* (Feng et al., 2016) and two chlorophyllous species of *Corallorhiza* (Barrett, Freudenstein, et al., 2014). A few photosynthesis genes, including *ccsA*, *cemA,* plus a few plastid-encoded photosystem genes were annotated as pseudogenes in these plants (Fig. 2). Functional data thus are urgently needed to confirm the pseudogenization of these genes in those heterotrophs.

ATP synthase genes (*atp* genes) are a clear exception to the rapid loss of genes encoding subunits of plastid thylakoid complexes along the transition to the nonphotosynthetic lifestyle (Fig. 2). In several holo-heterotrophic plants, including species of Orobanchaceae (Wicke et al., 2016), Orchidaceae (Barrett & Davis, 2012; Barrett, Freudenstein, et al., 2014; Feng et al., 2016), *Cuscuta* (Funk et al., 2007; McNeal et al., 2007), *Aneura mirabilis* (Aneuraceae) (Wickett et al., 2008), and *Petrosavia stellaris* (Petrosaviaceae) (Logacheva et al., 2014), *atp* genes are apparently retained with intact ORFs. This observation has led to the speculation of a prolonged or hidden secondary function of the thylakoid ATP synthase, including ATP synthesis from a source other than the photosynthetic proton gradient or the requirement of ATP hydrolysis (Wicke et al., 2013), which, for example, is needed for the twin-arginine protein translocator system (Kamikawa et al., 2015). More research of the role of plastids in nonphotosynthetic plants is needed to clarify whether an alternative function is causal to *atp* gene retention. The *rbcL* gene (large subunit of RuBisCO) was the first case in which a then unknown role of a primary photosynthesis-associated plastid gene contributed an explanation for an unexpectedly long retention. In addition to catalysing the first major step during carbon fixation, RuBisCO knowingly contributes to serine and glycine biosynthesis in the C2 pathway (Tolbert, 1997) and was shown later to also improve carbon efficiency without the Calvin cycle in greening seeds (Schwender, Goffman, Ohlrogge, & Shachar-Hill, 2004).

It is not implausible that parasitic plants might also help in revealing the role of *ycf1* and *ycf2*. Many holo-heterotrophs retain intact ORFs of both genes, corroborating earlier findings that these genes are essential to many plants (Drescher, Ruf, Calsa, Carrer, & Bock, 2000). For both these largest plastid ORFs, several functions have been proposed but continue to be the subject of a vivid scientific debate (*ycf1*: binding or docking plastid DNAs/mRNAs to the plastid envelope or thylakoid membrane—Boudreau et al., 1997; protein import as part of the inner translocon—Kikuchi et al., 2013; Nakai, 2015; but see Bölter & Soll, 2017; de Vries, Sousa, Bölter, Soll, & Gould, 2015 assembly of the plastid fatty acid synthase (ACCase)—Sjuts, Soll, & Bölter, 2017; *ycf2*: ftsH/CDC48-like protein involved in cell division, proteolysis, and/or protein transport—Wolfe, 1994).

To date, one of the greatest enigmas in heterotrophy-associated plastome degeneration is the series of losses of housekeeping genes (ribosomal proteins—*rpl*/*rps* genes, *infA*, *matK*, *clpP*, tRNAs, rRNAs), and of those genes whose products function in pathways other than photosynthesis (*accD*, *ycf1*, *ycf2*). Besides the fact that housekeeping gene loss, with the exception of PEP gene losses (see earlier), usually begins after the nonfunctionalization of photosynthesis-related genes, there is no clear signal from the set of conserved plastid genes *within* lineages, and there appears to be no definitive conserved set of essential genes *across* lineages. Interestingly, several tRNA genes appear to be lost already before the majority of ribosomal proteins (Fig. 2), indicating that tRNA import may be achieved more easily than import of ribosomal proteins. We also observe that species whose plastomes still retain pseudogene copies of photosynthesis genes are richer in ribosomal and tRNA genes, and that these plastomes often also have intact ORFs for *clpP*, *accD*, and *ycf1/2*, although some of those may be quite divergent. Their retention may be attributed to inefficient protein import, regulatory coupling of genes for biological processes, and the coordinated assembly and cotranslation of partnered proteins. The plastid-encoded L-glutamyl-tRNA (*trnE*), required for initiating tetrapyrrole biosynthesis, and the *accD* gene, needed for lipid biosynthesis, are often considered essential plastome genes (Barbrook, Howe, & Purton, 2006). However, loss of these genes from the plastomes of some holo-heterotrophs as well as in some photosynthetic plants (e.g. see Jansen & Ruhlman, 2012; Wicke et al., 2011 for reviews) suggests that current barriers of functional gene transfer or functional replacement can be overcome. Thus, current data imply that there is no such thing as *the* minimal plastid genome or the essential gene set common to all heterotrophs.

The pace at which plastomes of parasitic plants are reduced functionally and physically is poorly investigated. Combined evidence indicates that plastome degeneration is a highly lineage-specific process, perhaps a function of species-/lineage-specific recombination and/or replication error rates. In Orobanchaceae, and more specifically broomrapes (*Orobanche* spp.), the rate of gene loss was estimated to be about one gene per million years (Cusimano & Wicke, 2016), although loss through time seems to be no linear process (see Section 8). Functional complexes are lost rather rapidly around major lifestyle transitions or along the parasitic specialization, followed by one or more "stationary" phase(s) (Naumann et al., 2016; Wicke et al., 2016). To resolve paths and timing of reductive evolution more accurately, considerably more data from a much denser taxon sampling both within and across lineages are required, including a much higher intrageneric and intraspecific resolution per lineage (where possible). The latter will also allow addressing the question as to whether functional and physical reduction occur in bursts or proceed gradually over time.

7. EVOLUTION OF SUBSTITUTION RATES

Variation of nucleotide substitution rates exists between different lineages of plants, and among different classes of genes (Gaut, Yang, Takuno, & Eguiarte, 2011; Wicke & Schneeweiss, 2015). Differing rates of both synonymous and nonsynonymous divergence (Fig. 4) can be regarded as underlying causes of this variation, likely reflecting variation in selective constraints. To date, the molecular mechanisms underpinning substitution rate variation between lineages and among gene classes are not yet fully understood. However, location effects, gene length, selection on codon bias, and nucleotide composition that, in turn, affect mutation rate are known as good predictors of synonymous rate variation (e.g. Gaut et al., 2011; Morton, 1997; Wicke et al., 2013, 2014). Variation in

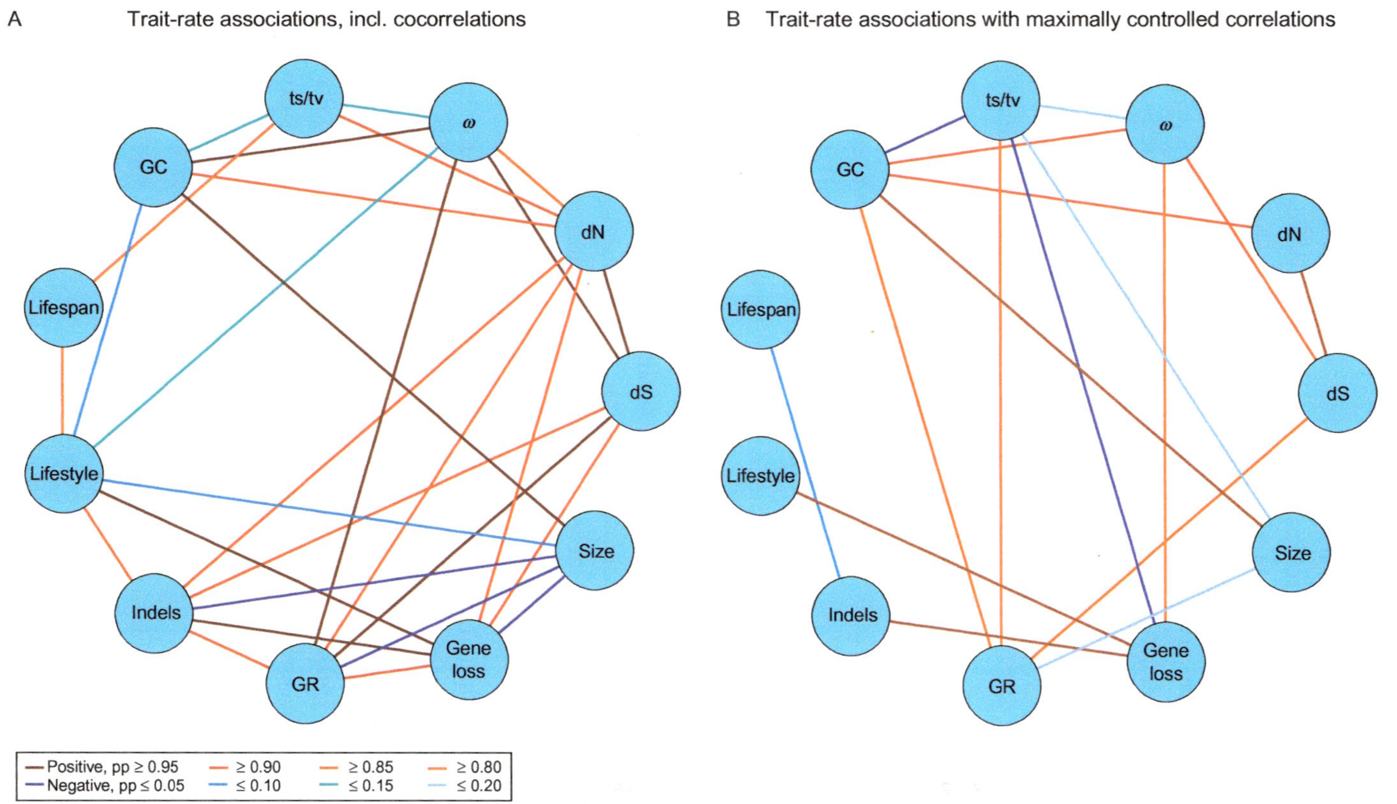

Fig. 4 Coevolutionary web of lifestyle, genetic, and molecular evolutionary changes. Based on the analyses of genetic, substitution, and indel rates as well as lifestyle traits of 20 Orobanchaceae genomes, (A) positive (*reddish*) and negative (*bluish*) associations between the various rates and traits are shown, whereby (B) control of potential cocorrelations among traits provides an overview of direct associations among the various molecular and lifestyle traits. ω, ratio of nonsynonymous to synonymous substitutions (dN/dS) *GR*, genomic rearrangements; *ts/tv*, ratio of transitions to transversions.

nonsynonymous substitution rates in ordinary green plants is still elusive, although it is often hypothesized to relate to a strong coevolution with synonymous substitutions rather than the relaxation of purifying selection. In autotrophic plants, such an elevation of the nucleotide substitution rate is often encountered in lineages with either severe departures from the conserved quadripartite architecture of plastomes (e.g. Guisinger, Kuehl, Boore, & Jansen, 2008; Sloan et al., 2014), or in plants that exhibit unusual lifestyles like carnivory (Wicke et al., 2014). Note that in some cases "localized hypermutation" in protein-coding genes of autotrophs are indeed the result of relaxation of selective constraints, e.g., in consequence of functional gene transfers (e.g. Magee et al., 2010; Rousseau-Gueutin et al., 2013).

Acceleration of substitution rates is a common trait of heterotrophic plant plastomes (e.g. dePamphilis, Young, & Wolfe, 1997; Nickrent, Blarer, Qiu, Vidal-Russell, & Anderson, 2004; Nickrent & Starr, 1994; Wolfe & dePamphilis, 1998; Wolfe, Morden, Ems, & Palmer, 1992; Young & dePamphilis, 2005). Although site, gene, and lineage effects, and any combination thereof, contribute to rate variation in plants in general (Gaut et al., 2011), selectional or mutational forces (site effects) might outweigh the contribution of others in plastomes of heterotrophs due to the relaxation of selective constraints on photosynthesis and photosynthesis-related genes. Relaxed purifying selection on the amino acid level is expected to result in higher nonsynonymous substitution rates (dN), whereas synonymous substitutions (dS) are not primarily affected. Studies in the Orobanchaceae and Orchidaceae indicate that sequence drift has accelerated early on along the evolution of heterotrophy in both these lineages (Barrett, Freudenstein, et al., 2014; Cusimano & Wicke, 2016; dePamphilis et al., 1997; Feng et al., 2016; Levy Karin, Wicke, Pupko, & Mayrose, 2017; Wicke et al., 2016). In Orobanchaceae, life history can be ruled out as sole causal factor of rate variation (Young & dePamphilis, 2005), implying that several mechanisms, which may include life history but also mutation rate, DNA repair efficiency, and perhaps speciation rate, jointly contribute to the evolution of molecular rates. A Bayesian cocorrelation analysis of molecular evolutionary rates and various genetic and lifestyle traits (Lartillot & Poujol, 2011) of 20 fully sequenced Orobanchaceae plastomes (Fan et al., 2016; Uribe-Convers et al., 2014; Wicke et al., 2016, 2013; Wolfe, Morden, Palmer, et al., 1992), carried out as recently described (Cusimano & Wicke, 2016), show that nonsynonymous and synonymous substitutions are tightly knit (Fig. 4).

Phylo-regression models recently revealed that lifestyle plus large-scale genomic features and the prevalence of indels all reflect evolutionary rate variation (dN, dS, and jointly) across trophic specializations in Orobanchaceae. Clearly, correlation is not causation. However, in the light of these data, it is tempting to speculate that rate acceleration in plastid genes is not only the result of the transition to a nonphotosynthetic lifestyle. Plastid genes are not involved in parasite/host interaction, so elevated dN and dS likely are not linked to the parasite/host arms race either, but perhaps predominantly to the relaxation of purifying selection in photosynthesis genes resulting from gaining the ability to utilize heterotrophically gained organic carbon. While these causes are experimentally and statistically difficult to untangle, we can speculate that the ability to withdraw nutrients from another plant provides such an extreme ecological advantage that it initiates a molecular–evolutionary feedback loop, in which substitutions and structural changes are tolerated to a greater extent than in nonheterotrophic plants. This feedback loop may then lead to the accumulation of deleterious mutations in the plastome, but also in distant genomic compartments harbouring "autotrophy and heterotrophy genes", thus eventually affecting the efficiency of the photosynthesis machinery, which, in turn, drives trophic specialization (Wicke et al., 2016). Naturally, this feedback loop is unlikely limited to plastid-encoded genes but may manifest in other genomic compartments as well because of the molecular coevolutionary web of genetic interactions between cellular components and metabolic pathways. Noteworthy here is the finding that, apart from substitution rate changes, microstructural mutations like short insertions and deletions (indels) or localized inversions increase in number in heterotrophs. Indel rates are thus a valuable additional proxy for relaxation of purifying selection (Wicke et al., 2016; farther: Wicke et al., 2014), because their origin might also relate to mechanisms underlying substitution rate changes.

8. MODELS OF PLASTOME DEGRADATION

Several conceptual models of plastome degradation have been postulated to describe heterotrophy-associated plastome degradation in plants in a simplified, idealized manner. Two of these explanatory models describe the series of physical and functional changes associated with the transition to heterotrophy (Barrett & Davis, 2012; Barrett, Freudenstein, et al., 2014; Naumann et al., 2016), and a third one integrates over the variation of molecular evolutionary rates and genetic changes during reductive plastome evolution (Wicke et al., 2016).

Barrett and Davis (2012) and Barrett, Freudenstein, et al. (2014) describe the reduction of plastomes as five stages of functional reduction, beginning (1) with the loss of *ndh* genes followed by (2) genes for the thylakoid photosynthesis complexes, (3) the plastid-encoded polymerase, (4) *atp* genes, and subsequently (5) the various housekeeping (RNA maturation, translation) and other genes, including four genes for nonbioenergetic functions (*accD*, *clpP*, *ycf1*, *ycf2*). The model of Barrett and Davis (2012) and Barrett, Freudenstein, et al. (2014) was slightly modified recently by Graham, Lam, and Merckx (2017), who provided an update by integrating the series of losses proposed in another study (Wicke et al., 2016, see later) and adding newer data of heterotrophs.

Naumann et al. (2016) suggested a "four-stage model of gene reduction" that, regarding the series of functional and physical losses, is in agreement with Barrett, Freudenstein, et al. (2014), Barrett, Specht, et al. (2014), and Graham et al. (2017) but also with a mechanistic model of plastome degradation by Wicke et al. (2016), see later. Inspired by observations in Orobanchaceae and Orchidaceae and several other lineages, Naumann et al. (2016) suggest that nonessential, photosynthesis-related genes are pseudogenized successively before their physical deletion in a "degradation stage I", whereby the order of gene losses often follows a recurring, convergent pattern. This phase is followed by a "stationary stage" that requires only nonphotosynthetic functions, in which also the rate of gene loss slows down. Further nonfunctionalization of plastid genes then depends on their functional replacement by imported proteins. This stage may be comparable to the extant state of autotrophic, nonparasitic plants, although heterotrophs often show accelerated rates of evolution in their retained plastomes (cf. mechanistic model below, but also see Section 6). "Degradation stage II" sets in as the last essential metabolic, nonbioenergetic genes (e.g. *accD* for lipid synthesis, L-glutamyl-tRNA for tetrapyrrole biosynthesis, perhaps *ycf1* and *ycf2* whose exact functions are still under debate) are functionally replaced by nuclear/cytosolic proteins—a process that renders the retention of the plastid translation apparatus unnecessary and thus the plastome itself dispensable. Accordingly, in the "absent stage" the plastome was completely lost, and ancient fragments, residing as dispersed copies in other parts of the genome, are the only remaining evidence for its past existence. The latter is observed in a few lineages that harbour secondary plastids (reviewed in Keeling, 2010), and it has been proposed for a few heterotrophic flowering plants as well (Molina et al., 2014; Nickrent, Ouyang, Duff, & dePamphilis, 1997).

Wicke et al. (2016) presented a mechanistic model that describes the coevolutionary web of lifestyle, genetic, and molecular evolutionary changes (Fig. 4) in a unified framework along the transition from autotrophy to a fully heterotrophic life history in plants. The basis for this evolutionary model includes all plastid genomes of heterotrophs sequenced until late 2016, as well as rigorous analyses of evolutionary rates and selectional pressures. The model by Wicke et al. (2016) recognizes five major stages of functional plastome degradation, with the first one setting in as plants gain the ability for heterotrophic carbon uptake and utilization. Heterotrophy first renders *ndh* genes dispensable, while more dramatic changes concur around the transition to obligate heterotrophy. These lifestyle changes relieve selection pressure on photosynthesis and some housekeeping functions (e.g. transcription—loss of PEP). These first phases of relaxations of functional constraints coincide with a steady increase of the rates of microstructural changes, nonsynonymous, and synonymous substitutions until a new rate equilibrium is reached. A similar shift of molecular evolutionary regimes occurs as selective constraints relax on functional complexes that were maintained for a prolonged period of time (e.g. *atp* genes, *rbcL*). The coevolutionary and causal relationships between the relaxation of selective pressures on photosynthesis, alternative or photosynthesis-unassociated functions, the plastid housekeeping machinery, and an increasing specialization on external carbon during this stage remain to be elucidated. However, at this phase, nonfunctionalization and physical reductions are considered to be accompanied by lifestyle-specific evolutionary rate shifts and a gradual reduction of the plastid GC content (Figs 3 and 4). Although the latter could not be shown to affect the rate of molecular evolution directly (Cusimano & Wicke, 2016; Wicke et al., 2016, 2013), low GC contents correlate with more structural rearrangements, including fragmental deletions, that in turn, are (co)correlated with substitution rates changes (Fig. 4).

Although we here are providing graphical summaries of all models (Fig. 5) and an updated compilation of the plastid gene contents of heterotrophs sequenced to date (Fig. 2), the commonalities among the various conceptual models, to us, outweigh their slight differences, which mainly pertain to the presumed order of functional losses. We therefore refrain from postulating a "merged" model of plastome degradation. Together, the models of Barrett, Freudenstein, et al. (2014) with their update by Graham et al. (2017) and that of Naumann et al. (2016), which all describe

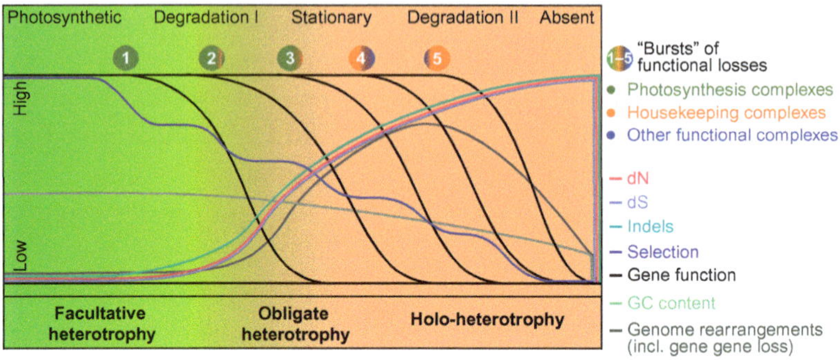

Fig. 5 Functional plastome degradation in relation to the evolution of molecular evolutionary rates. Functional reduction begins in photosynthetic heterotrophs with the loss of nonessential or stress-relevant genes (*ndh* genes) (1—first burst of functional losses). Heterotrophy-associated reductive plastome evolution proceeds by the loss of primary photosynthesis-related genes (*pet, psa, psb* genes) and the plastid-encoded polymerase (PEP) (2) during the "Degradation I"-stage around the boundary to a nonphotosynthetic lifestyle. Genes with a prolonged or alternative function such as *atp* genes and *rbcL* as well as nonessential housekeeping genes are lost after transitioning into a nonphotosynthetic, holo-heterotrophic lifestyle (3) before the rate of gene loss slows down in the stationary phase. When functional replacement of photosynthesis-unrelated metabolic genes (e.g. *accD, clpP, ycf1/2*) allows their deletion from the plastome at the border of the "Degradation II"-stage (4), also all other remaining housekeeping genes, including *trnE* (5) can be jettisoned then to reach the "Absent"-stage, i.e., the complete loss of a plastome. Alongside these functional reconfigurations as the heterotrophic lifestyle unfolds (*green to brown* background), genomic traits such as GC content (GC) gradually decrease as structural changes including gene deletions (GR) and evolutionary rates (dN, dS, indels) increase. In contrast to the steady elevation of nucleotide substitution and indel rates, selectional strength experiences several periods of relaxation and intensification. Figure modified from Wicke et al. (2016), incorporating the degeneration stage names suggested by Naumann et al. (2016). Refer to Fig. 2 for details on the order of gene losses.

the stages of plastome degradation, plus the mechanistic model by Wicke et al. (2016) that unifies evolutionary rate variation and the course of functional and physical plastome degeneration represent legitimate working hypotheses for future research. Testing and challenging these models by thoroughly designed studies that involve phylo-statistically powerful taxon and lineage samplings and/or extend to the nuclear-encoded photosynthesis genes offer exciting novel paths towards understanding heterotrophy-associated plastome degradation. These models thus provide both a basis and guidance for hypothesis-driven research of reductive plastome evolution in heterotrophs.

9. CONCLUSIONS AND FUTURE DIRECTIONS

Heterotrophic plants offer an exciting opportunity to understand general aspects of the genetics underlying plastome evolution, especially regarding the acceleration of molecular evolutionary rates, large- and small-scale genomic rearrangements, and the extent of intracellular DNA transfer, be it functional or nonfunctional. However, many aspects of reductive genome evolution remain unanswered to date. Several questions regarding the series and timing of functional and physical gene losses are still unclear, in part, because various data types are currently unavailable or because we still lack the statistical power regarding taxon sampling to resolve these issues. However, the currently available conceptual models of reductive plastome evolution provide excellent starting points for leaving the paths of descriptive science towards hypothesis-driven research. The research community should focus and collaborate to overcome technical issues regarding assembly and annotation problems, and to find solutions to minimize researcher biases in the categorization of genes based solely on DNA evidence. Plastid genomes may eventually be lost in plants, which undoubtedly represents the hardest part in this field because absence of evidence is no evidence of absence. However, we believe that, in a community effort, a convincing set of different data types will be generated eventually, which will allow corroborating (or falsifying) claims of lost plastomes.

On another path, comprehensive gene expression data from heterotrophic plants combined with protein evidence and ecophysiological measures of photosynthetic capacity would contribute valuable resources for photosynthesis and plastid research in general. Heterotrophs can be regarded as "natural mutants" that, in an explicit comparative-evolutionary framework, require no labour-intensive and time-consuming mutagenesis like green model plants to pyramid functional pathways. Unfortunately though, genetic knock-out experiments and plastid transformation are unavailable for parasites, with the exception of three Orobanchaceae (Fernandez-Aparicio, Rubiales, Bandaranayake, Yoder, & Westwood, 2011; Ishida, Yoshida, Ito, Namba, & Shirasu, 2011; Tomilov, Tomilova, & Yoder, 2007), and, moreover, most heterotrophs are hard to cultivate, if at all, and rarely fulfil their life cycle in less than 3 months. However, obtaining expression data is still possible through real-time, quantitative PCR experiments (e.g. Morden, Wolfe, dePamphilis, & Palmer, 1991; Naumann et al., 2016; Wolfe, Morden, Palmer, et al., 1992), and inexpensive protocols exist for the preservation of RNA-grade material directly from field collections. However, it must be kept in mind that for achieving high-quality and representative expression data from high-throughput sequencing, RNA libraries cannot be prepared on the basis of polyadenylated mRNA selection, because the plastid transcription/translation apparatus is a chimera of eukaryotic cytosolic features (e.g. poly-A-binding proteins), eubacterial components (e.g. Shine–Dalgarno interactions), and plastid innovations (e.g. regulatory step loops) (see Zerges, 2000 for a review).

Finally, we have shown here that the parasitic reduction syndrome equally affects haustorial parasites and mycoheterotrophic plants. The mode of organic carbon acquisition seems to play little or even no role for the course of plastome degeneration; it certainly does for several other metabolic traits though. Per se, the important aspect, maybe even the dominating trigger, appears to be the *heterotrophic* uptake of organic carbon that defines both lifestyles and unites haustorial parasites and mycoheterotrophs on the matter of reductive plastome evolution. We may assume that the ability to obtain organic carbon through sources other than own photosynthesis relaxes selective pressures on plastomes, diffusing into other genomic regions and functional pathways. Thus, the cause for the relaxation of purifying selection can be assumed to be the same for both groups of parasites, and it should be avoided to draw an artificial line between haustorial parasites and mycoheterophs regarding patterns of heterotrophy-associated plastome reduction. However, we acknowledge that it remains unclear to this date whether additional environmental factors contribute to the course and tempo of the degenerative process. Time certainly is an important contributor. Although divergence age estimates provide valuable evidence, we must not forget that the time since the transition to the obligate heterotrophic lifestyle likely represents the most significant predictor, which, unfortunately, is one that cannot be determined easily. In fact, there is no knowledge of how fast parasites specialize on the heterotrophic lifestyle after gaining the ability of take up and effectively utilize organic carbon from external sources. The prevalence of solely holo-heterotrophic lineages implies a rapid specialization process, perhaps because of the enormous evolutionary–ecological advantage the parasitic lifestyle provides in plants.

ACKNOWLEDGEMENTS

We would like to thank two anonymous reviewers for their comments and suggestions on an earlier version of this chapter. S.W. is a fellow of the *Emmy Noether*-program of the German Science Foundation (Deutsche Forschungsgemeinschaft, WI4507-3/1), whose support is gratefully acknowledged.

GLOSSARY

Achlorophyllous plant A nongreen plant, i.e., a plant with no (visible) chlorophyll that relies on the supply of organic carbon, nutrients, and water from another plant or a fungus

Autotrophic plant A "self-feeder" and primary producer of complex organic compounds (like carbohydrates) from simpler (inorganic) carbon sources (like CO_2) through photosynthesis, thereby converting light into chemical energy

Facultative heterotroph A plant with the ability to consume organic carbon from another plant or fungus but that can also fulfil its life cycle without ever connecting to another organism

Haustorial parasite A plant that retrieves all or some of its organic carbon, nutrients, and water from another plant via a physical connection, for which it develops a highly specialized, multifunctional organ called haustorium

Heterotrophic plant A plant that consumes organic carbon from another organism for energy production and biomolecule synthesis

Holo-heterotrophic plant A plant that completely relies on another plant or a fungus to take up and absorb organic carbon, nutrients, and water

Holoparasite A heterotrophic plant that obtains all of its organic carbon, nutrients, and water through a nonmutualistic interaction with another plant (or fungus)

Mycoheterotrophic plant A plant that retrieves all or some organic carbon, nutrients, and water from a mycorrhizal fungus

Obligate heterotroph or parasite A plant that depends on the heterotrophic consumption of organic carbon and/or nutrients and water during at least some developmental stage(s) to fulfil their life cycle

Parasitic plant Mostly used to refer to a haustorial parasite

Partial heterotroph A plant that can take up and utilize organic carbon heterotrophically in addition to assimilating CO_2 through own photosynthesis activity

REFERENCES

Ayliffe, M. A., Scott, N., & Timmis, J. N. (1998). Analysis of plastid DNA-like sequences within the nuclear genomes of higher plants. *Molecular Biology and Evolution*, *15*, 738–745.

Barbrook, A. C., Howe, C. J., & Purton, S. (2006). Why are plastid genomes retained in non-photosynthetic organisms? *Trends in Plant Science*, *11*, 101–108.

Barrett, C. F., & Davis, J. I. (2012). The plastid genome of the mycoheterotrophic *Corallorhiza striata* (Orchidaceae) is in the relatively early stages of degradation. *American Journal of Botany*, *99*, 1513–1523.

Barrett, C. F., Freudenstein, J. V., Li, J., Mayfield-Jones, D. R., Perez, L., Pires, J. C., et al. (2014). Investigating the path of plastid genome degradation in an early-transitional clade of heterotrophic orchids, and implications for heterotrophic angiosperms. *Molecular Evolution and Biology*, *31*, 3095–3112.

Barrett, C. F., Specht, C. D., Leebens-Mack, J., Stevenson, D. W., Zomlefer, W. B., & Davis, J. I. (2014). Resolving ancient radiations: Can complete plastid gene sets elucidate deep relationships among the tropical gingers (Zingiberales)? *Annals of Botany*, *113*, 119–133.

Bellot, S., Cusimano, N., Luo, S., Sun, G., Zarre, S., Gröger, A., et al. (2016). Assembled plastid and mitochondrial genomes, as well as nuclear genes, place the parasite family Cynomoriaceae in the Saxifragales. *Genome Biology and Evolution*, *8*, 2214–2230.

Bellot, S., & Renner, S. S. (2015). The plastomes of two species in the endoparasite genus Pilostyles (Apodanthaceae) each retain just five or six possibly functional genes. *Genome Biology and Evolution*, *8*, 189–201.

Blazier, J. C., Guisinger-Bellian, M. M., & Jansen, R. K. (2011). Recent loss of plastid-encoded *ndh* genes within *Erodium* (Geraniaceae). *Plant Molecular Biology*, *76*, 1–10.

Bock, R., & Timmis, J. N. (2008). Reconstructing evolution: Gene transfer from plastids to the nucleus. *BioEssays*, *30*, 556–566.

Bölter, B., & Soll, J. (2017). Ycf1/Tic214 is not essential for the accumulation of plastid proteins. *Molecular Plant*, *10*, 219–221.

Boudreau, E., Turmel, M., Goldschmidt-Clermont, M., Rochaix, J.-D., Sivan, S., Michaels, A., et al. (1997). A large open reading frame (orf1995) in the chloroplast DNA of *Chlamydomonas reinhardtii* encodes an essential protein. *Molecular Genetics and Genomics*, *253*, 649–653.

Braukmann, T. W. A., Broe, M. B., Stefanović, S., & Freudenstein, J. V. (2017). On the brink: The highly reduced plastomes of nonphotosynthetic Ericaceae. *New Phytologist*, *216*, 254–266. https://doi.org/10.1111/nph.14681.

Castandet, B., & Araya, A. (2011). RNA editing in plant organelles. Why make it easy? *Biochemistry Biokhimiia*, *76*, 924–931.

Cheng, J., Zeng, X., Ren, G., & Liu, Z. (2013). CGAP: A new comprehensive platform for the comparative analysis of chloroplast genomes. *BMC Bioinformatics*, *14*, 95.

Cronn, R., Knaus, B. J., Liston, A., Maughan, P. J., Parks, M., Syring, J. V., et al. (2012). Targeted enrichment strategies for next-generation plant biology. *American Journal of Botany*, *99*, 291–311. https://doi.org/10.3732/ajb.1100356.

Cronn, R., Liston, A., Parks, M., Gernandt, D. S., Shen, R., & Mockler, T. (2008). Multiplex sequencing of plant chloroplast genomes using Solexa sequencing-by-synthesis technology. *Nucleic Acids Research*, *36*, e122-e122. https://doi.org/10.1093/nar/gkn502.

Cusimano, N., & Wicke, S. (2016). Massive intracellular gene transfer during plastid genome reduction in nongreen Orobanchaceae. *New Phytologist*, *210*, 680–693.

de Vries, J., Sousa, F. L., Bölter, B., Soll, J., & Gould, S. B. (2015). YCF1: A green TIC? *The Plant Cell*, *27*, 1827–1833.

Delannoy, E., Fujii, S., des Francs, C. C., Brundrett, M., & Small, I. (2011). Rampant gene loss in the underground orchid *Rhizanthella gardneri* highlights evolutionary constraints on plastid genomes. *Molecular Evolution and Biology*, *28*, 2077–2086.

Delavault, P. M., Russo, N. M., Lusson, N. A., & Thalouarn, P. (1996). Organization of the reduced plastid genome of *Lathraea clandestina*, an achlorophyllous parasitic plant. *Physiologia Plantarum*, *96*, 674–682.

dePamphilis, C. W., & Palmer, J. D. (1990). Loss of photosynthetic and chlororespiratory genes from the plastid genome of a parasitic flowering plant. *Nature*, *348*, 337–339.

dePamphilis, C. W., Young, N. D., & Wolfe, A. D. (1997). Evolution of plastid gene *rps2* in a lineage of hemiparasitic and holoparasitic plants: Many losses of photosynthesis and complex patterns of rate variation. *Proceedings of the National Academy of Sciences of the United States of America*, *94*, 7367–7372.

Drescher, A., Ruf, S., Calsa, T., Carrer, H., & Bock, R. (2000). The two largest chloroplast genome-encoded open reading frames of higher plants are essential genes. *The Plant Journal*, *22*, 97–104.

Ems, S. C., Morden, C. W., Dixon, C. K., Wolfe, K. H., dePamphilis, C. W., & Palmer, J. D. (1995). Transcription, splicing and editing of plastid RNAs in the nonphotosynthetic plant *Epifagus virginiana*. *Plant Molecular Biology*, *29*, 721–733.

Eserman, L. A., Tiley, G. P., Jarret, R. L., Leebens-Mack, J. H., & Miller, R. E. (2014). Phylogenetics and diversification of morning glories (tribe Ipomoeeae, Convolvulaceae) based on whole plastome sequences. *American Journal of Botany*, *101*, 92–103.

Fan, W., Zhu, A., Kozaczek, M., Shah, N., Pabón-Mora, N., González, F., et al. (2016). Limited mitogenomic degradation in response to a parasitic lifestyle in Orobanchaceae. *Scientific Reports*, *6*, 36285.

Feng, Y.-L., Wicke, S., Li, J.-W., Han, Y., Lin, C.-S., Li, D.-Z., et al. (2016). Lineage-specific reductions of plastid genomes in an orchid tribe with partially and fully mycoheterotrophic species. *Genome Biology and Evolution*, *8*, 2164–2175.

Fernandez-Aparicio, M., Rubiales, D., Bandaranayake, P. C. G., Yoder, J., & Westwood, J. (2011). Transformation and regeneration of the holoparasitic plant *Phelipanche aegyptiaca*. *Plant Methods*, *7*, 36.

Funk, H., Berg, S., Krupinska, K., Maier, U., & Krause, K. (2007). Complete DNA sequences of the plastid genomes of two parasitic flowering plant species, *Cuscuta reflexa* and *Cuscuta gronovii*. *BMC Plant Biology*, *7*, 45.

Gaut, B., Yang, L., Takuno, S., & Eguiarte, L. E. (2011). The patterns and causes of variation in plant nucleotide substitution rates. *Annual Review of Ecology, Evolution, and Systematics*, *42*, 245–266.

Graham, S. W., Lam, V. K. Y., & Merckx, V. S. F. T. (2017). Plastomes on the edge: The evolutionary breakdown of mycoheterotroph plastid genomes. *The New Phytologist*, *214*, 48–55.

Guisinger, M. M., Kuehl, J. V., Boore, J. L., & Jansen, R. K. (2008). Genome-wide analyses of Geraniaceae plastid DNA reveal unprecedented patterns of increased nucleotide substitutions. *Proceedings of the National Academy of Sciences of the United States of America*, *105*, 18424–18429.

Henriquez, C. L., Arias, T., Pires, J. C., Croat, T. B., & Schaal, B. A. (2014). Phylogenomics of the plant family Araceae. *Molecular Phylogenetics and Evolution*, *75*, 91–102.

Huang, D. I., & Cronk, Q. C. B. (2015). Plann: A command-line application for annotating plastome sequences. *Applications in Plant Sciences*, *3*, 1500026.

Ishida, J. K., Yoshida, S., Ito, M., Namba, S., & Shirasu, K. (2011). *Agrobacterium rhizogenes*-mediated transformation of the parasitic plant *Phtheirospermum japonicum*. *PLoS One*, *6*, e25802.

Jansen, R. K., & Ruhlman, T. A. (2012). Plastid genomes of seed plants. In R. Bock & V. Knoop (Eds.), *Genomics of chloroplasts and mitochondria* (pp. 103–126). Dordrecht: Springer.

Kamikawa, R., Tanifuji, G., Ishikawa, S. A., Ishii, K.-I., Matsuno, Y., Onodera, N. T., et al. (2015). Proposal of a twin arginine translocator system-mediated constraint against loss of ATP synthase genes from nonphotosynthetic plastid genomes. *Molecular Evolution and Biology*, *32*, 2598–2604.

Keeling, P. J. (2010). The endosymbiotic origin, diversification and fate of plastids. *Philosophical Transactions of the Royal Society of London. Series B, Biological Sciences*, *365*, 729–748.

Kikuchi, S., Bedard, J., Hirano, M., Hirabayashi, Y., Oishi, M., Imai, M., et al. (2013). Uncovering the protein translocon at the chloroplast inner envelope membrane. *Science*, *339*, 571–574.

Lartillot, N., & Poujol, R. (2011). A phylogenetic model for investigating correlated evolution of substitution rates and continuous phenotypic characters. *Molecular Evolution and Biology*, *28*, 729–744.

Levy Karin, E., Wicke, S., Pupko, T., & Mayrose, I. (2017). An integrated model of phenotypic trait changes and site-specific sequence evolution. *Systematic Biology*, *66*, 917–933. https://doi.org/10.1093/sysbio/syx032.

Li, X., Zhang, T.-C., Qiao, Q., Ren, Z., Zhao, J., Yonezawa, T., et al. (2013). Complete chloroplast genome sequence of holoparasite *Cistanche deserticola* (Orobanchaceae) reveals gene loss and horizontal gene transfer from its host *Haloxylon ammodendron* (Chenopodiaceae). *PLoS One*, *8*, e58747.

Lim, G. S., Barrett, C. F., Pang, C.-C., & Davis, J. I. (2016). Drastic reduction of plastome size in the mycoheterotrophic *Thismia tentaculata* relative to that of its autotrophic relative *Tacca chantrieri*. *American Journal of Botany*, *103*, 1129–1137. https://doi.org/10.3732/ajb.1600042.

Lin, C.-P., Huang, J.-P., Wu, C.-S., Hsu, C.-Y., & Chaw, S.-M. (2010). Comparative chloroplast genomics reveals the evolution of Pinaceae genera and subfamilies. *Genome Biology and Evolution, 2*, 504–517.

Lin, C.-P., Ko, C.-Y., Kuo, C.-I., Liu, M.-S., Schafleitner, R., & Chen, L.-F. O. (2015). Transcriptional slippage and RNA editing increase the diversity of transcripts in chloroplasts: Insight from deep sequencing of *Vigna radiata* genome and transcriptome. *PLoS One, 10*, e0129396.

Lloyd, A. H., Rousseau-Gueutin, M., Timmis, J. N., Sheppard, A. E., & Ayliffe, M. A. (2012). Promiscuous organellar DNA. In R. Bock & V. Knoop (Eds.), *Genomics of chloroplasts and mitochondria* (pp. 201–221). Dordrecht: Springer Netherlands.

Logacheva, M. D., Schelkunov, M. I., Nuraliev, M. S., Samigullin, T. H., & Penin, A. A. (2014). The plastid genome of mycoheterotrophic monocot *Petrosavia stellaris* exhibits both gene losses and multiple rearrangements. *Genome Biology and Evolution, 6*, 238–246.

Logacheva, M. D., Schelkunov, M. I., Shtratnikova, V. Y., Matveeva, M. V., & Penin, A. A. (2016). Comparative analysis of plastid genomes of non-photosynthetic Ericaceae and their photosynthetic relatives. *Scientific Reports, 6* . 30042.

Lohan, A. J., & Wolfe, K. H. (1998). A subset of conserved tRNA genes in plastid DNA of nongreen plants. *Genetics, 150*, 425–433.

Magee, A. M., Aspinall, S., Rice, D. W., Cusack, B. P., Sémon, M., Perry, A. S., et al. (2010). Localized hypermutation and associated gene losses in legume chloroplast genomes. *Genome Research, 20*, 1700–1710.

Maréchal, A., & Brisson, N. (2010). Recombination and the maintenance of plant organelle genome stability. *New Phytologist, 186*, 299–317.

Martín, M., & Sabater, B. (2010). Plastid ndh genes in plant evolution. *Plant Physiology and Biochemistry, 48*, 636–645.

McKain, M. R., Hartsock, R. H., Wohl, M. M., & Kellogg, E. A. (2017). Verdant: Automated annotation, alignment and phylogenetic analysis of whole chloroplast genomes. *Bioinformatics, 33*, 130–132.

McNeal, J. R., Kuehl, J., Boore, J., & dePamphilis, C. W. (2007). Complete plastid genome sequences suggest strong selection for retention of photosynthetic genes in the parasitic plant genus *Cuscuta*. *BMC Plant Biology*, 7, 57.

Meurer, J., Lezhneva, L., Amann, K., Gödel, M., Bezhani, S., Sherameti, I., et al. (2002). A peptide chain release factor 2 affects the stability of UGA-containing transcripts in *Arabidopsis* chloroplasts. *The Plant Cell, 14*, 3255–3269.

Molina, J., Hazzouri, K. M., Nickrent, D., Geisler, M., Meyer, R. S., Pentony, M. M., et al. (2014). Possible loss of the chloroplast genome in the parasitic flowering plant *Rafflesia lagascae* (Rafflesiaceae). *Molecular Evolution and Biology, 31*, 793–803.

Moore, M., Dhingra, A., Soltis, P., Shaw, R., Farmerie, W., Folta, K., et al. (2006). Rapid and accurate pyrosequencing of angiosperm plastid genomes. *BMC Plant Biology, 6*, 17.

Morden, C. W., Wolfe, K. H., dePamphilis, C. W., & Palmer, J. D. (1991). Plastid translation and transcription genes in a nonphotosynthetic plant—Intact, missing and pseudogenes. *EMBO Journal, 10*, 3281–3288.

Morton, B. R. (1997). Rates of synonymous substitution do not indicate selective constraints on the codon use of the plant *psbA* gene. *Molecular Evolution and Biology, 14*, 412–419.

Nakai, M. (2015). YCF1: A green TIC: Response to the de Vries et al. Commentary. *The Plant Cell, 27*, 1834–1838.

Naumann, J., Der, J. P., Wafula, E. K., Jones, S. S., Wagner, S. T., Honaas, L. A., et al. (2016). Detecting and characterizing the highly divergent plastid genome of the nonphotosynthetic parasitic plant *Hydnora visseri* (Hydnoraceae). *Genome Biology and Evolution, 8*, 345–363.

Nickrent, D. L., Blarer, A., Qiu, Y.-L., Vidal-Russell, R., & Anderson, F. (2004). Phylogenetic inference in Rafflesiales: The influence of rate heterogeneity and horizontal gene transfer. *BMC Evolutionary Biology*, *4*, 40.

Nickrent, D. L., Ouyang, Y., Duff, R. J., & dePamphilis, C. W. (1997). Do nonasterid holoparasitic flowering plants have plastid genomes? *Plant Molecular Biology*, *34*, 717–729.

Nickrent, D. L., & Starr, E. M. (1994). High rates of nucleotide substitution in nuclear small-subunit (18S) rDNA from holoparasitic flowering plants. *Journal of Molecular Evolution*, *39*, 62–70.

Peltier, G., & Cournac, L. (2002). Chlororespiration. *Annual Review of Plant Biology*, *53*, 523–550.

Petersen, G., Cuenca, A., & Seberg, O. (2015). Plastome evolution in hemiparasitic mistletoes. *Genome Biology and Evolution*, *7*, 2520–2532.

Randle, C. P., & Wolfe, A. D. (2005). The evolution and expression of RBCL in holoparasitic sister-genera *Harveya* and *Hyobanche* (Orobanchaceae). *American Journal of Botany*, *92*, 1575–1585.

Richly, E., & Leister, D. (2004). NUPTs in sequenced eukaryotes and their genomic organization in relation to NUMTs. *Molecular Evolution and Biology*, *21*, 1972–1980.

Roquet, C., Coissac, É., Cruaud, C., Boleda, M., Boyer, F., Alberti, A., et al. (2016). Understanding the evolution of holoparasitic plants: The complete plastid genome of the holoparasite *Cytinus hypocistis* (Cytinaceae). *Annals of Botany*. https://doi.org/10.1093/aob/mcw135.

Rousseau-Gueutin, M., Ayliffe, M. A., & Timmis, J. N. (2011). Conservation of plastid sequences in the plant nuclear genome for millions of years facilitates endosymbiotic evolution. *Plant Physiology*, *157*, 2181–2193.

Rousseau-Gueutin, M., Huang, X., Higginson, E., Ayliffe, M. A., Day, A., & Timmis, J. N. (2013). Potential functional replacement of the plastidic *accD* gene by recent transfers to the nucleus in some angiosperm lineages. *Plant Physiology*, *16*, 1918–1929.

Samigullin, T. H., Logacheva, M. D., Penin, A. A., & Vallejo-Roman, C. M. (2016). Complete plastid genome of the recent holoparasite *Lathraea squamaria* reveals earliest stages of plastome reduction in Orobanchaceae. *PLoS One*, *11*, e0150718.

Schelkunov, M. I., Shtratnikova, V. Y., Nuraliev, M. S., Selosse, M.-A., Penin, A. A., & Logacheva, M. D. (2015). Exploring the limits for reduction of plastid genomes: A case study of the mycoheterotrophic orchids *Epipogium aphyllum* and *Epipogium roseum*. *Genome Biology and Evolution*, *7*, 1179–1191.

Schwender, J., Goffman, F., Ohlrogge, J. B., & Shachar-Hill, Y. (2004). Rubisco without the Calvin cycle improves the carbon efficiency of developing green seeds. *Nature*, *432*, 779–782.

Sjuts, I., Soll, J., & Bölter, B. (2017). Import of soluble proteins into chloroplasts and potential regulatory mechanisms. *Frontiers in Plant Sciences*, *8*, 168.

Sloan, D. B., Triant, D. A., Forrester, N. J., Bergner, L. M., Wu, M., & Taylor, D. R. (2014). A recurring syndrome of accelerated plastid genome evolution in the angiosperm tribe Sileneae (Caryophyllaceae). *Molecular Phylogenetics and Evolution*, *72*, 82–89.

Song, Y., Yu, W.-B., Tan, Y., Liu, B., Yao, X., Jin, J., et al. (2017). Evolutionary comparisons of the chloroplast genome in Lauraceae and insights into loss events in the Magnoliids. *Genome Biology and Evolution*, *9*, 2354–2364.

Straub, S. C. K., Cronn, R. C., Edwards, C., Fishbein, M., & Liston, A. (2013). Horizontal transfer of DNA from the mitochondrial to the plastid genome and its subsequent evolution in milkweeds (Apocynaceae). *Genome Biology and Evolution*, *5*, 1872–1885.

Straub, S. C. K., Parks, M., Weitemier, K., Fishbein, M., Cronn, R. C., & Liston, A. (2012). Navigating the tip of the genomic iceberg: Next-generation sequencing for plant systematics. *American Journal of Botany*, *99*, 349–364. https://doi.org/10.3732/ajb.1100335.

Su, H.-J., & Hu, J.-M. (2016). The complete chloroplast genome of hemiparasitic flowering plant *Schoepfia jasminodora*. *Mitochondrial DNA*, *1*, 767–769.

Su, H.-J., Hu, J.-M., Anderson, F. E., Der, J. P., & Nickrent, D. L. (2015). Phylogenetic relationships of Santalales with insights into the origins of holoparasitic Balanophoraceae. *Taxon*, *64*, 491–506.

The Angiosperm Phylogeny Group. (2016). An update of the angiosperm phylogeny group classification for the orders and families of flowering plants: APG IV. *Botanical Journal of the Linnean Society*, *181*, 1–20.

Tolbert, N. E. (1997). The C2 oxidative photosynthetic carbon cycle. *Annual Review of Plant Physiology and Plant Molecular Biology*, *48*, 1–25.

Tomilov, A. A., Tomilova, N. B., & Yoder, J. I. (2007). *Agrobacterium tumefaciens* and *Agrobacterium rhizogenes* transformed roots of the parasitic plant *Triphysaria versicolor* retain parasitic competence. *Planta*, *225*, 1059–1071.

Uribe-Convers, S., Duke, J. R., Moore, M. J., & Tank, D. C. (2014). A long PCR–based approach for DNA enrichment prior to next-generation sequencing for systematic studies. *Applications in Plant Sciences*, *2*, 1300063.

Wakasugi, T., Tsudzuki, J., Ito, S., Nakashima, K., Tsudzuki, T., & Sugiura, M. (1994). Loss of all ndh genes as determined by sequencing the entire chloroplast genome of the black pine Pinus Thunbergii. *Proceedings of the National Academy of Sciences of the United States of America*, *91*, 9794–9798.

Wicke, S., Müller, K. F., dePamphilis, C. W., Quandt, D., Bellot, S., & Schneeweiss, G. M. (2016). Mechanistic model of evolutionary rate variation en route to a non-photosynthetic lifestyle in plants. *Proceedings of the National Academy of Sciences of the United States of America*, *113*, 9045–9050.

Wicke, S., Müller, K. F., dePamphilis, C. W., Quandt, D., Wickett, N. J., Zhang, Y., et al. (2013). Mechanisms of functional and physical genome reduction in photosynthetic and non-photosynthetic parasitic plants of the broomrape family. *The Plant Cell*, *25*, 3711–3725.

Wicke, S., Schäferhoff, B., dePamphilis, C. W., & Müller, K. F. (2014). Disproportional plastome-wide increase of substitution rates and relaxed purifying selection in genes of carnivorous Lentibulariaceae. *Molecular Evolution and Biology*, *31*, 529–545.

Wicke, S., & Schneeweiss, G. M. (2015). Next generation organellar genomics: Potentials and pitfalls of high-throughput technologies for molecular evolutionary studies and plant systematics. In E. Hörandl & M. Appelhans (Eds.), *Next generation sequencing in plant systematics, Regnum Vegetabile* (pp. 9–50). Koenigstein: Koeltz Scientific Books.

Wicke, S., Schneeweiss, G. M., dePamphilis, C. W., Müller, K. F., & Quandt, D. (2011). The evolution of the plastid chromosome in land plants: Gene content, gene order, gene function. *Plant Molecular Biology*, *76*, 273–297.

Wickett, N. J., Zhang, Y., Hansen, S. K., Roper, J. M., Kuehl, J. V., Plock, S. A., et al. (2008). Functional gene losses occur with minimal size reduction in the plastid genome of the parasitic liverwort *Aneura mirabilis*. *Molecular Biology and Evolution*, *25*, 393–401.

Wolfe, K. H. (1994). Similarity between putative ATP-binding sites in land plant plastid ORF2280 proteins and the FtsH/CDC48 family of ATPases. *Current Genetics*, *25*, 379–383.

Wolfe, A. D., & dePamphilis, C. W. (1998). The effect of relaxed functional constraints on the photosynthetic gene *rbcL* in photosynthetic and nonphotosynthetic parasitic plants. *Molecular Biology and Evolution*, *15*, 1243–1258.

Wolfe, K. H., Morden, C. W., Ems, S. C., & Palmer, J. D. (1992). Rapid evolution of the plastid translational apparatus in a nonphotosynthetic plant: Loss or accelerated sequence evolution of tRNA and ribosomal protein genes. *Journal of Molecular Evolution*, *35*, 304–317.

Wolfe, K. H., Morden, C. W., & Palmer, J. D. (1992). Function and evolution of a minimal plastid genome from a nonphotosynthetic parasitic plant. *Proceedings of the National Academy of Sciences of the United States of America*, *89*, 10648–10652.

Wu, C.-S., Lai, Y.-T., Lin, C.-P., Wang, Y.-N., & Chaw, S.-M. (2009). Evolution of reduced and compact chloroplast genomes (cpDNAs) in gnetophytes: Selection toward a lower-cost strategy. *Molecular Phylogenetics and Evolution*, *52*, 115–124.

Wu, C.-S., Wang, T.-J., Wu, C.-W., Wang, Y.-N., & Chaw, S.-M. (2017). Plastome evolution in the sole hemiparasitic genus Laurel dodder (*Cassytha*) and insights into the plastid phylogenomics of Lauraceae. *Genome Biology and Evolution*, *9*, 2604–2614.

Wyman, S. K., Boore, J. L., & Jansen, R. K. (2004). Automatic annotation of organellar genomes with DOGMA. *Bioinformatics*, *20*, 3252–3255.

Young, N. D., & dePamphilis, C. W. (2005). Rate variation in parasitic plants: Correlated and uncorrelated patterns among plastid genes of different function. *BMC Evolutionary Biology*, *5*, 16.

Yu, J., Wang, C., & Gong, X. (2017). Degeneration of photosynthetic capacity in mixotrophic plants, *Chimaphila japonica* and *Pyrola decorata* (Ericaceae). *Plant Diversity*, *39*, 80–88.

Zerges, W. (2000). Translation in chloroplasts. *Biochimie*, *82*, 583–601.

SUBJECT INDEX

Note: Page numbers followed by "*f*" indicate figures, and "*t*" indicate tables.

AUTHOR INDEX

Note: Page numbers followed by "*f*" indicate figures, "*t*" indicate tables, and "*np*" indicate footnotes.

B

C

E

F

G

I

J

K

L

N

O

P

Q

R

S

T

W

X

Y

Z

Kolibri 360 GmbH
Ferdinand-Jühlke-Straße 7
99095 Erfurt, Deutschland
produktsicherheit@kolibri360.de